Perspectives in Mathematical Sciences I

Probability and Statistics

Statistical Science and Interdisciplinary Research

Series Editor: Sankar K. Pal *(Indian Statistical Institute)*

Description:
In conjunction with the Platinum Jubilee celebrations of the Indian Statistical Institute, a series of books will be produced to cover various topics, such as Statistics and Mathematics, Computer Science, Machine Intelligence, Econometrics, other Physical Sciences, and Social and Natural Sciences. This series of edited volumes in the mentioned disciplines culminate mostly out of significant events — conferences, workshops and lectures — held at the ten branches and centers of ISI to commemorate the long history of the institute.

Vol. 1 Mathematical Programming and Game Theory for Decision Making
edited by S. K. Neogy, R. B. Bapat, A. K. Das & T. Parthasarathy (Indian Statistical Institute, India)

Vol. 2 Advances in Intelligent Information Processing: Tools and Applications
edited by B. Chandra & C. A. Murthy (Indian Statistical Institute, India)

Vol. 3 Algorithms, Architectures and Information Systems Security
edited by Bhargab B. Bhattacharya, Susmita Sur-Kolay, Subhas C. Nandy & Aditya Bagchi (Indian Statistical Institute, India)

Vol. 4 Advances in Multivariate Statistical Methods
edited by A. SenGupta (Indian Statistical Institute, India)

Vol. 5 New and Enduring Themes in Development Economics
edited by B. Dutta, T. Ray & E. Somanathan
(Indian Statistical Institute, India)

Vol. 6 Modeling, Computation and Optimization
edited by S. K. Neogy, A. K. Das and R. B. Bapat (Indian Statistical Institute, India)

Vol. 7 Perspectives in Mathematical Sciences I: Probability and Statistics
edited by N. S. N. Sastry, T. S. S. R. K. Rao, M. Delampady and B. Rajeev (Indian Statistical Institute, India)

Platinum Jubilee Series

Statistical Science and
Interdisciplinary Research — Vol. 7

Perspectives in Mathematical Sciences I

Probability and Statistics

Editors

N. S. Narasimha Sastry

T. S. S. R. K. Rao

Mohan Delampady

B. Rajeev

Indian Statistical Institute, India

Series Editor: ***Sankar K. Pal***

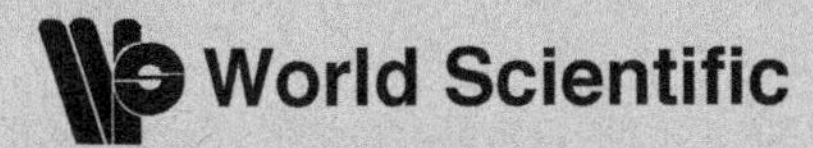

NEW JERSEY · LONDON · SINGAPORE · BEIJING · SHANGHAI · HONG KONG · TAIPEI · CHENNAI

Published by

World Scientific Publishing Co. Pte. Ltd.
5 Toh Tuck Link, Singapore 596224
USA office: 27 Warren Street, Suite 401-402, Hackensack, NJ 07601
UK office: 57 Shelton Street, Covent Garden, London WC2H 9HE

British Library Cataloguing-in-Publication Data
A catalogue record for this book is available from the British Library.

Statistical Science and Interdisciplinary Research — Vol. 7
PERSPECTIVES IN MATHEMATICAL SCIENCES I
Probability and Statistics

ISBN-13 978-981-4273-62-6
ISBN-10 981-4273-62-7

Printed in Singapore.

Foreword

The Indian Statistical Institute (ISI) was established on 17th December, 1931 by a great visionary Professor Prasanta Chandra Mahalanobis to promote research in the theory and applications of statistics as a new scientific discipline in India. In 1959, Pandit Jawaharlal Nehru, the then Prime Minister of India introduced the ISI Act in the parliament and designated it as an *Institution of National Importance* because of its remarkable achievements in statistical work as well as its contribution to economic planning.

Today, the Indian Statistical Institute occupies a prestigious position in the academic firmament. It has been a haven for bright and talented academics working in a number of disciplines. Its research faculty has done India proud in the arenas of Statistics, Mathematics, Economics, Computer Science, among others. Over seventy five years, it has grown into a massive banyan tree, like the institute emblem. The Institute now serves the nation as a unified and monolithic organization from different places, namely Kolkata, the Headquarters, Delhi, Bangalore, and Chennai, three centers, a network of five SQC-OR Units located at Mumbai, Pune, Baroda, Hyderabad and Coimbatore, and a branch (field station) at Giridih.

The platinum jubilee celebrations of ISI have been launched by Honorable Prime Minister Prof. Manmohan Singh on December 24, 2006, and the Government of India has declared 29th June as the "Statistics Day" to commemorate the birthday of Professor Mahalanobis nationally.

Professor Mahalanobis, was a great believer in interdisciplinary research, because he thought that this will promote the development of not only Statistics, but also the other natural and social sciences. To promote interdisciplinary research, major strides were made in the areas of computer science, statistical quality control, economics, biological and social sciences, physical and earth sciences.

The Institute's motto of "unity in diversity" has been the guiding principle of all its activities since its inception. It highlights the unifying role of statistics in relation to various scientific activities.

In tune with this hallowed tradition, a comprehensive academic programme, involving Nobel Laureates, Fellows of the Royal Society, Abel prize winner and other dignitaries, has been implemented throughout the Platinum Jubilee year, highlighting the emerging areas of ongoing frontline research in its various scientific divisions, centers, and outlying units. It includes international and national-level seminars, symposia, conferences and workshops, as well as series of special lectures. As an outcome of these events, the Institute is bringing out a series of comprehensive volumes in different subjects under the title *Statistical Science and Interdisciplinary Research*, published by the World Scientific Press, Singapore.

The present volume titled *Perspectives in Mathematical Sciences I: Probability and Statistics* is the seventh one in the series. The volume consists of eleven chapters, written by eminent probabilists and statisticians from different parts of the world. These chapters provide a current perspective of different areas of research, emphasizing the major challenging issues. They deal mainly with statistical inference, both frequentist and Bayesian, with applications of the methodology that will be of use to practitioners. I believe the state-of-the art studies presented in this book will be very useful to both researchers as well as practitioners.

Thanks to the contributors for their excellent research contributions, and to the volume editors Profs. N. S. Narasimha Sastry, T. S. S. R. K. Rao, M. Delampady and B. Rajeev for their sincere effort in bringing out the volume nicely in time. Initial design of the cover by Mr. Indranil Dutta is acknowledged. Sincere efforts by Prof. Dilip Saha and Dr. Barun Mukhopadhyay for editorial assistance are appreciated. Thanks are also due to World Scientific for their initiative in publishing the series and being a part of the Platinum Jubilee endeavor of the Institute.

December 2008
Kolkata

Sankar K. Pal
Series Editor and
Director

Preface

Indian Statistical Institute, a premier research institute founded by Professor Prasanta Chandra Mahalanobis in Calcutta in 1931, celebrated its platinum jubilee during the year 2006-07. On this occasion, the institute organized several conferences and symposia in various scientific disciplines in which the institute has been active.

From the beginning, research and training in probability, statistics and related mathematical areas including mathematical computing have been some of the main activities of the institute. Over the years, the contributions from the scientists of the institute have had a major impact on these areas.

As a part of these celebrations, the Division of Theoretical Statistics and Mathematics of the institute decided to invite distinguished mathematical scientists to contribute articles, giving "a perspective of their discipline, emphasizing the current major issues". A conference entitled "Perspectives in Mathematical Sciences" was also organized at the Bangalore Centre of the institute during February 4-8, 2008.

The articles submitted by the speakers at the conference, along with the invited articles, are brought together here in two volumes (Part I and Part II).

Part I consists of articles in Probability and Statistics. Articles in Statistics are mainly on statistical inference, both frequentist and Bayesian, for problems of current interest. These articles also contain applications illustrating the methodologies discussed. The articles on probability are based on different "probability models" arising in various contexts (machine learning, quantum probability, probability measures on Lie groups, economic phenomena modelled on iterated random systems, "measure free martingales", and interacting particle systems) and represent active areas of research in probability and related fields.

Part II consists of articles in Algebraic Geometry, Algebraic Number Theory, Functional Analysis and Operator Theory, Scattering Theory, von Neumann Algebras, Discrete Mathematics, Permutation Groups, Lie Theory and Super Symmetry.

All the authors have taken care to make their exposition fairly self-contained. It is our hope that these articles will be valuable to researchers at various levels.

The editorial committee thanks all the authors for writing the articles and sending them in time, the speakers at the conference for their talks and various scientists who have kindly refereed these articles. Thanks are also due to the National Board for Higher Mathematics, India, for providing partial support to the conference. Finally, we thank Ms. Asha Lata for her help in compiling these volumes.

October 16, 2008

N. S. Narasimha Sastry
T. S. S. R. K. Rao
Mohan Delampady
B. Rajeev

Contents

Chapter 1

Entropy and Martingale

K. B. Athreya[1] and M. G. Nadkarni[2]

[1]*Department of Mathematics,*
Iowa State University, Ames, Iowa, 50011, USA
and
I. M. I, Department of Mathematics,
Indian Institute of Science, Bangalore, 560012, India
kba@iastate.edu

[2]*Department of Mathematics*
University of Mumbai, Kalina, Mumbai, 400098, India
mgnadkarni@gmail.com

1.1. Introduction

This article discusses the concepts of relative entropy of a probability measure with respect to a dominating measure and that of measure free martingales. There is considerable literature on the concepts of relative entropy and standard martingales, both separately and on connection between the two. This paper draws from results established in [1] (unpublished notes) and [6]. In [1] the concept of relative entropy and its maximization subject to a finite as well as infinite number of linear constraints is discussed. In [6] the notion of measure free martingale of a sequence $\{f_n\}_{n=1}^{\infty}$ of real valued functions with the restriction that each f_n takes only finitely many distinct values is introduced. Here is an outline of the paper.

In section 1.2 the concepts of relative entropy and Gibbs-Boltzmann measures, and a few results on the maximization of relative entropy and the weak convergence of of the Gibbs-Boltzmann measures are presented. We also settle in the negative a problem posed in [6]. In section 1.3 the notion of measure free martingale is generalized from the case of finitely many valued sequence $\{f_n\}_{n=1}^{\infty}$ to the general case where each f_n is allowed to be a Borel function taking possibly uncountably many values. It is shown that every martingale is a measure free martingale, and conversely that every measure free martingale admits a finitely additive measure on a certain algebra under which it is a martingale. Conditions under which such a measure is countably additive are given. Last section is devoted to an ab initio discussion of the existence of an equivalent martingale measure and the uniqueness of such a measure if they are chosen to maximize certain relative entropies.

1.2. Relative Entropy and Gibbs-Boltzmann Measures

1.2.1. *Entropy Maximization Results*

Let $(\Omega, \mathcal{B}, \mu)$ be a measure space. A $\mathcal{B}$- measurable function $f : \Omega \to [0, \infty)$ is called a probability density function (p.d.f) with respect to μ if $\int_\Omega f(\omega)\mu(d\omega) = 1$. Then $\nu_f(A) = \int_A f(\omega)\mu(d\omega), A \in \mathcal{B}$, is a probability measure dominated by μ. The relative entropy of ν_f with respect to μ is defined as

$$H(f, \mu) = -\int_\Omega f(\omega) \log f(\omega)\mu(d\omega) \tag{1}$$

provided the integral on the right hand side is well defined, although it may possibly be infinite. In particular, if μ is a finite measure, this holds since the function $h(x) = x \log x$ is bounded on $(0, 1)$ and hence $\int_\Omega(-f(\omega) \log f(\omega))^+ \mu(d\omega) < \infty$. This does allow for the possibility that $H(f, \mu)$ could be $-\infty$ when $\mu(\Omega)$ is finite. We will show below that if $\mu(\Omega)$ is finite and positive then $H(f, \mu) \leq \log \mu(\Omega)$ for all p.d.f. f with respect to μ. In particular if $\mu(\Omega) = 1$, $H(f, \mu) \leq 0$.

We recall here for the benefit of the reader that a $\mathcal{B}$ measurable non-negative real valued function f always has a well defined integral with respect to μ. It is denoted by $\int_\Omega f(\omega)\mu(d\omega)$. The integral may be finite or infinite. A real valued $\mathcal{B}$ measurable function f can be written as $f = f_+ - f_-$, where, for each $\omega \in \Omega$,

$$f_+(\omega) = \max\{0, f(\omega)\}, f_-(\omega) = -\min\{0, f(\omega)\}.$$

If at least one of f_+, f_- has a finite integral, then we say that f has a well defined integral with respect to μ and write

$$\int_\Omega f(\omega)\mu(d\omega) = \int_\Omega f_+(\omega)\mu(d\omega) - \int_\Omega f_-(\omega)\mu(d\omega).$$

Now note the simple fact from calculus. The function $\phi(x) = x - 1 - \log x$ on $(0, \infty)$ has a unique minimum at $x = 1$ and $\phi(1) = 0$. Thus for all $x > 0$, $\log x \leq x - 1$ with equality holding if and only if $x = 1$. So if f_1 and f_2 are two probability density functions on $(\Omega, \mathcal{B}, \mu)$, then for all ω,

$$f_1(\omega) \log f_2(\omega) - f_1(\omega) \log f_1(\omega) \leq f_2(\omega) - f_1(\omega), \tag{2}$$

with equality holding if and only if $f_1(\omega) = f_2(\omega)$. Assume now that $f_1(\omega) \log f_1(\omega)$, $f_1(\omega) \log f_2(\omega)$ have definite integrals with respect to μ and that one of them is finite. On integrating the two sides of (2) we get

$$\begin{aligned}
&\int_\Omega f_1(\omega) \log f_2(\omega) \mu(d\omega) - \int_\Omega f_1(\omega) \log f_1(\omega) \mu(d\omega) \\
&\leq \int_\Omega f_2(\omega) \mu(d\omega) - \int_\Omega f_1(\omega) \mu(d\omega) \\
&= 1 - 1 = 0.
\end{aligned}$$

The middle inequality becomes an equality if and only if equality holds in (2) a.e. with respect to μ. We have proved:

Proposition 2.1. *Let $(\Omega, \mathcal{B}, \mu)$ be a measure space and let f_1, f_2 be two probability density functions on $(\Omega, \mathcal{B}, \mu)$. Assume that the functions $f_1 \log f_1, f_1 \log f_2$ have definite integrals with respect to μ and that one of them is finite. Then*

$$H(f_1, \mu) = -\int_\Omega f_1(\omega) \log f_1(\omega) \mu(d\omega) \leq -\int_\Omega f_1(\omega) \log f_2(\omega) \mu(d\omega), \tag{3}$$

with equality holding if and only if $f_1(\omega) = f_2(\omega)$, a.e. μ.

Note that if $\mu(\Omega)$ is finite and positive and if we set $f_2(\omega) = (\mu(\Omega))^{-1}$, for all ω, then the right hand side of (3) becomes $\log \mu(\Omega)$ and we conclude that relative entropy of $H(f_1, \mu)$ is well defined and at most $\log \mu(\Omega)$.

Let f_0 be a probability density function on $(\Omega, \mathcal{B}, \mu)$ such that $\lambda \equiv H(f_0, \mu)$ is finite and let

$$\mathcal{F}_\lambda = \{f : f \text{ a p.d.f. wrt } \mu \text{ and } -\int_\Omega f(\omega) \log f_0(\omega) \mu(d\omega) = \lambda\}. \tag{4}$$

From Proposition 2.1 it follows that for any $f \in \mathcal{F}_\lambda$,

$$H(f,\mu) = -\int_\Omega f(\omega)\log f(\omega)\mu(d\omega) \leq -\int_\Omega f(\omega)\log f_0(\omega)\mu(d\omega) = \lambda$$
$$= -\int_\Omega f_0(\omega)\log f_0(\omega)\mu(d\omega),$$

with equality holding if and only if $f = f_0$, *a.e.* μ. We summarize this as:

Theorem 2.1. *Let $f_0, \lambda, \mathcal{F}_\lambda$, be as in (4) above. Then*

$$\sup\{H(f,\mu) : f \in \mathcal{F}_\lambda\} = H(f_0,\mu)$$

and f_0 is the unique maximiser.

Theorem 2.1. says that any probability density function f_0 with respect to μ with finite entropy relative to μ appears as the unique solution to an entropy maximizing problem in an appropriate class of probability density functions. Of course, this assertion has some meaning only if $\mathcal{F}_\lambda$ does not consist of f_0 alone. A useful reformulation of this result is as follows:

Theorem 2.2. *Let $h : \Omega \to \mathbb{R}$ be a $\mathcal{B}$ measurable function. Let c and λ be real numbers such that*

$$\psi(c) = \int_\Omega e^{ch(\omega)}\mu(d\omega) < \infty, \qquad \int_\Omega \mid h(\omega) \mid e^{ch(\omega)}\mu(d\omega) < \infty, \text{ and}$$

$$\lambda = \frac{\int_\Omega h(\omega)e^{ch(\omega)}\mu(d\omega)}{\psi(c)}.$$

Let $f_0 = \frac{e^{ch}}{\psi(c)}$ and let

$$\mathcal{F}_\lambda = \{f : f \text{ a p.d.f. wrt } \mu, \text{ and } \int_\Omega f(\omega)h(\omega)\mu(d\omega) = \lambda\}.$$

Then

$$\sup\{H(f,\mu) : f \in \mathcal{F}_\lambda\} = H(f_0,\mu),$$

and f_0 is the unique maximiser.

Here are some sample examples of the above considerations.

Example 1. Let $\Omega = \{1, 2, \cdots, N\}, N < \infty$, μ counting measure on Ω, $h = 1, \lambda = 1$. Then $\mathcal{F}_\lambda = \{(p_i)_{i=1}^N, p_i \geq 0, \sum_{i=1}^N p_i = 1\}$. Then

$$f_0(j) = \frac{1}{N}, j = 1, 2, \cdots, N,$$

the uniform distribution on $\{1, 2, \cdots, N\}$, maximizes the relative entropy of the class $\mathcal{F}_\lambda$ with respect to μ.

Example 2. Let $\Omega = \mathbb{N}$, the natural numbers $\{1, 2, \cdots\}$, $\mu =$ the counting measure on Ω, $h(j) = j, j \in \mathbb{N}$. Fix $\lambda, 1 \leq \lambda < \infty$ and let

$$\mathcal{F}_\lambda = \{(p_j)_{j=1}^\infty : \forall\ j,\ p_j \geq 0, \sum_{j=1}^\infty p_j = 1, \sum_{j=1}^\infty j p_j = \lambda\}.$$

Then $f_0(j) = (1-p)p^{j-1}, j = 1, 2, \cdots$, where $p = 1 - \frac{1}{\lambda}$, maximizes the relative entropy of the class $\mathcal{F}_\lambda$ with respect to μ.

Example 3. Let $\Omega = \mathbb{R}$, $\mu =$ Lebesgue measure on $\mathbb{R}$, $h(x) = x^2, 0 < \lambda < \infty$. Set $\mathcal{F}_\lambda = \{f : f \geq 0, \int_\mathbb{R} f(x)dx = 1, \int_\mathbb{R} x^2 f(x)dx = \lambda\}$. Then

$$f_0(x) = \frac{1}{\sqrt{2\pi\lambda}} e^{-\frac{x^2}{2\lambda}},$$

i.e., the Gaussian distribution with mean zero and variance λ, maximizes the relative entropy of the class $\mathcal{F}_\lambda$ with respect to the Lebesgue measure.

These examples are well known (see [5]) and the usual method is by the use of Lagrange's multipliers. The present method extends to the case of arbitrary number of constraints (see [1], [8]).

Definition 2.1. Let $(\Omega, \mathcal{B}, \mu)$ be a measure space. Let $h : \Omega \to \mathbb{R}$ be $\mathcal{B}$ measurable and let c be a real number. Let $\psi(c) = \int_\Omega e^{ch(\omega)} \mu(d\omega)$ be finite. Let

$$\nu_{(\mu,c,h)}(A) = \frac{\int_A e^{ch(\omega)} \mu(d\omega)}{\psi(c)}, A \in \mathcal{B}.$$

The probability measure $\nu_{(\mu,c,h)}$ is called the *Gibbs-Boltzmann measure* corresponding to (μ, c, h).

Example 4. (Spin system on N states.) Let $\Omega = \{-1, 1\}^N$, N a positive integer, and let $V : \Omega \to \mathbb{R}$, $0 < \beta_T < \infty$ be given. Let μ denote the counting measure on Ω. The measure

$$\nu_{(\mu,\beta_T,V)}(A) = \frac{\sum_{\omega \in A} e^{-\beta_T V(\omega)}}{\sum_{\omega \in \Omega} e^{-\beta_T V(\omega)}}, \qquad A \subset \Omega, \qquad (5)$$

is called the *Gibbs distribution* with potential function V and temperature constant β_T for the spin system of N states. The denominator on the right side of (5) is known as the *partition function.*

1.2.2. *Weak Convergence of Gibbs-Boltzmann Distribution*

Let Ω be a Polish space, i.e., a complete separable metric space and let $\mathcal{B}$ denote its Borel σ-algebra. Recall that a sequence $(\mu_n)_{n=1}^{\infty}$ of probability measures on $(\Omega, \mathcal{B})$ converges weakly to a probability measure μ on $(\Omega, \mathcal{B})$, if

$$\int_{\Omega} f(\omega)\mu_n(d\omega) \to \int_{\Omega} f(\omega)\mu(d\omega)$$

for every continuous bounded function $f : \Omega \to \mathbb{R}$. Now let $(\mu_n)_{n=1}^{\infty}$ be a sequence of probability measures on $(\Omega, \mathcal{B})$, $(h_n)_{n=1}^{\infty}$ a sequence of $\mathcal{B}$ measurable functions from Ω to $\mathbb{R}$ and $(c_n)_{n=1}^{\infty}$ a sequence of real numbers. Assume that for each $n \geq 1$, $\int_{\Omega} e^{c_n h_n(\omega)}\mu(d\omega) < \infty$. For each $n \geq 1$, let $\nu_{(\mu_n, c_n, h_n)}$ be the Gibbs-Boltzmann measure corresponding (μ_n, c_n, h_n) as in definition 2.1. An important problem is to find conditions on $(\mu_n, h_n, c_n)_{n=1}^{\infty}$ so that $(\nu_{(\mu_n, c_n, h_n)})_{n=1}^{\infty}$ converges weakly. We address this question in a somewhat special context. We start with some preliminaries.

Let $C \subset \mathbb{R}$ be a compact subset and μ a probability measure on Borel subsets of $\mathbb{R}$ with support contained in C. For $c \in \mathbb{R}$, let

$$\psi(c) = \int_C e^{cx}\mu(dx).$$

Since C is bounded and μ is a probability measure, the function ψ is well defined and infinitely differentiable on $\mathbb{R}$. For any $k \geq 1$,

$$\psi^{(k)}(c) = \int_C e^{cx}x^k\mu(dx).$$

Note that the function $f_c(x) = \frac{e^{cx}}{\psi(c)}$ is a probability density function with respect to μ with mean $\phi(c) = \frac{\psi'(c)}{\psi(c)}$.

Proposition 2.2.

(i) *ϕ is infinitely differentiable and $\phi'(c) > 0$ for all $c \in \mathbb{R}$, provided μ is not supported on a single point. If μ is a Dirac measure, i.e., if μ is supported on a single point, then $\phi'(c) = 0$ for all c.*

(ii) $\lim_{c\to-\infty}\phi(c) = \inf\{x : \mu(-\infty, x) > 0\} \equiv a$,
(iii) $\lim_{c\to+\infty}\phi(c) = \sup\{x : \mu[x, \infty) > 0\} \equiv b$,
(iv) *for any* α, $a < \alpha < b$, *there is a unique* c *such that* $\phi(c) = \alpha$.

Proof: If μ is a Dirac measure then the claims are trivially true, so we assume that μ is not a Dirac measure. Since ψ is infinitely differentiable and $\psi(c) > 0$ for all c, ϕ is also infinitely differentiable. Moreover,

$$\phi'(c) = \frac{(\int_C x^2 e^{cx}\mu(dx))\psi(c) - (\psi'(c))^2}{(\psi(c))^2}$$

can be seen as the variance of a non-constant random variable X_c whose distribution is absolutely continuous with respect to μ with probability density function $f_c(x) = \frac{e^{cx}}{\psi(c)}$. (Note that X_c is non-constant since μ is not concentrated at a single point and f_c is positive on the support of μ.) Thus $\phi'(c) =$ variance of $X_c > 0$, for all c. This proves (i).

Although a direct verification of (ii) is possible we will give a slightly different proof. We will show that as $c \to -\infty$, the random variable X_c converges in distribution to the constant function a so that $\phi(c)$ which is the expected value of X_c converges to a. Note that by definition of a, for all $\epsilon > 0$, $\mu([a, a+\epsilon)) > 0$ while $\mu((-\infty, a)) = 0$. Also $\mu((b, \infty)) = 0$, whence

$$P(X_c > a+\epsilon) = \frac{\int_{(a+\epsilon,b]} e^{cx}\mu(dx)}{\psi(c)} = \frac{\int_{(a+\epsilon,b]} e^{c(x-a)}\mu(dx)}{\int_{[a,b]} e^{c(x-a)}\mu(dx)},$$

For $c < 0$, and $0 < \epsilon < \frac{b-a}{2}$,

$$P(X_c > a+\epsilon) \le \frac{e^{c\epsilon}\mu((a+\epsilon, b])}{e^{c\frac{\epsilon}{2}}\mu([a, a+\frac{\epsilon}{2}))}$$

$$= e^{c\frac{\epsilon}{2}}\frac{\mu((a+\epsilon, b])}{\mu([a, a+\frac{\epsilon}{2}))} \to 0 \quad \text{as} \quad c \to -\infty.$$

Also, since $\mu((-\infty, a)) = 0$, $P(X_c < a) = 0$. So, $X_c \to a$ in distribution as $c \to -\infty$, whence $\phi(c) \to a$ as $c \to -\infty$. This proves (ii). Proof of (iii) is similar. Finally (iv) follows from the intermediate value theorem since ϕ is strictly increasing and continuous with range (a, b). This completes the proof of Proposition 2.2.

Proposition 2.2 also appears at the beginning of the theory of large deviations (see [10]) thus giving a glimpse of the natural connection between large deviation theory and entropy theory. (See Varadhan's interview, p31, [11].)

The requirement that μ have compact support can be relaxed in the above proposition. The following is a result under a relaxed condition.

Let μ be a measure on $\mathbb{R}$. Let $I = \{c : \int_{\mathbb{R}} e^{cx}\mu(dx) < \infty\}$. It can be shown that I is a connected subset of $\mathbb{R}$ which can be empty, a singleton, or an interval that is half open, open, closed, finite, semi-finite or all of $\mathbb{R}$ (see [2]). Suppose I has a non-empty interior I^0. Then in I^0 the function $\psi(c) = \int_{\mathbb{R}} e^{cx}\mu(dx)$ is infinitely differentiable with $\psi^{(k)}(c) = \int_{\mathbb{R}} e^{cx}x^k\mu(dx)$. Further $\phi(c) = \frac{\psi'(c)}{\psi(c)}$ satisfies

$$\phi'(c) = \frac{\psi''(c) - (\psi'(c))^2}{(\psi(c))^2},$$

which is positive, being equal to the variance of a random variable with probability density function $\frac{e^{cx}}{\psi(c)}$ with respect to μ. Thus, for any α satisfying $\inf_{c\in I^0}\phi(c) < \alpha < \sup_{c\in I^0}\phi(c)$, there is a unique c_0 in I^0 such that $\phi(c_0) = \alpha$.

Let μ be a probability measure on $\mathbb{R}$. Note that a real number λ is the mean $\int_{\mathbb{R}} x\nu(dx)$ of a probability measure ν absolutely continuous with respect to μ if and only if $\mu(\{x : x \le \lambda\}), \mu(\{x : x \ge \lambda\})$ are both positive.

As a corollary of Proposition 2.2 we have:

Corollary 2.1. *Let the closed support of μ be a compact set C. Let α be such that $\mu\{x : x \le \alpha\}, \mu\{x : x \ge \alpha\}$ are both positive. Let*

$$\mathcal{F}_\alpha = \{f : f \text{ a pdf}, \int_C xf(x)\mu(dx) = \alpha\}.$$

Then there exists a unique probability density function g with respect to μ such that

$$H(g,\mu) = \max\{H(f,\mu) : f \in \mathcal{F}_\alpha\}.$$

If $\alpha = \inf C$ or if $\alpha = \sup C$, then μ necessarily assigns positive mass to α and $g = \frac{1}{\mu(\{\alpha\})} \times 1_{\{\alpha\}}$. Let $\inf C < \alpha < \sup C$. Then there is a unique c such that with $g = f_c = \frac{e^{cx}}{\int_C e^{cx}\mu(dx)}$ one has $\alpha = \int_C xf_c(x)\mu(dx)$ and

$$H(g,\mu) = H(f_c,\mu) = \max\{H(f,\mu) : f \in \mathcal{F}_\alpha\}.$$

Keeping in mind the notation of the above corollary, and the fact that α uniquely determines c, we write $\nu_{\alpha,\mu}$ to denote the probability measure $f_c d\mu$, i.e., the measure with probability density function f_c with respect to μ. It is also the Gibbs-Boltzmann measure $\nu_{\mu,c,h}$ with $h(x) \equiv x$. We are now ready to state the result on the weak convergence of Gibbs-Boltzmann measures.

Theorem 2.3. *Let C be a compact subset of $\mathbb{R}$. Let $(\mu_n)_{n=1}^{\infty}$ be a sequence of probability measures such that (i) support of each μ_n is contained in C, (ii) $\mu_n \to \mu$ weakly. Let*

$$a = \inf\{x : \mu((-\infty, x)) > 0\}, b = \sup\{x : \mu((x, \infty)) > 0\}.$$

Let $a < \alpha < b$. Then for all large n, ν_{α,μ_n} is well defined and $\nu_{\alpha,\mu_n} \to \nu_{\alpha,\mu}$ weakly.

Proof: Since $\mu_n \to \mu$ weakly, and $a < \alpha < b$, for n large, $\mu_n(-\infty, \alpha) > 0$, $\mu_n(\alpha, \infty) > 0$. So, by Proposition 2.2 it follows that there is a unique c_n such that with $f_{c_n} = \frac{e^{c_n x}}{\int_C e^{c_n x} \mu_n(dx)}$,

$$\int_C x f_{c_n}(x) \mu_n(dx) = \alpha.$$

Thus ν_{α,μ_n} is well defined for all large n. Next we claim that c_n's are bounded. If c_n's are not bounded, then there is a subsequence of $(c_n)_{n=1}^{\infty}$ which diverges to $-\infty$ or to $+\infty$. Suppose a subsequence of $(c_n)_{n=1}^{\infty}$ diverges to $-\infty$. Note that for all $\epsilon > 0$, and $c_n < 0$,

$$\nu_{\alpha,\mu_n}[a+\epsilon, \infty) = \frac{\int_{[a+\epsilon,\infty)} e^{c_n x} \mu_n(dx)}{\int_{\mathbb{R}} e^{c_n x} \mu_n(dx)} \le \frac{e^{c_n \epsilon} \mu_n([a+\epsilon, \infty)}{e^{c_n \frac{\epsilon}{2}} \mu_n([a, a+\frac{\epsilon}{2}))}. \quad (6)$$

Since $\mu_n \to \mu$ weakly, for each $\epsilon > 0$, $\liminf_{n\to\infty} \mu_n([a, a+\epsilon)) > 0$. Therefore, over the subsequence in question, $\nu_{\alpha,\mu_n}([a+\epsilon, \infty)) \to 0$ by (6), and since $\nu_{\alpha,\mu_n}((-\infty, a)) = 0$ we see that $(\nu_{\alpha,\mu_n})_{n=1}^{\infty}$ converges weakly to Dirac measure at a. Since C is compact, this implies that $\int_C x \nu_{x,\mu_n}(dx) \to a$ as $n \to \infty$, contradicting the fact that $\int_C x \nu_{\alpha,\mu_n}(dx) = \alpha > a$, for all n. Similarly, $(c_n)_{n=1}^{\infty}$ is bounded above. So $(c_n)_{n=1}^{\infty}$ is a bounded sequence, which in fact converges as we see below. Let a subsequence $(c_{n_k})_{k=1}^{\infty}$ converge to a real number c. Then, since $\mu_n \to \mu$ weakly, and since all μ_n have support contained in C, a compact set, we see that

$$\int_C e^{c_{n_k} x} \mu_{n_k}(dx) \to \int_C e^{cx} \mu(dx),$$

and

$$\int_C xe^{c_{n_k}x}\mu_{n_k}(dx) \to \int_C xe^{cx}\mu(dx),$$

as $k \to \infty$, whence, $\alpha = \phi(c_{n_k}) \to \phi(c)$, so that $\phi(c) = \alpha$. Again, by Proposition 2.2., c is uniquely determined by α, so that all subsequential limits of $(c_n)_{n=1}^{\infty}$ are the same. So $c_n \to c$ as $n \to \infty$. Clearly then, since C is compact, $f_{c_n} \to f_c$ uniformly on C, so that $\nu_{\alpha,\mu_n} \to \nu_{\alpha,\mu}$ weakly, thus proving Theorem 2.3.

This theorem allows us to answer a question raised in [6]. Let C be a compact subset of $\mathbb{R}$. For each $\epsilon > 0$ let $C_\epsilon = \{x_{\epsilon,1} < x_{\epsilon,2} \cdots < x_{\epsilon,k_\epsilon}\}$ be a ϵ-net in C, i.e., for all $x \in C$, there is a $x_{\epsilon,j}$ such that $| x - x_{\epsilon,j} |< \epsilon$. Fix α such that $\inf C < \alpha < \sup C$. Then for small enough ϵ it must hold that $x_{\epsilon,1} < \alpha < x_{\epsilon,k_\epsilon}$. Let μ_ϵ be the uniform distribution on C_ϵ. Let $\nu_{\mu_\epsilon,\alpha}$ be the Gibbs-Boltzmann distribution on C_ϵ corresponding to μ_ϵ and α. The problem raised in [6] was whether $\nu_{\alpha,\mu_\epsilon}$ converges to a unique limit as $\epsilon \to 0$. Theorem 2.3. above answers this in the negative. Take $C = [0, 1]$, and let $(x_n)_{n=1}^{\infty}$ be a sequence of points in C which become dense in C and such that if μ_n denotes the uniform distribution on the first n points of the sequence, then the sequence $(\mu_n)_{n=1}^{\infty}$ has no unique weak limit. By Theorem 2.3, the associated Gibbs-Boltzmann distributions $(\nu_{\epsilon,\mu_n})_{n=1}^{\infty}$ will also not have a unique weak limit.

(Here is a way of constructing such a sequence $(x_n)_{n=1}^{\infty}$. Let μ_1 and μ_2 be two different probability measures on $[0, 1]$ both equivalent to Lebesgue measure. Let $(X_n)_{n=1}^{\infty}$, $(Y_n)_{n=1}^{\infty}$ be two sequences of points in $[0, 1]$ such that the sequence of empirical distributions based on $(X_n)_{n=1}^{\infty}$ converges weakly to μ_1 and that based on $(Y_n)_{n=1}^{\infty}$ converges weakly to μ_2. Let $(Z_i)_{i=1}^{\infty}$ be defined as follows:

$$Z_i = X_i, 1 \le i \le n_1, Z_i = Y_i, n_1 < i \le n_2, Z_i = X_i, n_2 < i \le n_3, \cdots.$$

One can choose $n_1 < n_2 < n_3 < \cdots$ in such a way that the empirical distribution of $(Z_i)_{i=1}^{\infty}$ converges to μ_1 over the sequence $(n_{2k+1})_{k=1}^{\infty}$ and to μ_2 over the sequence $(n_{2k})_{k=1}^{\infty}$. Since μ_1, μ_2 are equivalent to the Lebesgue measure on $[0, 1]$, the sequence $(Z_n)_{n=1}^{\infty}$ is dense in $[0, 1]$.)

Remarks. See [1] for some further applications of the above discussion. The quantity $H(f, \mu)$ or its negative has been known in statistical literature as Kullback-Leibler information. In financial mathematics, the quantity $-H(f, \mu)$ is called the relative entropy with respect to μ, so then one deals with f minimizing the relative entropy.

1.2.3. *Relative Entropy and Conditioning*

Let $(\Omega, \mathcal{B}, \mu)$ be a probability space, where $(\Omega, \mathcal{B})$ is a standard Borel space (see [7], [9]). For a given countable collection $\{B_n\}_{n=1}^{\infty}$ in $\mathcal{B}$ let $\mathbb{Q}$, $\mathcal{C}$, be the partition and the σ-algebra generated by it. (By partition generated by $\{B_n\}_{n=1}^{\infty}$ we mean the collection:

$$\mathbb{Q} = \{q : q = \cap_{i=1}^{\infty} B_i^{\epsilon_i}\},$$

where $\epsilon_i = 0$ or 1, and $B_i^0 = B_i, B_i^1 = \Omega - B_i$.)

For any subset A of Ω, by saturation of A with respect to $\mathbb{Q}$ we mean the the union of elements of $\mathbb{Q}$ which have non-empty intersection with A.

It is known that $\mathcal{C} = \{C \in \mathcal{B} : C$ a union of elements in $\mathbb{Q}\}$. We regard $\mathcal{C}$ also as a σ-algebra on $\mathbb{Q}$. Note that $\mathcal{C}$-measurable functions are the functions which are $\mathcal{B}$ measurable and which are constant on elements of $\mathbb{Q}$. A $\mathcal{C}$-measurable function on Ω is therefore also a $\mathcal{C}$-measurable function on $\mathbb{Q}$, and if f is such a function, we regard it both as a function on Ω and on $\mathbb{Q}$. We write $f(q)$ to denote the constant value of such a function on $q, q \in \mathbb{Q}$. In addition to $\mathcal{C}$, we also need a larger σ-algebra, denoted by $\mathcal{A}$, generated by analytic subsets $\mathbb{Q}$ (see [9]).

Since $(\Omega, \mathcal{B})$ is a standard Borel space for any probability measure μ on $(\Omega, \mathcal{B})$ there exists a regular conditional probability given $\mathcal{C}$ (equivalently disintegration of μ with respect to $\mathbb{Q}$). This means that there is a function $\mu(\cdot, \cdot)$ on $\mathcal{B} \times \mathbb{Q}$, such that

1) $\mu(\cdot, q)$ is a probability measure $\mathcal{B}$,

2) $\mu(q, q) = 1$,

3) for each $A \in \mathcal{B}$, $\mu(A, \cdot)$ is measurable with respect to $\mathcal{A}$ and

$$\mu(A) = \int_{\Omega} \mu(A \cap q(\omega), \omega)\mu(d\omega) = \int_{\mathbb{Q}} \mu(A, q)\mu \mid_{\mathcal{C}} (dq),$$

where $q(\omega)$ is the element of $\mathbb{Q}$ containing ω,

4) if $\mu'(\cdot, \cdot)$ is another such function then $\mu(\cdot, \omega) = \mu'(\cdot, \omega)$ for a.e. ω (with respect to μ).

Note that sets in $\mathcal{A}$ are measurable with respect to every probability measure on $\mathcal{B}$. Further, we can say that there is a μ-null set N which is a union of elements of $\mathbb{Q}$ and such that (i) $\mathcal{C}\mid_{\Omega-N}$ is a standard Borel structure on $\Omega - N$, (ii) for each $A \in \mathcal{C}\mid_{\Omega-N}$, $\mu(A,\cdot)$ is measurable with respect to this Borel structure.

The function $\mu(\cdot,\cdot)$ is called conditional probability distribution of μ with respect to $\mathbb{Q}$, or, the disintegration of μ with respect to the partition $\mathbb{Q}$. If f is a $\mathcal{B}$ measurable function with finite integral with respect to μ, then the function $h(\omega) = \int_q f(y)\mu(dy,q), \omega \in q$ is called the conditional expectation of f with respect to $\mathbb{Q}$ (or with respect to $\mathcal{C}$) and denoted by $E_\mu(f \mid \mathbb{Q})$ or $E_\mu(f \mid \mathcal{C})$. If $\mathbb{Q}$ is the partition induced by a measurable function g, $E_\mu(f \mid \mathbb{Q})$ is called the conditional expectation of f given g and written $E_\mu(f \mid g)$. (See [9], p. 209)

The measure μ is completely determined by $\mu(\cdot,\cdot)$ together with the restriction of μ to $\mathcal{C}$, denoted by $\mu\mid_{\mathcal{C}}$. We note also that if ν' is any probability measure on $\mathcal{C}$ then $\nu \equiv \int_{\mathbb{Q}} \mu(\cdot,q)\nu'(dq)$ is a measure on $\mathcal{B}$ having the same conditional distribution (or disintegration) with respect to $\mathbb{Q}$ as that of μ.

Let μ and ν be two probability measures on $\mathcal{B}$, with ν absolutely continuous with respect to μ. Let $\mu(\cdot,q), \nu(\cdot,q), q \in \mathbb{Q}$ be the disintegration of μ and ν with respect to the partition $\mathbb{Q}$. Then, for a.e. ω with respect to μ,

$$\frac{d\nu}{d\mu}(\omega) = \frac{d\nu(\cdot,q)}{d\mu(\cdot,q)} \cdot \frac{d\nu\mid_{\mathcal{C}}}{d\mu\mid_{\mathcal{C}}}(\omega), \text{if } \omega \in q.$$

A calculation using this identity shows that

$$H(\frac{d\nu}{d\mu},\mu) = \int_\Omega H(\frac{d\nu(\cdot,q)}{d\mu(\cdot,q)},\mu(\cdot,q))\nu\mid_{\mathcal{C}}(dq) + H(\frac{d\nu\mid_{\mathcal{C}}}{d\mu\mid_{\mathcal{C}}},\mu\mid_{\mathcal{C}}). \tag{7}$$

Assume now that f is a real valued function having finite expectation with respect to μ, and let g be a real valued function on $\mathbb{Q}$ for which there is a probability measure ν, absolutely continuous with respect to μ, such that for all $q \in \mathbb{Q}$,

$$\int_q f(\omega)\nu(d\omega,q) = g(q). \tag{8}$$

(Note that g is necessarily $\mathcal{C}$ measurable.).

Theorem 2.4. *If ν_0 is a probability measure which maximizes $H(\frac{d\nu}{d\mu},\mu)$ among all probability measure ν, absolutely continuous with respect to μ and satisfying (8), then for a.e. q (with respect to $\mu\mid_{\mathcal{C}}$), $\nu_0(\cdot,q)$ maximizes the relative entropy $H(\frac{d\lambda}{d\mu(\cdot,q)},\mu(\cdot,q))$ as λ ranges over all probability measures on q absolutely continuous with respect to $\mu(\cdot,q)$ and satisfying*

$$\int_q f(\omega)\lambda(d\omega)=g(q), q\in\mathbb{Q}.$$

Proof: Assume in order to arrive at a contradiction, that the theorem is false. Then there is a set $E\subset\mathbb{Q}$ of positive $\mu\mid_{\mathcal{C}}$ measure and a transition probability $\lambda(\cdot,\cdot)$ on $\mathcal{B}\times E$ such that for each $q\in E$,

(i) $\lambda(q,q)=1$,

(ii) $\lambda(\cdot,q)$ is absolutely continuous with respect to $\mu(\cdot,q)$,

(iii) $H(\frac{d\nu_0(\cdot,q)}{d\mu(\cdot,q)},\mu(\cdot,q))<H(\frac{d\lambda(\cdot,q)}{d\mu(\cdot,q)},\mu(\cdot,q))$, and

(iv) $\int_q f(\omega)\lambda(d\omega,q)=g(q)$.

The existence of such an E and the transition probability $\lambda(\cdot,\cdot)$ is easy to see if the partition $\mathbb{Q}$ is finite or countable. In the general case the proof relies on some non-trivial measure theory. Define a new transition probability on $\mathcal{B}\times\mathbb{Q}$ as follows: For all $A\in\mathcal{B}$,

$$T(A,q)=\nu_0(A,q) \text{ if } q\in\mathbb{Q}-E, T(A,q)=\lambda(A,q) \text{ if } q\in E.$$

The measure T defined on $\mathcal{B}$ by

$$T(A)=\int_A T(A,q)\nu_0\mid_{\mathcal{C}}(dq),$$

is absolutely continuous with respect to μ, $T(\cdot,\cdot)$ is its disintegration with respect to $\mathbb{Q}$, $T\mid_{\mathcal{C}}=\nu_0\mid_{\mathcal{C}}$ and for each $q\in\mathbb{Q}$

$$\int_q f(\omega)T(d\omega,q)=g(q).$$

Finally, by formula (7), and in view of the values of T on E,

$$H(\frac{dT}{d\mu},\mu)=\int_{\mathbb{Q}} H(\frac{dT(\cdot,q)}{d\mu(\cdot,q)},\mu(\cdot,q))\nu_0\mid_{\mathcal{C}}(dq)+H(\frac{d\nu_0\mid_{\mathcal{C}}}{d\mu\mid_{\mathcal{C}}},\mu\mid_{\mathcal{C}})$$

is strictly bigger than $H(\frac{d\nu_0}{d\mu},\mu)$ which is equal to

$$H(\frac{d\nu_0}{d\mu},\mu) = \int_{\mathbb{Q}} H(\frac{d\nu_0(\cdot,q)}{d\mu(\cdot,q)},\mu(\cdot,q))\nu_0 \mid_{\mathcal{C}} (dq) + H(\frac{d\nu_0 \mid_{\mathcal{C}}}{d\mu \mid_{\mathcal{C}}},\mu \mid_{\mathcal{C}}).$$

This contradicts the maximality of $H(\frac{d\nu_0}{d\mu},\mu)$, and proves the theorem.

Note that the measures $\nu_0(\cdot,q), q \in \mathbb{Q}$, remain unchanged even if ν_0 maximizes $H(\frac{d\nu}{d\mu},\mu)$ under the additional constraint that $\int_\Omega f(\omega)\nu(d\omega) = \alpha$ for some fixed α. However, $\nu_0 \mid_{\mathcal{C}}$ need not maximize $H(\frac{dm}{d\mu|_{\mathcal{C}}},\mu \mid_{\mathcal{C}})$ among all probability measures m on $\mathcal{C}$ satisfying $\int_{\mathbb{Q}} g(q)m(dq) = \alpha$.

1.3. Measure Free Martingales, Weak Martingales, Martingales

1.3.1. *Finite Range Case*

In this section we will discuss the notion of measure free martingales, and and its relation to the usual martingale. In [6] the simpler case of measure free martingale, where functions assume only finitely many values, was introduced and we recall it below.

Let Ω be a non-empty set. Let $(f_n)_{n=1}^\infty$ be a sequence of real valued functions such that each f_n has finite range, say $(x_{n1}, x_{n2}, \cdots, x_{nk_n})$, and these values are assumed on subsets $\Omega_{n1}, \Omega_{n2}, \cdots, \Omega_{nk_n}$. These sets form a partition of Ω which we denote by $\mathbb{P}_n$. We denote by $\mathbb{Q}_n$ the partition generated by $\mathbb{P}_1, \mathbb{P}_2, \cdots, \mathbb{P}_n$ and the algebra generated by $\mathbb{Q}_n$ is denoted by $\mathcal{A}_n$. Let $\mathcal{A}_\infty$ denote the algebra $\cup_{n=1}^\infty \mathcal{A}_n$.

Define $\mathcal{A}_n$ measurable functions m_n, M_n as follows: for $Q \in \mathbb{Q}_n$ and $\omega \in Q$,

$$m_n(\omega) = \min_{x\in Q} f_{n+1}(x),$$

$$M_n(\omega) = \max_{x\in Q} f_{n+1}(x).$$

Definition 3.1. The sequence $(f_n, \mathcal{A}_n)_{n=1}^\infty$ is said to be a measure free martingale or probability free martingale if

$$m_n(\omega) \le f_n(\omega) \le M_n(\omega), \forall \omega \in \Omega,\ n \ge 1.$$

Clearly, for each $Q \in \mathbb{Q}_n$, the function f_n is constant on Q. We denote this constant by $f_n(Q)$. With this notation, it is easy to see that

$(f_n, \mathcal{A}_n)_{n=1}^{\infty}$ is a measure free martingale or probability free martingale if and only if for each n and for each $Q \in \mathbb{Q}_n$, $f_n(Q)$ lies between the minimum and the maximum values of $f_{n+1}(Q')$ as Q' runs over $Q \cap \mathbb{Q}_{n+1}$. It is easy to see that if there is a probability measure on $\mathcal{A}_\infty$ with respect to which $(f_n, \mathcal{A}_n)_{n=1}^{\infty}$ is a martingale then $(f_n, \mathcal{A}_n)_{n=1}^{\infty}$ is also a measure free martingale. Indeed let P be such a measure. Then, for any Q in $\mathbb{Q}_n$, $f_n(Q)$, is equal to

$$\frac{1}{P(Q)} \sum_{\{Q' \in \mathbb{Q}_{n+1}, Q' \subseteq Q\}} f_{n+1}(Q')P(Q'),$$

so that $f_n(Q)$ lies between the minimum and the maximum values $f_{n+1}(Q')$, $Q' \in Q \cap \mathbb{Q}_{n+1}$. In [6], the following converse is proved.

Theorem 3.1. *Given a measure free martingale $(f_n, \mathcal{A}_n)_{n=1}^{\infty}$, there exists for each $n \geq 0$, a measure P_n on $\mathcal{A}_n$ such that*

$$P_{n+1} |_{\mathcal{A}_n} = P_n, \quad E_{n+1}(f_{n+1} \mid \mathcal{A}_n) = f_n,$$

where E_{n+1} denotes the conditional expectation with respect to the probability measure P_{n+1}. There is a finitely additive probability measure P on the algebra $\mathcal{A}_\infty$ such that, for each n, $P |_{\mathcal{A}_n} = P_n$. Moreover such a P is unique if certain naturally occurring entropies are maximized.

1.3.2. *The General Case*

In the rest of this section we will dispense with the requirement that the functions f_n assume only finitely many values.

Let $(\Omega, \mathcal{B})$ be a standard Borel space, and let $(f_n)_{n=1}^{\infty}$ be a sequence of real valued Borel functions on Ω. For each n, let $\mathbb{P}_n = \{f_n^{-1}(\{\omega\}) : \omega \in \mathbb{R}\}$ denote the partition of Ω generated by f_n, and let $\mathbb{Q}_n$ denote the partition generated $f_1, f_2, \cdots, f_n$, i.e., $\mathbb{Q}_n$ is the superposition of $\mathbb{P}_1, \mathbb{P}_2, \cdots, \mathbb{P}_n$. Let $\mathcal{B}_n$ be the σ-algebra generated by $f_1, f_2, \cdots, f_n$. Since $(\Omega, \mathcal{B})$ is a standard Borel space, for each n, $\mathcal{B}_n$ is the collection of sets in $\mathcal{B}$ which can be written as a union of elements in $\mathbb{Q}_n$. For $q \in \mathbb{Q}_n$, f_n is constant on q and we denote this value by $f_n(q)$. The algebra $\cup_{n=1}^{\infty} \mathcal{B}_n$ will be denoted by $\mathcal{B}_\infty$ and we will assume that it generates $\mathcal{B}$. Note that we have changed the notation slightly. In section 1.3.1 above we denoted an element in $\mathbb{Q}_n$ by Q, while from now we will use the lower case q.

Definition 3.2. Let $(f_n)_{n=1}^{\infty}$ be a sequence $\mathcal{B}$ measurable real valued functions on Ω.

(i) Say that sequence $(f_n)_{n=1}^{\infty}$ is a measure free martingale if for each n and for each $q \in \mathbb{Q}_n$, $f_n(q)$ is in the convex hull of the values assumed by f_{n+1} on q (note that f_{n+1} need not be constant on $q \in \mathbb{Q}_n$.)

(ii) Let ν_∞ be a finitely additive probability measure on $\mathcal{B}_\infty$ such that for each $n < \infty$, its restriction to $\mathcal{B}_n$ is countably additive. The sequence $(f_n)_{n=1}^{\infty}$ is called a weak martingale with respect to ν_∞ if for each n, $E_{\nu_{n+1}}(f_{n+1} \mid f_n) = f_n$ *a.e.* ν_{n+1}.

(iii) A weak martingale $(f_n)_{n=1}^{\infty}$ with respect to ν_∞ is a martingale if ν_∞ is countably additive, in which case the countably additive extension of ν_∞ to $\mathcal{B}$ is denoted by ν, and we call $(f_n)_{n=1}^{\infty}$ a martingale with respect to ν.

Clearly every martingale is a weak martingale, and if $(f_n)_{n=1}^{\infty}$ is a weak martingale with respect to ν_∞, then for each n we can modify f_n on a ν_n null set so that the resulting new sequence of functions is a measure free martingale. Indeed if $\nu_n(\cdot,\cdot)$ is the conditional probability distribution of ν_{n+1} given $\mathbb{Q}_n$, then

$$f_n(q) = \int_q f_{n+1}(\omega)\nu_n(d\omega, q),\ \nu_n \ a.\ e.\ q \in \mathbb{Q}_n. \tag{9}$$

At those q's where the equality in (9) holds $f_n(q)$ lies between the infimum and the supremum of the values assumed by f_{n+1} on q. On the other q's we modify f_{n+1} by simply setting $f_{n+1}(\omega) = f_n(q), \omega \in q$. The modified sequence $(f_n)_{n=1}^{\infty}$ is the required measure free martingale. In the converse direction we show that every measure free martingale admits a finitely additive measure on $\mathcal{B}_\infty$ under which it is a weak martingale.

Proposition 3.1. *Let $(\Omega, \mathcal{B})$ be a standard Borel space and let $\mathbb{Q}$, $\mathcal{C}$ be the partition and the σ-algebra generated by a countable collection of sets in $\mathcal{B}$. Let $\mathcal{A}$ be the σ-algebra generated by analytic subsets of Ω which are unions of elements of $\mathbb{Q}$. Let f and g be respectively $\mathcal{B}$ and $\mathcal{C}$ measurable real valued functions on Ω such that for each $q \in \mathbb{Q}$, $g(q)$ is in the convex hull of the values assumed by f on q. Then there exists a transition probability $\nu(\cdot,\cdot)$ on $\mathcal{B} \times \mathbb{Q}$ such that for each $A \in \mathcal{B}$, the function $\nu(A,\cdot)$ is $\mathcal{A}$ measurable, while $\nu(\cdot, q)$ is a probability measure on $\mathcal{B}$ supported on at most two points of q satisfying*

$$g(q) = \int_q f(\omega)\nu(d\omega, q), \forall q \in \mathbb{Q}.$$

Proof: The sets

$$S_1 = \{\omega \in \Omega : f(\omega) \leq g(\omega)\}, S_2 = \{\omega \in \Omega : g(\omega) \leq f(\omega)\}$$

are in $\mathcal{B}$. For each $q \in \mathbb{Q}$, since $g(q)$ is in the convex hull of the values assumed by f on q, both S_1 and S_2 have non-empty intersection with q. By von-Neumann selection theorem (see [9], p 199) there exist coanalytic sets $C_1 \subset S_1, C_2 \subset S_2$ which intersect each $q \in \mathbb{Q}$ in exactly one point. For each $q \in \mathbb{Q}$, let

$$\omega_1(q) = S_1 \cap q, \omega_2(q) = S_2 \cap q.$$

Then

$$f(\omega_1(q)) \leq g(q) \leq f(\omega_2(q)),$$

so that the middle real number $g(q)$ is a unique convex combination of $f(\omega_1(q)), f(\omega_2(q))$. If $f(\omega_1(q)) = f(\omega_2(q)) = g(q)$ write $p_1(q) = 1, p_2(q) = 0$, otherwise write

$$p_1(q) = \frac{f(\omega_2(q)) - g(q)}{f(\omega_2(q)) - f(\omega_1(q))}, p_2(q) = \frac{g(q) - f(\omega_1(q))}{f(\omega_2(q)) - f(\omega_1(q))}.$$

Then

$$p_1(q)f(\omega_1(q)) + p_2(q)f(\omega_2(q)) = g(q).$$

For each $q \in \mathbb{Q}$, let $\nu(\cdot, q)$ be the probability measure on q with masses $p_1(q), p_2(q)$ at $\omega_1(q), \omega_2(q)$ respectively. The sets C_1, C_2 are co-analytic and functions $f \mid_{C_1}, f \mid_{C_2}$ are $\mathcal{B} \mid_{C_1}, \mathcal{B} \mid_{C_2}$ measurable respectively, whence $p_1(\cdot), p_2(\cdot)$ are $\mathcal{A}$ measurable. For any $A \in \mathcal{B}$, and $q \in \mathbb{Q}$,

$$\nu(A, q) = p_1(q)1_A(\omega_1(q)) + p_2(q)1_A(\omega_2(q)),$$

whence, for each $A \in \mathcal{B}$, $\nu(A, \cdot)$ is $\mathcal{A}$ measurable. The proposition is proved.

Theorem 3.2. *A measure free martingale admits a finitely additive measure under which it is weak martingale.*

Proof: Let $(f_n)_{n=1}^{\infty}$ be a measure free martingale. Let $\mathbb{Q}_1, \mathbb{Q}_2, \mathcal{B}_1, \mathcal{B}_2$ be the partition and the σ-algebra generated by f_1, f_2 respectively. Let ν_1 be a probability measure on $\mathcal{B}_1$. Since, for each $q \in \mathbb{Q}_1$, $f_1(q)$ is in the convex hull of the values assumed by f_2 on q, by the Proposition 3.1, there is a transition probability $\nu_1(\cdot,\cdot)$ on $\mathcal{B} \times \mathbb{Q}_1$ such that for each $q \in \mathbb{Q}_1$, $\nu_1(q,q) = 1$, and

$$\int_q f_2(\omega)\nu_1(d\omega, q) = f_1(q).$$

For any $A \in \mathcal{B}_2$, define

$$\nu_2(A) = \int_{\mathbb{Q}_1} \nu_1(A, q)\nu_1(dq).$$

Then ν_2 is a countably additive measure on $\mathcal{B}_2$,

$$\nu_2 \mid_{\mathcal{B}_1} = \nu_1, \quad E_{\nu_2}(f_2 \mid f_1) = f_1.$$

Having defined ν_2, it is now clear how to construct $\nu_3, \nu_4, \cdots, \nu_n, \cdots$ such that for each n,

$$\nu_{n+1} \mid_{\mathcal{B}_n} = \nu_n$$

and

$$E_{\nu_{n+1}}(f_{n+1} \mid f_n) = f_n.$$

The finitely additive measure ν_∞ defined on $\mathcal{B}_\infty$ by

$$\nu_\infty(A) = \nu_n(A), A \in \mathcal{B}_n$$

satisfies, for each n,

$$\nu_\infty \mid_{\mathcal{B}_n} = \nu_n$$

and $(f_n)_{n=1}^\infty$ is thus a a weak martingale with respect to ν_∞. This proves the theorem.

It is natural to ask the question as to when is ν_∞ countably additive. There is an answer to this. The refining system of partitions $(\mathbb{Q}_n)_{n=1}^\infty$ as well as the associated σ-algebras $(\mathcal{B}_n)_{n=1}^\infty$ is called a filtration. It is said to be complete if for every decreasing sequence $(q_n)_{n=1}^\infty$, of non-empty elements, $q_n \in \mathbb{Q}_n$, their intersection $\cap_{n=1}^\infty q_n$ is non-empty. We have:

Theorem 3.3. *If the filtration $(\mathbb{Q}_n)_{n=1}^\infty$ associated with the measure free martingale is complete, then ν_∞ is countably additive.*

This is indeed a consequence of the Kolmogorov consistency theorem formulated in terms of filtrations which is as follows. Let the filtration $(\mathbb{Q}_n)_{n=1}^{\infty}$ arise from a sequence of Borel functions $(f_n)_{n=1}^{\infty}$, not necessarily a measure free martingale. For $1 \leq i \leq j$, we define the natural projection map $\Pi_{ij} : \mathbb{Q}_j \to \mathbb{Q}_i$ by

$$\Pi_{ij}(q) = r, \quad \text{if } q \subset r, q \in \mathbb{Q}_j, r \in \mathbb{Q}_i.$$

For each i let $\Pi_i : \Omega \to \mathbb{Q}_i$ be defined by

$$\Pi_i(\omega) = q \in \mathbb{Q}_i, \text{if } \omega \in q.$$

If $q \in \mathbb{Q}_n$ and $i \leq n$, then f_i is constant on q, and we write $f_i(q)$ to denote this constant value. For each n let τ_n be the smallest topology on $\mathbb{Q}_n$ which makes the map $q \to (f_1(q), f_2(q), \cdots, f_n(q)), q \in \mathbb{Q}_n$, continuous. We note that maps $\Pi_{ij}, i \leq j$ are continuous. The topology τ_n generates the σ-algebra $\mathcal{B}_n$. Any probability measure μ on $\mathcal{B}_n$ is compact approximable with respect to this topology, i.e., given $B \in \mathcal{B}_n$ and $\epsilon > 0$, there is a set $C \subset B$, C compact w.r.t. τ_n, such that $\mu(B - C) < \epsilon$. For each n, let P_n be a countably additive probability measure on $\mathcal{B}_n$. Assume that $P_{n+1}|_{\mathcal{B}_n} = P_n$. Define P_∞ on $\cup_{n=1}^{\infty} \mathcal{B}_n$ by $P(A) = P_n(A)$, if $A \in \mathcal{B}_n$. P is obviously finitely additive.

We have the Kolmogorov consistency theorem in our setting, arrived at after a discussion with B. V. Rao, and as pointed out by Rajeeva Karandikar, it is proved also in [7].

Theorem 3.4. *If the the filtration $(\mathbb{Q}_n)_{n=1}^{\infty}$ is complete, then P_∞ is countably additive.*

Proof: For simplicity, write P for P_∞. If P is not countably additive, then there exists a decreasing sequence $(A_n)_{n=1}^{\infty}$ in $\cup_{n=1}^{\infty} \mathcal{B}_n$, with $\cap_{n=1}^{\infty} A_n = \emptyset$, such that for all n, $P(A_n) > a > 0$ for some positive real a. Without loss of generality we can assume that for each n, A_n is in $\mathcal{B}_n$. We can choose a set $C_1 \subset A_1$, $C_1 \in \mathcal{B}_1$, C_1 compact with respect to the topology τ_1, such that

$$P(A_1 - C_1) = P_1(A_1 - C_1) < \frac{a}{4}.$$

Note that

$$A_2 \cap C_1 \in \mathcal{B}_2, P(A_2 \cap (A_1 - C_1) \leq P(A_1 - C_1) < \frac{a}{4},$$

whence

$$P(A_2 \cap C_1) > \frac{3}{4}a.$$

We next choose $C_2 \subset A_2 \cap C_1$, $C_2 \in \mathcal{B}_2$, C_2 compact in the topology τ_2, such that $P(A_2 \cap C_1 - C_2) < \frac{a}{4^2}$. Then $P(A_2 - C_2) \leq \frac{a}{4} + \frac{a}{4^2}$. Note that

$$A_3 \cap C_2 \in \mathcal{B}_3, \quad P(A_3 \cap C_2) > a - (\frac{a}{4} + \frac{a}{4^2}).$$

Proceeding thus we get a decreasing sequence $(C_n)_{n=1}^{\infty}$ such that for all n, $C_n \in \mathcal{B}_n$, $C_n \subset A_n$, C_n compact in the topology on τ_n, and

$$P(A_n - C_n) < \frac{a}{4} + \frac{a}{4^2} + \cdots + \frac{a}{4^n}.$$

Clearly each C_n is non-empty. For each n choose an element q_n in C_n. Since C_n is compact the sequence $(\Pi_{nj}q_j)_{j=n}^{\infty}$ has a subsequence converging to a point in C_n. By Cantor's diagonal procedure it is possible to choose the sequence $(q_n)_{n=1}^{\infty}$ in such a way that for each i, $(\Pi_{ij}q_j)_{j=i}^{\infty}$ is convergent in the topology τ_i to an element p_i in $\mathbb{Q}_i$. By continuity of the map Π_{ij} we have $\Pi_{ij}p_j = p_i$ if $i \leq j$, i.e., if $i \leq j$ then $p_j \subset p_i$. By completeness of the filtration we conclude that $\cap_{i=1}^{\infty} p_i \neq \emptyset$. But

$$\cap_{i=1}^{\infty} p_i \subset \cap_{i=1}^{\infty} C_i \subset \cap_{i=1}^{\infty} A_i = \emptyset.$$

The contradiction proves the theorem.

Remark. (i) The requirement that the filtration be complete has been crucial in the above discussion. Here is an example due to S. M. Srivastava of a filtration on the real line which is not complete, but the quotient topologies are locally compact second countable. Let $\Omega = \mathbb{R}$, and let, for $n \geq 1$, $\mathbb{Q}_n = \{\{r\}, r < n, [n, \infty)\}$, i.e., the nth partition $\mathbb{Q}_n$ consists of all singletons less than n together with the interval $[n, \infty)$). Clearly $(\mathbb{Q}_n)_{n=1}^{\infty}$ is a filtration. For each n, the quotient topology on $\mathbb{Q}_n$ is isomorphic to the usual topology on $\mathbb{R}$. The set $C_n = [n, \infty)$ is compact in the quotient topology, $C_{n+1} \subset C_n$, but $\cap_{n=1}^{\infty} C_n$ is empty, so the filtration is not complete.

(ii) S. Bochner ([3]) has formulated and proved the Kolmogorov consistency theorem for projective families. One can derive the above version by proper identification of our sets and maps as a projective system, once topologies τ_n are described.

(iii) The totality of finitely additive measure on $\mathcal{B}_{\infty}$ which render the measure free martingale $f_n, n = 1, 2, 3, \cdots$ into a weak martingale is a convex

set whose extreme points are precisely those ν_∞ for which ν_1 is a point mass, and for which, for each n, the disintegration $\nu_n(\cdot,\cdot)$ of ν_{n+1} with respect to $\mathbb{Q}_n$ has the property that for each $q \in \mathbb{Q}_n$, $\nu(\cdot, q)$ is supported on at most two points.
(iv) We note that measure free martingales have some nice properties not shared by the usual martingales. If $f_n, n = 1, 2, 3, \cdots$ is a measure free martingale, then $[f_n], n = 1, 2, 3, \cdots$, where $[x]$ means the integral part of x, and $min\{f_n, K\}, k = 1, 2, 3, \cdots$, where K is fixed real number are also measure free martingale. In other words, measure free martingales are closed under discretization and truncation.

1.4. Equivalent Martingale Measures

Let $(\Omega, \mathcal{B})$ be a standard Borel space and let μ be a probability measure on $\mathcal{B}$. Let $(f_n)_{n=1}^\infty$ be a sequence of Borel measurable real valued functions on Ω, not necessarily a measure free martingale. In this section we discuss conditions, necessary as well as sufficient, for there to exist a measure ν, having the same null sets as μ, and which renders the sequence $(f_n)_{n=1}^\infty$ a martingale. Clearly, if such ν exists then we can modify $(f_n)_{n=1}^\infty$ on a ν-null set, which is therefore also μ-null, so that the new sequence of functions is a measure free martingale. Thus a necessary condition for the existence of a ν, equivalent to μ, under which $(f_n)_{n=1}^\infty$ is a martingale is that $(f_n)_{n=1}^\infty$ admit a modification on a μ-null set so that the resulting sequence is a measure free martingale.

Again assume that such a ν exists. For each n, let μ_n, ν_n respectively be the restriction of μ, ν to $\mathcal{B}_n$. Let $\mu_n(\cdot,\cdot), \nu_n(\cdot,\cdot)$ denote the disintegration of μ_{n+1}, ν_{n+1} with respect to the partition $\mathbb{Q}_n$. Since μ_{n+1} and ν_{n+1} have the same null sets, for μ_n almost every $q \in \mathbb{Q}_n$, $\mu_n(\cdot, q)$ and $\nu_n(\cdot, q)$ have the same null sets, and since

$$\int_q f_{n+1}(\omega)\nu_n(d\omega, q) = f_n(q),$$

it follows that for μ_n a.e. q

$$\mu_n(\{\omega \in q : f_{n+1}(\omega) \le f_n(q)\}, q) > 0, \mu_n(\{\omega \in q : f_{n+1}(\omega) \ge f_n(q)\}, q) > 0. \tag{11}$$

Thus, if a ν equivalent to μ under which $(f_n)_{n=1}^\infty$ is a martingale exists, then for each n, for μ_n a.e. $q \in \mathbb{Q}_n$, (11) holds.

Fix n, fix a $q \in \mathbb{Q}_n$, and write $m = \mu_n(\cdot, q)$. Assume that

$$\int_q | f_{n+1}(\omega) | m(d\omega) < \infty,$$

and that

$$m(\{\omega \in q : f_{n+1}(\omega) < f_n(q)\}) > 0, m(\{\omega \in q : f_{n+1}(\omega) > f_n(q)\}) > 0.$$

Write

$$E = \{\omega \in q : f_{n+1}(\omega) < f_n(q)\}, F = \{\omega \in q : f_{n+1}(\omega) > f_n(q)\},$$

$$m(E) = \alpha, m(F) = \beta,$$

$$c = \frac{\int_E f_{n+1}(\omega) m(d\omega)}{\alpha}, d = \frac{\int_F f_{n+1}(\omega) m(d\omega)}{\beta}.$$

Note that $c < f_n(q) < d$ and with $a = \frac{d - f_n(q)}{d-c}, b = \frac{f_n(q) - c}{d-c}$, we have

$$a + b = 1, ac + bd = f_n(q).$$

Define $\nu_n'(\cdot, \cdot)$ as follows:

$$\frac{d\nu_n'(\cdot, q)}{dm} = (\frac{a}{\alpha} 1_E + \frac{b}{\beta} 1_F).$$

Note here that q ranges over $\mathbb{Q}_n$, so that m will vary with it. Further, $a, \alpha, b, \beta, m = \mu_n(\cdot, q)$ are measurable functions of q, so that $\nu_n'(\cdot, \cdot)$ is a transition probability. If $m(\{\omega \in q : f_{n+1}(\omega) = f_n(q)\}) = 0$, then $\nu_n'(\cdot, q)$ and m are equivalent, and

$$\int_q f_{n+1}(\omega) \nu_n'(d\omega, q) = ac + bd = f_n(q).$$

In any case, if we set $s = \{\omega \in q : f_{n+1}(\omega) = f_n(q)\}$, then $s \in \mathbb{Q}_{n+1}$, so that we can speak of Dirac measure $\delta_{\{s\}}$ at s, and consider the measure $\nu_n(\cdot, q)$ defined by:

$$\nu_n(\cdot, q) = (1 - m(s)) \nu_n'(\cdot, q) + \delta_{\{s\}} m(s)$$

The measure $\nu_n(\cdot, q)$ and m have the same null sets, $\nu_n(\cdot, q)$ is measurable in q, and

$$\int_q f_{n+1}(\omega) \nu_n(d\omega, q) = f_n(q)$$

We define inductively,

$$\nu_1 = \mu_1, \nu_2 = \int_\Omega \nu_1(\cdot, q)\nu_1(dq), \cdots, \nu_{n+1}(\cdot) = \int_\Omega \nu_n(\cdot, q)\nu_n(dq).$$

Then for each n, ν_n is a probability measure on $\mathcal{B}_n$, equivalent to μ_n,

$$\nu_{n+1} \mid_{\mathcal{B}_n} = \nu_n, \quad E_n(f_{n+1} \mid \mathcal{B}_n) = f_n,$$

where E_n is the conditional expectation operator in $(\Omega, \mathcal{B}_{n+1}, \nu_{n+1})$.

If the filtration $(\mathbb{Q}_n)_{n=1}^\infty$ is complete, the naturally defined measure ν_∞ on $\mathcal{B}_\infty$ has a countably additive extension, say ν, to all of $\mathcal{B}$. However, in general, ν need not be equivalent to μ (see example 4.1. below). If there are positive constants A and B such that for all n, $A < \frac{d\nu_n}{\mu_n} < B$, then clearly the ν_∞ defined on $\mathcal{B}_\infty$ has an extension to $\mathcal{B}$ which has the same null sets as μ. We have proved:

Theorem 4.1.

(a) *Let $(f_n)_{n=1}^\infty$ be a sequence of Borel measurable functions on $(\Omega, \mathcal{B}, \mu)$ and let $\mathbb{Q}_n, \mathcal{B}_n, \mu_n, \mu_n(\cdot,\cdot)$ be as above. If there is a probability measure ν equivalent to μ with respect to which $(f_n)_{n=1}^\infty$ is a martingale, then $(f_n)_{n=1}^\infty$ can be modified on a μ-null set so that the resulting sequence of functions is a measure free martingale. Further, for each n, for almost every $q \in \mathbb{Q}_n$, the sets $\{\omega \in q : f_{n+1}(\omega) \leq f_n(q)\}, \{\omega \in q : f_n(q) \geq f_{n+1}(\omega)\}$ have positive $\mu_n(\cdot, q)$- measure.*

(b) *If for every n, and for almost every $q \in \mathbb{Q}_n$, the sets $\{\omega \in q : f_{n+1}(\omega) \leq f_n(q)\}, \{\omega \in q : f_n(q) \leq f_{n+1}(\omega)\}$ have positive $\mu_n(\cdot, q)$-measure, then for each n we have a ν_n equivalent to μ_n such that $\nu_{n+1} \mid_{\mathcal{B}_n} = \nu_n$, $E_{n+1}(f_{n+1} \mid \mathcal{B}_n) = f_n$, where E_{n+1} stands for the conditional expectation operator on $(\Omega, \mathcal{B}_{n+1}, \nu_{n+1})$ with respect to $\mathcal{B}_n$.*

(c) *Finally, if there are positive constants A and B such that for all n, $A < \frac{d\nu_n}{d\mu_n} < B$, then the naturally defined ν_∞ on $\mathcal{B}_\infty$ has an extension ν to $\mathcal{B}$ which has the same null sets as μ and $(f_n)_{n=1}^\infty$ is a martingale with respect to ν.*

The equivalent martingale measure ν in the above theorem is obtained by a rather naive modification of μ. Indeed, $\mu_n(\cdot, q)$ is only rescaled over the sets $\{\omega \in q : f_{n+1}(\omega) < f_n(q)\}, \{\omega \in q : f_n(q) > f_{n+1}(\omega)\}$, so that if $\mu_n(\cdot, q)$ is not 'well distributed ' on these sets, then this persists with $\nu_n(\cdot, q)$. This circumstance can be changed if we assume that for each n and for each $q \in \mathbb{Q}_n$, f_{n+1} is bounded on q, in addition to the requirement that sets $\{\omega \in q : f_{n+1}(\omega) \leq f_n(q)\}, \{\omega \in q : f_n(q) \geq f_{n+1}(\omega)\}$ have positive $\mu_n(\cdot, q)$ measure. We know from entropy considerations of section 1.1 that there exists a unique $c_n = c_n(q)$ such that

$$\frac{\int_q f_{n+1}(\omega) e^{c_n(q) f_{n+1}(\omega)} \mu_n(d\omega, q)}{\int_q e^{c_n(q) f_{n+1}(\omega)} \mu_n(d\omega, q)} = f_n(q).$$

The function $q \to c_n(q)$ is $\mathcal{B}_n$ measurable. We set, for each n and for each $q \in \mathbb{Q}_n$,

$$d\nu_n(\cdot, q) = \frac{e^{c_n(q) f_{n+1}(\omega)}}{\int_q e^{c_n(q) f_{n+1}(\omega)} \mu_n(d\omega, q)} \cdot d\mu_n(\cdot, q),$$

$$\nu_1 = \mu_1, \nu_{n+1} = \int_\Omega \nu_n(\cdot, q) \nu_n(dq).$$

We change notation and write $\nu_n = B_n$. The natural finitely additive measure B_∞ on $\mathcal{B}_\infty$ renders $(f_n)_{n=1}^\infty$ into a weak martingale. If we assume that there are positive constants A, C such that for all n, $A < \frac{dB_n}{d\mu_n} < C$, then B_∞ extends to a countably additive measure B on $\mathcal{B}$, B equivalent to μ, and, $(f_n)_{n=1}^\infty$ is a martingale with respect to B.

We may summarise this as

Theorem 4.2. *If for each n, for μ_n a.e. $q \in \mathbb{Q}_n$,*

(i) *f_{n+1} is bounded on q,*

(ii) $\mu(\{\omega \in q : f_{n+1}(\omega) \leq f_n(q)\}, q) > 0, \mu(\{\omega \in q : f_{n+1}(\omega) \geq f_n(q)\}, q) > 0$,

then there exists a unique finitely additive measure B_∞ on $\mathcal{B}_\infty$ such that for each n,

(a) *the restriction B_n of B_∞ to $\mathcal{B}_n$ is countably additive and equivalent to μ_n,*

(b) *B_{n+1} maximizes the relative entropy $H(\frac{d\lambda_{n+1}}{d\mu_{n+1}}, \mu_{n+1})$ among all finitely additive probability measures λ on $\mathcal{B}_\infty$ which render $(f_n)_{n=1}^\infty$ into a weak martingale and such that $\lambda_1 = \mu_1$, λ_n equivalent to μ_n for all n,*

(c) *if there exist constants A, B such that for each n, $A < \frac{dB_n}{d\mu_n} < B$, then B_∞ extends to a countably additive measure measure B on $\mathcal{B}$, B and ν are equivalent, and $(f_n)_{n=1}^\infty$ is a martingale under B.*

Remarks. 1) The measure B, however, need not maximize the relative entropy $H(\frac{d\lambda}{d\mu}, \mu)$ among all measures λ on $\mathcal{B}$ equivalent to μ and under which $(f_n)_{n=1}^\infty$ is a martingale.
2) One may call B a Boltzmann measure equivalent to μ and the associated sequence $(f_n)_{n=1}^\infty$ a Boltzmann martingale.

We now give a more general condition for the existence of a martingale measure equivalent to a given μ than the one given in Theorem 4.1. (c). Let $(f_n)_{n=1}^\infty$ be a sequence of measurable functions on $(\Omega, \mathcal{B}, \mu)$ for which there exists a measure ν_∞ on $\mathcal{B}_\infty$ such that (i) $(f_n)_{n=1}^\infty$ is a weak martingale with respect to ν_∞, (ii) for each n, μ_n and ν_n are equivalent, (iii) $\mu_1 = \nu_1$.

Now the Radon-Nikodym derivative $\frac{d\nu_n}{d\mu_n}$ is computed as follows:

$$\frac{d\nu_{n+1}}{d\mu_{n+1}}(\omega) = \frac{d\nu_n(\cdot, q_n)}{d\mu_n(\cdot, q_n)}(\omega)\frac{d\nu_n}{d\mu_n}(\omega), \omega \in q_n \in \mathbb{Q}_n.$$

So, on iteration, we have:

$$\frac{d\nu_{n+1}}{\mu_{n+1}}(\omega) = (\Pi_{i=1}^n \frac{d\nu_i(\cdot, q_i)}{d\mu_i(\cdot, q_i)}(\omega)) \times \frac{d\nu_1}{d\mu_1}, \omega \in q_i \in \mathbb{Q}_i, i = 1, 2, \cdots, n.$$

Since we have chosen $\nu_1 = \mu_1$, $\frac{d\nu_1}{d\mu_1}(\omega) = 1$ for all ω, so that

$$\frac{d\nu_{n+1}}{\mu_{n+1}}(\omega) = \Pi_{i=1}^n \frac{d\nu_i(\cdot, q_i)}{d\mu_i(\cdot, q_i)}(\omega), \omega \in q_i \in \mathbb{Q}_i, i = 1, 2, \cdots, n.$$

Further,

$$E_\mu(\frac{d\nu_{n+1}}{d\mu_{n+1}} \mid \mathcal{B}_n)(\omega) = \frac{d\nu_n}{d\mu_n}(\omega)$$

where E_μ denotes the conditional expectation operator with respect to μ.

Indeed for any set $A \in \mathcal{B}_n$,

$$\begin{aligned}\int_A \frac{d\nu_{n+1}}{d\mu_{n+1}}(\omega)\mu(d\omega) &= \int_A \frac{d\nu_n}{d\mu_n}(\omega)\frac{d\nu_n(\omega,q)}{d\mu_n(\omega,q)}d\mu \\ &= \int_A \frac{d\nu_n}{d\mu_n}(\omega)\frac{d\nu_n(\omega,q)}{d\mu_n(\omega,q)}d\mu_{n+1} \\ &= \int_A \frac{d\nu_n}{d\mu_n}(\omega)(\int_q \frac{d\nu_n(\omega,q)}{d\mu_n(\omega,q)}\mu_n(d\omega,q))\mu_n(d\omega) \\ &= \int_A \frac{d\nu_n}{d\mu_n}(\omega)\mu_n(d\omega) \\ &= \int_A \frac{d\nu_n}{d\mu_n}(\omega)\mu(d\omega).\end{aligned}$$

The sequence of functions $(g_n = \frac{d\nu_n}{d\mu_n})_{n=1}^{\infty}$ is therefore a martingale of non-negative functions on the probability space $(\Omega, \mathcal{B}, \mu)$, so converges μ a.e. to a function g. If $\int_\Omega g(\omega)\mu(d\omega) = 1$, or if the sequence $(g_n)_{n=1}^{\infty}$ is uniformly integrable, then, by martingale convergence theorem (see [4], p. 319), for each n, $g_n = E_\mu(g \mid \mathcal{B}_n)$, equivalently, for each n, $gd\mu \mid_{\mathcal{B}_n} = d\nu_n$ and so ν_∞ is countably additive and extends to a measure ν on $\mathcal{B}$, with $\frac{d\nu}{d\mu} = g$. Further $(f_n)_{n=1}^{\infty}$ is a martingale with respect to it. If, in addition, $g > 0$ a.e. μ, then ν is equivalent to μ and $(f_n)_{n=1}^{\infty}$ is a martingale with respect to it.

We have proved:

Theorem 4.3.

(a) *The sequence* $(g_n = \frac{d\nu_n}{d\mu_n})_{n=1}^{\infty}$ *of Radon-Nikodym derivatives is a martingale of non-negative functions and converges* μ *a.e. to a function* g.

(b) *If* $\int_\Omega g(\omega)\mu(d\omega) = 1$ *or if the* g_n*'s are uniformly integrable with respect to* μ, *then* $(f_n)_{n=1}^{\infty}$ *is a martingale with respect to* ν *given by* $d\nu = gd\mu$. *If in addition* $g > 0$ *a.e.* μ, ν *is equivalent to* μ.

(c) *If* $\sum_{n=1}^{\infty} \mid 1 - \frac{d\nu_n(\cdot,q)}{d\mu_n(\cdot,q)} \mid < \infty$ a.s. μ *then* $g > 0$ *a.e.* μ.

Write

$$H(\frac{d\nu_{n+1}}{d\mu_{n+1}}, \mu_{n+1}) \mid \mathbb{Q}_n) = \int_{\mathbb{Q}_n} H(\frac{d\nu_n(\cdot,q)}{d\mu_n(\cdot,q)}, \mu_n(\cdot,q))\nu_n(dq).$$

We know from formula (7) of section 1.2.3 that

$$H(\frac{d\nu_{n+1}}{d\mu_{n+1}}, \mu_{n+1}) = H(\frac{d\nu_{n+1}}{d\mu_{n+1}}, \mu_{n+1}) \mid \mathbb{Q}_n) + H(\frac{d\nu_n}{d\mu_n}, \mu_n).$$

Iterating we get

$$H(\frac{d\nu_{n+1}}{d\mu_{n+1}}, \mu_{n+1}) = \sum_{i=1}^{n+1} H(\frac{d\nu_i}{d\mu_i}, \mu_i) \mid \mathbb{Q}_{i-1}).$$

(Here, when $i = 1$, $\mathbb{Q}_{i-1} = \mathbb{Q}_0$ which we take to be the trivial partition $\{\emptyset, \Omega\}$.)

Since $(g_n = \frac{d\nu_n}{d\mu_n})_{n=1}^{\infty}$ is a martingale with respect to μ, the sequence $(g_n \log^+ g_n)_{n=1}^{\infty}$ is a submartingale provided, for each n, $E_\mu(g_n \log^+ g_n)$ is finite ([4], p. 296). We assume that this is the case. Then

$$E_\mu(g_n \log^+ g_n) \leq E_\mu(g_{n+1} \log^+ g_{n+1}), n = 1, 2, \cdots,$$

so that $\lim_{n\to\infty} E_\mu(g_n \log^+ g_n)$ exists, which may be finite or infinite. We assume that this limit is finite, say c. Since $(g_n \log^+ g_n)_{n=1}^{\infty}$ is a submartingale of non-negative functions it has a limit which is indeed $g \log^+ g$, since $(g_n)_{n=1}^{\infty}$ has limit g, μ a.e. Moreover by Fatou's lemma, $E_\mu(g \log^+ g) \leq c$. Assume that the sequence $(g_n)_{n=1}^{\infty}$ is uniformly integrable so that this sequence together with g forms a martingale. From martingale theory ([4], p. 296) the sequence $(g_n \log^+ g_n)_{n=1}^{\infty}$ together with the function $g \log^+ g$ is a submartingale of non-negative functions, so, again from martingale theory ([4], p. 325) we conclude that

$$\lim_{n\to\infty} E_\mu(g_n \log g_n^+) = E_\mu(g \log^+ g).$$

Since $g_n \log^- g_n$ remains bounded independent of n (which is the case at ω where $g_n(\omega) \leq 1$), we also have

$$\lim_{n\to\infty} \int_\Omega g_n(\omega) \log^- g_n(\omega) \mu_n(d\omega) = \int_\Omega g \log^- (\omega) \mu(d\omega).$$

Thus we have

$$\lim_{n\to\infty} H(\frac{d\nu_n}{d\mu_n}, \mu_n) = H(\frac{d\nu}{d\mu}, \mu).$$

We have proved:

Theorem 4.4.

(a) *Assume that the martingale $(g_n = \frac{d\nu_n}{d\mu_n})_{n=1}^{\infty}$ is uniformly integrable and that $\int_\Omega g_n \log^+ g_n \mu_n(d\omega) \leq c$, for some real number c. Then $\lim_{n\to\infty} H(\frac{d\nu_n}{d\mu_n}, \mu_n)$ exists and we have:*

$$\lim_{n\to\infty} H(\frac{d\nu_n}{d\mu_n}, \mu_n) = \sum_{i=1}^{\infty} H(\frac{d\nu_i}{d\mu_i}, \mu_i) \mid \mathbb{Q}_{i-1}) = H(\frac{d\nu}{d\mu}, \mu).$$

(b) *In addition to the hypothesis and notations of Theorem 4.2, assume that the martingale* $(\frac{dB_n}{d\mu_n})_{n=1}^{\infty}$ *is uniformly integrable and that the sequence of relative entropies* $(H(\frac{dB_n}{d\mu_n}, \mu_n))_{n=1}^{\infty}$ *remains bounded. Then* $\lim_{n\to\infty} H(\frac{dB_n}{d\mu_n}, \mu_n)$ *exists and we have:*

$$\lim_{n\to\infty} H(\frac{dB_n}{d\mu_n}, \mu_n) = \sum_{n=1}^{\infty} H(\frac{dB_n}{d\mu_n}, \mu_n) \mid \mathbb{Q}_{n-1}) = H(\frac{dB}{d\mu}, \mu).$$

Among all ν *absolutely continuous with respect to* μ *under which* $(f_n)_{n=1}^{\infty}$ *is a martingale and* $\nu_1 = \mu_1$, B *is the unique one which maximizes, for each* n, *the relative entropy entropy* $H(\frac{d\nu_n}{d\mu_n}, \mu_n)$.

Example 4.1. Consider $\mathbb{R}^2$ together with the measure $\mu = \sigma \times \sigma$ where σ is the normal distribution with mean zero and variance one. Let $\mathbb{Q}_1$ be the partition $\{x\} \times \mathbb{R}, x \in \mathbb{R}$. Let $f_i, i = 1, 2$ be the co-ordinate maps. The partition of $\mathbb{R}^2$ given by f_1 is $\mathbb{Q}_1$. The distribution ν on $\mathbb{R}^2$ equivalent to μ, satisfying $E_\nu(f_2 \mid f_1) = E_\nu(f_2 \mid \mathbb{Q}_1) = f_1$, and maximizing the relative entropy with respect to μ is the bivariate distribution of $(f_1, f_1 + f_2)$, where f_1, f_2 are independent with normal distribution of mean zero and variance one.

More generally, let $\mathbb{R}^n$ be given the measure $\mu = \sigma^n$, the n-fold product of σ. Let $f_1, f_2, \cdots, f_n$ be the co-ordinate random variables. Then

$$\mathbb{Q}_i = \{\{\omega_1, \omega_2, \cdots, \omega_i\} \times \mathbb{R}^{n-i} : (\omega_1 . \omega_2, \cdots, \omega_i) \in \mathbb{R}^i\}$$

is the partition of $\mathbb{R}^n$ given by the functions $f_1, f_2, \cdots, f_i$. Let ν_n be the measure induced on $\mathbb{R}^n$ by the vector random variable $(f_1, f_1 + f_2, \cdots, f_1 + f_2 + \cdots + f_n)$ where $(f_1, f_2, \cdots . f_n)$ has distribution $\mu = \sigma^n$. Then, among all probability measures λ on $\mathbb{R}^n$ equivalent to μ and satisfying $E_\lambda(f_{i+1} \mid f_i) = f_i, 1 \le i \le n-1$, ν_n is the unique one which simultaneously maximizes the relative entropies

$$-\int_{\mathbb{R}^i} \frac{d\lambda_i}{d\mu_i}(\omega) \log \frac{d\lambda_i}{d\mu_i}(\omega) \mu_i(d\omega),$$

$1 \le i \le n$, where μ_i, λ_i are respectively the measures μ and λ restricted to the the σ-algebra $\mathcal{B}_i$ generated $f_1, f_2, \cdots, f_i$.

Finally, let (i) $\Omega = \mathbb{R}^{\mathbb{N}}$, where $\mathbb{N}$ is the set of natural numbers, (ii) $\mu =$ countable product of σ with itself, (iii) for each n, $f_n =$ projection on the

nth co-ordinate space. The partition $\mathbb{Q}_n$ of Ω generated by $f_1, f_2, \cdots, f_n$ is the collection

$$\{\{(\omega_1, \omega_2, \cdots, \omega_n)\} \times \mathbb{R}^{\{n+1, n+2, \cdots\}} : (\omega_1, \omega_2, \cdots, \omega_n) \in \mathbb{R}^n.\}$$

For each n, let ν_n denote the measure on $\mathcal{B}_n$, induced by the martingale $f_1, f_1+f_2, \cdots, \sum_{i=1}^{n} f_i)$, where $f_1, f_2, \cdots, f_n$ are independent random variable, each with distribution σ. Let ν_∞ be the measure on the algebra $\cup_{n=1}^{\infty} \mathcal{B}_n$ whose restriction to each $\mathcal{B}_n$ is ν_n. Then among all probability measures λ_∞ on $\mathcal{B}_\infty$ which satisfies (a) for each n, μ_n and λ_n are equivalent, (b) for each n, $E_{\lambda_n}(f_{n+1} \mid f_n) = f_n$, (c) $\lambda_1 = \sigma$, the measure ν_∞ is the unique one which maximizes simultaneously the relative entropies $H(\frac{d\lambda_n}{d\mu_n}, \mu_n)$, $n = 1, 2, \cdots$. The extension ν of ν_∞ to the Borel σ-algebra of $\mathbb{R}^{\mathbb{N}}$ is, however, singular to μ.

References

[1] Athreya, K. B. (1994). *Entropy Maximization*, Preprint Series No 1231, April 1994, Institute for Mathematics and Applications, University of Minnesota.

[2] Athreya, K. B. and Lahiri, S. N. (2006). *Measure Theory and Probability Theory*, Springer.

[3] Bochner, S. (1960). *Harmonic Analysis and The Theory of Probability*, Dover.

[4] Doob, J. L. (1955). *Stochastic Processes*, John Wiley.

[5] Durret, S. (1991). *Probability: Theory and Examples*, Wadsworth and Brooks-Cole, Pacific Grove, CA.

[6] Karandikar, R. L. and Nadkarni, M. G. (2005). Measure Free Martingales. *Proc. Indian Acad. Sci (Math. Sci)*. **15** 111–116.

[7] Parthasarathy, K. R. (1967). *Probability Measures on Metric Spaces*. Academic Press.

[8] Rao, C. R. (1973). *Linear Statistical Inference and its Applications*, Second Edition. John Wiley.

[9] Srivastava, S. M. (1998). *A course on Borel Sets*. Graduate Texts in Mathematics, 180, Springer-Verlag, New York.

[10] Stroock, D. (1993). *Probability Theory, An Analytic View*. Cambridge University Press.

[11] Bhatia, R. (2008). A Conversation with S. R. S. Varadhan, Varadhan's Interview by Rajendra Bhatia. *Mathematical Intellingencer*. **30** 24–42.

Chapter 2

Marginal Quantiles: Asymptotics for Functions of Order Statistics

G. Jogesh Babu

Department of Statistics,
326 Joab L. Thomas Building,
The Pennsylvania State University,
*University Park, PA 16802-2111, USA**

Methods for quantile estimation based on massive streaming data are reviewed. Marginal quantiles help in the exploration of massive multivariate data. Asymptotic properties of the joint distribution of marginal sample quantiles of multivariate data are also reviewed. The results include weak convergence to Gaussian random elements. Asymptotics for the mean of functions of order statistics are also presented. Application of the latter result to regression analysis under partial or complete loss of association among the multivariate data is described.

2.1. Introduction

Data depth provides an ordering of all points from the center outward. Contours of depth are often used to reveal the shape and structure of multivariate data set. The depth of a point x in a one-dimensional data set $\{x_1, x_2, \cdots, x_n\}$ can be defined as the minimum of the number of data points on one side of x (cf. [10]).

Several multidimensional depth measures $D_n(x; x_1, \cdots, x_n)$ for $x \in R^k$ were considered by many that satisfy certain mathematical conditions. If the data is from a spherical or elliptic distribution, the depth contours are generally required to converge to spherical or elliptic shapes. In this paper we concentrate on marginal quantiles. They help in describing percentile contours, which lead to a description of the densities and the multivariate distributions.

This approach is useful in quickly exploring massive datasets that are

*Research supported in part by NSF grant AST-0707833.

becoming more and more common in diverse fields such as Internet traffic, large sky surveys etc. For example, several ongoing sky surveys such as the Two Micron All Sky Survey and the Sloan Digital Sky Survey are providing maps of the sky at infrared and optical wavelengths, respectively generating data sets measured in the tens of Terabytes. These surveys are creating catalogs of objects (stars, galaxies, quasars, etc.) numbering in billions, with up to a hundred measured numbers for each object. Yet, this is just a fore-taste of the much larger datasets to come from surveys such as Large Synoptic Survey Telescope. This great opportunity comes with a commensurate technological challenge: how to optimally store, manage, combine, analyze and explore these vast amounts of complex information, and to do it quickly and efficiently? It is difficult even to compute a median of massive one dimensional data. As the multidimensional case is much more complex, marginal quantiles can be used to study the structure.

In this review article we start with description of estimation methods for quantiles and density for massive streaming data. Then describe asymptotic properties of joint distribution of marginal sample quantiles of multivariate data. We conclude with recent work on asymptotics for the mean of functions of order statistics and their applications to regression analysis under partial or complete loss of association among the multivariate data.

2.1.1. *Streaming Data*

As described above, massive streaming datasets are becoming more and more common. The data is in the form of a continuous stream with no fixed size. Finding trends in these massive size data is very important. One cannot wait till all the data is in and stored for retrieval for statistical analysis. Even to compute median from a stored billion data points is not feasible. In this case one can think of the data as a streaming data and use low storage methods to continually update the estimate of median and other quantiles ([2] and [7]). Simultaneous estimation of multiple quantiles would aid in density estimation.

Consider the problem of estimation of p-th quantile based on a very large dataset with n points of which a fixed number, say m, points can be placed into memory for sorting and ranking. Initially, each of these points is given a weight and a score based on p. Now a new point from the dataset is put in the array and all the points in the existing array above it will have their ranks increased by 1. The weights and scores are updated for these $m + 1$ points. The point with the largest score will then be dropped from the array, and the process is repeated. Once all the data points are run through the procedure, the data point with rank closest to np will be taken as an estimate of the p-th quantile. See [7] for the details.

Methods for estimation of several quantiles simultaneously are needed for the density estimation when the data is streaming. The method developed by [8] uses estimated ranks, assigned weights, and a scoring function that determines the most attractive candidate data points for estimates of the quantiles. The method uses a small fixed storage and its computation time is $O(n)$. Simulation studies show that the estimates are as accurate as the sample quantiles.

While the estimated quantiles are useful and informative on their own, it is often more useful to have information about the density as well. The probability density function can give a more intuitive picture of such characteristics as the skewness of the distribution or the number of modes. Any of the many standard curve fitting methods can now be employed to obtain an estimate of the cumulative distribution function.

The procedure is also useful in the approximation of the unknown underlying cumulative distribution function by fitting a cubic spline through the estimates obtained by this extension. The derivative of this spline fit provides an estimate of the probability density function.

The concept of convex hull peeling is useful in developing procedures for median in 2 or more dimensions. The convex hull of a dataset is the minimal convex set of points that contains the entire dataset. The convex hull based multivariate median is obtained by successively peeling outer layers until the dataset cannot be peeled any further. The centroid of the resulting set is taken as the multivariate median. Similarly, a multivariate interquartile range is obtained by successively peeling convex hull surfaces until approximately 50% of the data is contained within a hull. This hull is then taken as the multivariate interquartile range. See [5] for a nice overview. This procedure requires assumptions on the shape of the density. To avoid this one could use joint distribution of marginal quantiles to find the multidimensional structure.

2.2. Marginal Quantiles

Babu & Rao (cf. [3]) derived asymptotic results on marginal quantiles and quantile processes. They also developed tests of significance for population medians based on the joint distribution of marginal sample quantiles. Joint asymptotic distribution of the sample medians was developed by [9]; see also [6], where they assume the existence of the multivariate density. On the other hand Babu & Rao work with a much weaker assumption, the existence of densities of univariate marginals alone.

2.2.1. *Joint Distribution of Marginal Quantiles*

Let $\mathbf{F}$ denote a k-dimensional distribution function and let F_j denote the j-th marginal distribution function. The quantile functions of the marginals are defined by:

$$F_j^{-1}(u) = \inf\{x : F_j(x) \geq u\}, \text{ for } 0 < u < 1.$$

Thus $F_j^{-1}(u)$ is u-th quantile of the jth marginal.

Let $\mathbf{X}_1, \ldots, \mathbf{X}_n$ be independent random vectors with common distribution $\mathbf{F}$, where $\mathbf{X}_i = (X_{i1}, \ldots, X_{ik})$. Hence F_j is the distribution of X_{ij}. To get the joint distribution of sample quantiles, let $0 < q_1, \ldots, q_k < 1$. Let δ_j denote the density of F_j at $F_j^{-1}(q_j)$ and let $\hat{\theta}_j$ denote the q_j-th sample quantile based on the j-th coordinates $X_{1j}, \ldots, X_{nj}$ of the sample. [3] obtained the following theorem.

Theorem 2.1. *Let F_j be twice continuously differentiable in a neighborhood of $F_j^{-1}(q_j)$ and $\delta_j > 0$. Then the asymptotic distribution of*

$$\sqrt{n}(\hat{\theta}_1 - F_1^{-1}(q_1), \ldots, \hat{\theta}_k - F_k^{-1}(q_k))$$

is k-variate Gaussian distribution with mean vector zero and variance-covariance matrix Σ given by

$$\Sigma = \begin{pmatrix} q_1(1-q_1)\delta_1^{-2} & \sigma_{12} & \cdots & \sigma_{1k} \\ \vdots & \vdots & \cdots & \vdots \\ \sigma_{k1} & \sigma_{k2} & \cdots & q_k(1-q_k)\delta_k^{-2} \end{pmatrix},$$

where for $i \neq j$, $\sigma_{ij} = (F_{ij}(F_i^{-1}(q_i), F_j^{-1}(q_j)) - q_i q_j)/(\delta_i \delta_j)$.

The proof uses Bahadur's representation of the sample quantiles (see [4]).

In practice σ_{ij} can be directly estimated using bootstrap method,

$$\widehat{\sigma_{ij}} = E^*(n(\theta_i^* - \hat{\theta}_i)(\theta_j^* - \hat{\theta}_j)),$$

where θ_j^* denotes the bootstrapped marginal sample quantile and E^* denotes the expectation under the bootstrap distribution function. An advantage of the bootstrap procedure is that it avoids density estimation altogether.

2.2.2. *Weak Convergence of Quantile Process*

We now describe the weak limits of the entire marginal quantile processes. For $(q_1, \ldots, q_k) \in (0,1)^k$, define the sample quantile process,

$$Z_n(q_1, \ldots, q_k) = \sqrt{n}\left(\delta_1(\hat{\theta}_1 - F_1^{-1}(q_1)), \ldots, \delta_k(\hat{\theta}_k - F_k^{-1}(q_k))\right).$$

The following theorem is from Section 4 of [3].

Theorem 2.2. *Suppose for $j = 1, \ldots, k$, the marginal d.f. F_j is twice differentiable on (a_j, b_j), where*

$$-\infty \leq a_j = \sup\{x : F_j(x) = 0\}$$

$$\infty \geq b_j = \inf\{x : F_j(x) = 1\}.$$

Further suppose that the first two derivatives F_j' and F_j'' of F_j satisfy the conditions

$$F_j' \neq 0 \text{ on } (a_j, b_j),$$

$$\max_{1 \leq j \leq k} \sup_{a_j < x < b_j} F_j(x)(1 - F_j(x)) \frac{|F_j''(x)|}{(F_j'(x))^2} < \infty,$$

and F_j' is non-decreasing (non-increasing) on an interval to the right of a_j (to the left of b_j). Then $Z_n(q_1, \ldots, q_k)$ converges weekly to a Gaussian random element $(W_1, \ldots, W_k)$ on $C[0,1]^k$.

Thus, each marginal of Z_n converges weakly to a Brownian bridge. The covariance of the limiting Gaussian random element is given by

$$E(W_i(t)W_j(s)) = P(F_i(X_{1i}) \leq t, F_j(X_{1j}) \leq s) - ts.$$

2.3. Regression under Lost Association

[11] developed a method of estimation of linear regression coefficients when the association among the paired data is partially or completely lost. He considered the simple linear regression problem

$$Y_i = \alpha + \beta U_i + \epsilon_i,$$

where U_i are independent identically distributed (i.i.d.) with mean μ and standard deviation σ_U, the residual errors ϵ_i are i.i.d with mean zero and standard deviation σ_ϵ. Further, $\{U_i\}$ and $\{\epsilon_i\}$ are assumed to be independent sequences. If Π_n denotes the set of all permutations of $\{1,\ldots,n\}$, then it is natural to find estimators $\hat{\alpha}, \hat{\beta}$ of α, β that minimize

$$h(\alpha,\beta) = \min_{\pi\in\Pi_n} \sum_{i=1}^{n} (Y_{\pi(i)} - \alpha - \beta U_i)^2.$$

[11] has shown that the permutation that minimizes h is free from α, β.

The main difficulty is the computational complexity. As there are $n!$ permutations, conceivably it requires that many computations. [11] has shown that $\hat{\beta}$ depends only on two permutations. In particular, he has shown that

$$\frac{1}{n}\sum_{i=1}^{n} U_{(i)}Y_{(i)} \text{ and } \frac{1}{n}\sum_{i=1}^{n} U_{(i)}Y_{(n-i+1)}$$

appear in the definition of $\hat{\beta}$. Hence the results on their limits are needed to obtain the asymptotics for $\hat{\beta}$. This would aid in the estimation of the bias of $\hat{\beta}$. Further testing of hypothesis or obtaining confidence intervals for $\hat{\beta}$ require limiting distribution of

$$\frac{1}{\sqrt{n}}\sum_{i=1}^{n} U_{(i)}Y_{(i)}.$$

See the Example 2.2 in the last section.

In the next section we present some work in progress on the strong law of large numbers and central limit theorems for means of general functions of order statistics. These results would aid in establishing

$$\hat{\beta} \xrightarrow{a.e} \beta_1 = sign(\beta)\sqrt{\beta^2 + \sigma_\epsilon^2\sigma_U^{-1}}.$$

2.4. Mean of Functions of Order Statistics

This section is based on the current research by [1]. We present some recent results on strong law of large numbers and the central limit theorem for the means of functions of order statistics. Let $\mathbf{X}_i$ and X_{ij} be as in Section 2.2.1. Let $X_{n:i}^{(j)}$ denote the i-th order statistic of $\{X_{1j},\ldots,X_{nj}\}$. Suppose ϕ is a measurable function on $\mathbb{R}^k$ and the function γ defined by

$$\gamma(u) = \phi(F_1^{-1}(u),\ldots,F_k^{-1}(u)),\ 0 < u < 1,$$

is integrable on $(0,1)$.

Theorem 2.3. *Suppose F_j are continuous, $\phi(F_1^{-1}(u_1), \ldots, F_k^{-1}(u_k))$ is continuous in the neighborhood of the diagonal $u_1 = u, \ldots, u_k = u, 0 < u < 1$, and for some A and $0 < c_0 < 1/2$,*

$$|\phi(F_1^{-1}(u_1), \ldots, F_k^{-1}(u_k))| \leq A\left(1 + \sum_{j=1}^{k} |\gamma(u_j)|\right),$$

whenever $(u_1, \ldots, u_k) \in (0, c_0)^k \cup (1 - c_0, 1)^k$. Then

$$\frac{1}{n} \sum_{i=1}^{n} \phi(X_{n:i}^{(1)}, \ldots, X_{n:i}^{(k)}) \xrightarrow{a.e.} \int_0^1 \gamma(y)dy.$$

For example, in the two dimensional case,

$$\phi(F_1^{-1}(u), F_2^{-1}(v)) = \min(u, v)^{-\alpha}(1 - \max(u, v))^{-\alpha}$$

with $0 < \alpha < \frac{1}{2}$ satisfies the conditions of Theorem 2.3.

To establish asymptotic normality, we require

$$\lim_{u \to 0+} \sqrt{u}(|\gamma(u)| + |\gamma(1 - u)|) = 0,$$

square integrability of partial derivatives ψ_j,

$$\psi_j(u) = \frac{\partial \phi(F_1^{-1}(u_1), \ldots, F_k^{-1}(u_k))}{\partial u_j}|_{(u, \ldots, u)},$$

and some smoothness conditions on ψ_j and $\phi(F_1^{-1}(u_1), \ldots, F_k^{-1}(u_k))$, in addition to the conditions of Theorem 2.3.

Theorem 2.4. *Assume for any pair $(1 \leq j \neq r \leq k)$, the joint distribution $F_{j,r}$ of (X_{ij}, X_{ir}) is continuous. Under regularity assumptions that include the conditions mentioned above, we have*

$$\frac{1}{\sqrt{n}} \sum_{i=1}^{n} (\phi(X_{n:i}^{(1)}, \ldots, X_{n:i}^{(k)}) - \int_0^1 \gamma(y)dy) \xrightarrow{dist} N(0, \sigma^2),$$

where

$$\begin{aligned}\sigma^2 =& 2 \sum_{1 \leq j \neq r \leq k} \int_0^1 \int_0^1 [F_{j,r}(F_j^{-1}(x), F_r^{-1}(y)) - xy]\psi_j(x)\psi_j(y)dxdy \\ &+ 2 \sum_{j=1}^{k} \int_0^1 \int_0^y x(1 - y)\psi_j(x)\psi_j(y)dxdy.\end{aligned}$$

Details are in [1].

2.5. Examples

The above results are illustrated with two of examples.

Example 2.1. Let $\mathbf{X}_i$ be as in Section 2.2.1. Let the marginals X_{ij} be uniformly distributed and let $\phi(u_1, \ldots, u_k) = u_1^{a_1} \cdots u_k^{a_k}$, for some $a_j \geq 1$. Then

$$\frac{1}{n}\sum_{i=1}^{n}(X_{n:i}^{(1)})^{a_1}\cdots(X_{n:i}^{(k)})^{a_k} \xrightarrow{a.e.} \frac{1}{a_1+\cdots+a_k+1},$$

$$\frac{1}{\sqrt{n}}\sum_{i=1}^{n}\left[(X_{n:i}^{(1)})^{a_1}\cdots(X_{n:i}^{(k)})^{a_k} - \frac{1}{a_1+\cdots+a_k+1}\right] \xrightarrow{dist.} N(0,\sigma^2),$$

where

$$\sigma^2 = 2\sum_{1\leq j<r\leq k} a_j a_r E(X_{1j}X_{1r})^M + \frac{(2M-3)(M^2-2)}{2M+1}\sum_{j=1}^{k} a_j^2,$$

and $M = a_1 + \cdots + a_k$.

Note that the limit in this example does not depend on the joint distribution of X_{1j}. In particular, if $a_1 = a_2 = 1$, we obtain that both $\frac{1}{n}\sum_{i=1}^{n} X_{n:i}^{(1)}X_{n:i}^{(2)}$ and $\frac{1}{n}\sum_{i=1}^{n}\left(X_{n:i}^{(1)}\right)^2$ converge to the same limit $E(X_{11}^2) = \frac{1}{3}$ a.e.

Example 2.2. (*Regression with lost associations.*) Let $\{(X_i, Y_i), 1 \leq i \leq n\}$ be i.i.d. bivariate normal random vectors with correlation ρ, means μ_1, μ_2, and standard deviations σ_1, σ_2. Let the marginal distributions of X_1 and Y_1 be denoted by F and G. Clearly,

$$G^{-1}(F(x)) = \mu_2 + \frac{\sigma_2}{\sigma_1}(x - \mu_1).$$

Then

$$\frac{1}{n}\sum_{i=1}^{n} X_{n:i}Y_{n:i} \xrightarrow{a.e.} \int_0^1 F^{-1}(u)G^{-1}(u)\,du = E(X_1G^{-1}(F(X_1)))$$
$$= \mu_1\mu_2 + \sigma_1\sigma_2,$$

and

$$\frac{1}{\sqrt{n}}\sum_{i=1}^{n}(X_{n:i}Y_{n:i} - \mu_1\mu_2 - \sigma_1\sigma_2) \xrightarrow{dist.} N(0,\sigma^2),$$

where

$$\sigma^2 = \mu_1^2\sigma_2^2 + \mu_2^2\sigma_1^2 + (1+\rho^2)\sigma_1^2\sigma_2^2 + 2\rho\,\mu_1\mu_2\,\sigma_1\sigma_2.$$

Regression under broken samples are considered in Section 2.3, where it is indicated that the regression coefficients depend only on

$$\frac{1}{n}\sum_{i=1}^{n} X_{n:i}Y_{n:i} \text{ and } \frac{1}{n}\sum_{i=1}^{n} X_{n:i}Y_{n:(n-i+1)}.$$

Acknowledgment

I would like to thank an anonymous referee for the comments which helped in improving the presentation.

References

[1] Babu, G. J., Bai, Z. D., Choi, K.-P. and Mangalam, V. (2008). Limit theorems for functions of marginal quantiles. Submitted.

[2] Babu, G. J. and McDermott, J. P. (2002). Statistical methodology for massive datasets and model selection. In *Astronomical Data Analysis II* (Jean-Luc Starck and Fionn D. Murtagh, Eds.), Proceedings of SPIE. **4847** 228–237.

[3] Babu, G. J. and Rao, C. R. (1988). Joint asymptotic distribution of marginal quantiles and quantile functions in samples from a multivariate population. *J. Multivariate Anal.* **27** 15–23.

[4] Bahadur, R. R. (1966). A note on quantiles in large samples. *Ann. Math. Statist.* **37** 577–580.

[5] Barnett, V. (1981). *Interpreting Multivariate Data.* Wiley, New York.

[6] Kuan, K. S. and Ali, M. M. (1980). Asymptotic distribution of quantiles from a multivariate distribution. *Multivariate Statistical Analysis* (Proceedings of Conference at Dalhousie University), Halifax, Nova Scotia, 1979, 109–120. North-Holland, Amsterdam.

[7] Liechty, J. C., Lin, D. K. J. and McDermott, J. P. (2003). Single-pass low-storage arbitrary quantile estimation for massive datasets. *Statistics and Computing.* **13** 91–100.

[8] McDermott, J. P., Babu, G. J., Liechty, J. C. and Lin, D. K. J. (2007). Data skeletons: simultaneous estimation of multiple quantiles for massive streaming datasets with applications to density estimation. *Statistics and Computing.* **17** 311–321.

[9] Mood, A. M. (1941). On the joint distribution of the medians in samples from a multivariate population. *Ann. Math. Statist.* **12** 268–278.

[10] Tukey, J. W. (1975). Mathematics and the picturing of data. In *Proceedings of the International Congress of Mathematicians*, Canadian Mathematical Congress, Montreal, **2** 523–531.

[11] Mangalam, V. (2006). The regression under lost association. Private Communication.

Chapter 3

Statistics on Manifolds with Applications to Shape Spaces

Rabi Bhattacharya[1] and Abhishek Bhattacharya[2]

[1] *Department of Mathematics, University of Arizona, Tucson, Arizona 85721, USA*
rabi@math.arizona.edu
*http://math.arizona.edu/~rabi/**

[2] *Department of Statistical Science, Duke University, Durham, North Carolina 27708, USA*
ab216@stat.duke.edu
http://stat.duke.edu/~ab216/

This article provides an exposition of recent developments on the analysis of *landmark based shapes* in which a k-ad, i.e., a set of k points or landmarks on an object or a scene, are observed in 2D or 3D, for purposes of identification, discrimination, or diagnostics. Depending on the way the data are collected or recorded, the appropriate shape of an object is the maximal invariant specified by the space of orbits under a group G of transformations. All these spaces are manifolds, often with natural Riemannian structures. The statistical analysis based on Riemannian structures is said to be *intrinsic.* In other cases, proper distances are sought via an *equivariant embedding* of the manifold M in a vector space E, and the corresponding statistical analysis is called *extrinsic.*

3.1. Introduction

Statistical analysis of a probability measure Q on a differentiable manifold M has diverse applications in directional and axial statistics, morphometrics, medical diagnostics and machine vision. In this article, we are mostly concerned with the analysis of landmark based data, in which each observation consists of $k > m$ points in m-dimension, representing k locations on an object, called a k-ad. The choice of landmarks is generally made with expert help in the particular field of application. The objects of study can

*Research supported in part by NSF Grant DMS 0806011.

be anything for which two k-ads are equivalent modulo a group of transformations appropriate for the particular problem depending on the method of recording of the observations. For example, one may look at k-ads modulo size and Euclidean rigid body motions of translation and rotation. The analysis of shapes under this invariance was pioneered by [27, 28] and [13]. Bookstein's approach is primarily registration-based requiring two or three landmarks to be brought into a standard position by translation, rotation and scaling of the k-ad. For these shapes, we would prefer Kendall's more invariant view of a shape identified with the orbit under rotation (in m-dimension) of the k-ad centered at the origin and scaled to have unit size. The resulting shape space is denoted Σ_k^m. A fairly comprehensive account of parametric inference on these manifolds, with many references to the literature, may be found in [21]. The nonparametric methodology pursued here, along with the geometric and other mathematical issues that accompany it, stems from the earlier work of [9–11].

Recently there has been much emphasis on the statistical analysis of other notions of shapes of k-ads, namely, *affine shapes* invariant under affine transformations, and *projective shapes* invariant under projective transformations. Reconstruction of a scene from two (or more) aerial photographs taken from a plane is one of the research problems in affine shape analysis. Potential applications of projective shape analysis include face recognition and robotics – for robots to visually recognize a scene ([36], [1]).

Examples of analysis with real data suggest that appropriate nonparametric methods are more powerful than their parametric counterparts in the literature, for distributions that occur in applications ([7]).

There is a large literature on registration via landmarks in functional data analysis (see, e.g., [12], [43], [37]), in which proper alignments of curves are necessary for purposes of statistical analysis. However this subject is not closely related to the topics considered in the present article.

The article is organized as follows. Section 3.2 provides a brief expository description of the geometries of the manifolds that arise in shape analysis. Section 3.3 introduces the basic notion of the *Fréchet mean* as the unique minimizer of the *Fréchet function* $F(p)$, which is used here to nonparametrically discriminate different distributions. Section 3.4 outlines the asymptotic theory for *extrinsic mean*, namely, the unique minimizer of the Fréchet function $F(p) = \int_M \rho^2(p,x)Q(dx)$ where ρ is the distance inherited by the manifold M from an equivariant embedding J. In Section 3.5, we describe the corresponding asymptotic theory for *intrinsic means* on Riemannian manifolds, where ρ is the geodesic distance. In Section 3.6, we apply the theory of extrinsic and intrinsic analysis to some manifolds including the shape spaces of interest. Finally, Section 3.7 illustrates the theory with three applications to real data.

3.2. Geometry of Shape Manifolds

Many differentiable manifolds M naturally occur as submanifolds, or surfaces or hypersurfaces, of a Euclidean space. One example of this is the sphere $S^d = \{p \in \mathbb{R}^{d+1} \colon \|p\| = 1\}$. The shape spaces of interest here are not of this type. They are generally quotients of a Riemannian manifold N under the action of a transformation group. A number of them are quotient spaces of $N = S^d$ under the action of a compact group G, i.e., the elements of the space are orbits in S^d traced out by the application of G. Among important examples of this kind are axial spaces and Kendall's shape spaces. In some cases the *action of the group is free*, i.e., $gp = p$ only holds for the identity element $g = e$. Then the elements of the orbit $O_p = \{gp \colon g \in G\}$ are in one-one correspondence with elements of G, and one can identify the orbit with the group. The orbit inherits the differential structure of the Lie group G. The tangent space T_pN at a point p may then be decomposed into a *vertical subspace* of dimension that of the group G along the orbit space to which p belongs, and a *horizontal* one which is orthogonal to it. The projection π, $\pi(p) = O_p$ is a *Riemannian submersion* of N onto the quotient space N/G. In other words, $\langle d\pi(v), d\pi(w)\rangle_{\pi(p)} = \langle v, w\rangle_p$ for horizontal vectors $v, w \in T_pN$, where $d\pi : T_pN \to T_{\pi(p)}N/G$ denotes the differential, or Jacobian, of the projection π. With this metric tensor, N/G has the natural structure of a Riemannian manifold. The intrinsic analysis proposed for these spaces is based on this Riemannian structure (See Section 3.5).

Often it is simpler both mathematically and computationally to carry out an extrinsic analysis, by embedding M in some Euclidean space $E^k \approx \mathbb{R}^k$, with the distance induced from that of E^k. This is also pursued when an appropriate Riemannian structure on M is not in sight. Among the possible embeddings, one seeks out *equivariant embeddings* which preserve many of the geometric features of M.

Definition 3.1. For a Lie group H acting on a manifold M, an embedding $J : M \to \mathbb{R}^k$ is *H-equivariant* if there exists a group homomorphism $\phi : H \to GL(k, \mathbb{R})$ such that

$$J(hp) = \phi(h)J(p) \; \forall p \in M, \; \forall h \in H. \tag{2.1}$$

Here $GL(k, \mathbb{R})$ is the *general linear group* of all $k \times k$ non-singular matrices.

3.2.1. *The Real Projective Space* $\mathbb{R}P^d$

This is the axial space comprising axes or lines through the origin in $\mathbb{R}^{d+1}$. Thus elements of $\mathbb{R}P^d$ may be represented as equivalence classes

$$[x] = [x^1 : x^2 : \dots x^{m+1}] = \{\lambda x : \lambda \neq 0\},\ x \in \mathbb{R}^{d+1} \setminus \{0\}. \tag{2.2}$$

One may also identify $\mathbb{R}P^d$ with S^d/G, with G comprising the identity map and the antipodal map $p \mapsto -p$. Its structure as a d-dimensional manifold (with quotient topology) and its Riemannian structure both derive from this identification. Among applications are observations on galaxies, on axes of crystals, or on the line of a geological fissure ([42], [35], [22], [3], [29]).

3.2.2. *Kendall's (Direct Similarity) Shape Spaces* Σ_m^k

Kendall's shape spaces are quotient spaces S^d/G, under the action of the *special orthogonal group* $G = SO(m)$ of $m \times m$ orthogonal matrices with determinant $+1$. For the important case $m = 2$, consider the space of all planar k-ads $(z_1, z_2, \dots, z_k)$ $(z_j = (x_j, y_j))$, $k > 2$, excluding those with k identical points. The set of all centered and normed k-ads, say $u = (u_1, u_2, \dots, u_k)$ comprise a unit sphere in a $(2k-2)$-dimensional vector space and is, therefore, a $(2k-3)$-dimensional sphere S^{2k-3}, called the *preshape sphere.* The group $G = SO(2)$ acts on the sphere by rotating each landmark by the same angle. The orbit under G of a point u in the preshape sphere can thus be seen to be a circle S^1, so that Kendall's planer shape space Σ_2^k can be viewed as the quotient space $S^{2k-3}/G \sim S^{2k-3}/S^1$, a $(2k-4)$-dimensional compact manifold. An algebraically simpler representation of Σ_2^k is given by the complex projective space $\mathbb{C}P^{k-2}$, described in Section 3.6.4. For many applications in archaeology, astronomy, morphometrics, medical diagnosis, etc., see [14, 15], [29], [21], [10, 11], [7] and [39].

3.2.3. *Reflection (Similarity) Shape Spaces* $R\Sigma_m^k$

Consider now the *reflection shape* of a k-ad as defined in Section 3.2.2, but with $SO(m)$ replaced by the larger *orthogonal group* $O(m)$ of all $m \times m$ orthogonal matrices (with determinants either $+1$ or -1). The reflection shape space $R\Sigma_m^k$ is the space of orbits of the elements u of the preshape sphere whose columns span $\mathbb{R}^m$.

3.2.4. *Affine Shape Spaces* $A\Sigma_m^k$

The *affine shape* of a k-ad in $\mathbb{R}^m$ may be defined as the orbit of this k-ad under the group of all *affine transformations* $x \mapsto F(x) = Ax + b$, where A is an arbitrary $m \times m$ non-singular matrix and b is an arbitrary point in $\mathbb{R}^m$. Note that two k-ads $x = (x_1, \ldots, x_k)$ and $y = (y_1, \ldots, y_k)$, (x_j, y_j $\in \mathbb{R}^m$ for all j) have the same affine shape if and only if the centered k-ads $u = (u_1, u_2, \ldots, u_k) = (x_1 - \bar{x}, \ldots, x_k - \bar{x})$ and $v = (v_1, v_2, \ldots, v_k) = (y_1 - \bar{y}, \ldots, y_k - \bar{y})$ are related by a transformation $Au \doteq (Au_1, \ldots, Au_k) = v$. The centered k-ads lie in a linear subspace of $\mathbb{R}^m$ of dimension $m(k-1)$. Assume $k > m + 1$. The affine shape space is then defined as the quotient space $H(m,k)/GL(m,R)$, where $H(m,k)$ consists of all centered k-ads whose landmarks span $\mathbb{R}^m$, and $GL(m,\mathbb{R})$ is the general linear group on $\mathbb{R}^m$ (of all $m \times m$ nonsingular matrices) which has the relative topology (and distance) of $\mathbb{R}^{m^2}$ and is a manifold of dimension m^2. It follows that $A\Sigma_m^k$ is a manifold of dimension $m(k-1) - m^2$. For $u, v \in H(m,k)$, since $Au = v$ iff $u'A' = v'$, and as A varies $u'A'$ generates the linear subspace L of $H(m,k)$ spanned by the m rows of u. The affine shape of u, (or of x), is identified with this subspace. Thus $A\Sigma_m^k$ may be identified with the set of all m dimensional subspaces of $\mathbb{R}^{k-1}$, namely, the *Grassmannian* $G_m(k-1)$-a result of [40] (Also see [16], pp. 63-64, 362-363). Affine shape spaces arise in certain problems of bioinformatics, cartography, machine vision and pattern recognition ([4, 5], [38], [40]).

3.2.5. *Projective Shape Spaces* $P\Sigma_m^k$

For purposes of machine vision, if images are taken from a great distance, such as a scene on the ground photographed from an airplane, affine shape analysis is appropriate. Otherwise, *projective shape* is a more appropriate choice. If one thinks of images or photographs obtained through a central projection (a pinhole camera is an example of this), a ray is received as a point on the image plane (e.g., the film of the camera). Since axes in 3D comprise the projective space $\mathbb{R}P^2$, k-ads in this view are valued in $\mathbb{R}P^2$. Note that for a 3D k-ad to represent a k-ad in $\mathbb{R}P^2$, the corresponding axes must all be distinct. To have invariance with regard to camera angles, one may first look at the original noncollinear (centered) 3D k-ad u and achieve affine invariance by its affine shape (i.e., by the equivalence class Au, $A \in GL(3,\mathbb{R})$), and finally take the corresponding equivalence class of axes in $\mathbb{R}P^2$ to define the projective shape of the k-ad as the equivalence class, or orbit, with respect to projective transformations on $\mathbb{R}P^2$. A projective shape (of a k-ad) is singular if the k axes lie on a vector plane ($\mathbb{R}P^1$). For $k > 4$, the space of all non-singular shapes is the 2D projective shape space, denoted $P_0\Sigma_2^k$.

In general, a projective (general linear) transformation α on $\mathbb{R}P^m$ is defined in terms of an $(m+1)\times(m+1)$ nonsingular matrix $A \in GL(m+1, \mathbb{R})$ by

$$\alpha([x]) = \alpha([x^1 : \ldots : x^{m+1}]) = [A(x^1, \ldots, x^{m+1})'], \tag{2.3}$$

where $x = (x^1, \ldots, x^{m+1}) \in \mathbb{R}^{m+1} \setminus \{0\}$. The group of all projective transformations on $\mathbb{R}P^m$ is denoted by $PGL(m)$. Now consider a k-ad $(y_1, \ldots, y_k)$ in $\mathbb{R}P^m$, say $y_j = [x_j]$ $(j = 1, \ldots, k)$, $k > m+2$. The projective shape of this k-ad is its orbit under $PGL(m)$, i.e., $\{(\alpha y_1, \ldots, \alpha y_k) \colon \alpha \in PGL(m)\}$. To exclude singular shapes, define a k-ad $(y_1, \ldots, y_k) = ([x_1], \ldots, [x_k])$ to be in *general position* if the linear span of $\{y_1, \ldots, y_k\}$ is $\mathbb{R}P^m$, i.e., if the linear span of the set of k representative points $\{x_1, \ldots, x_k\}$ in $\mathbb{R}^{m+1}$ is $\mathbb{R}^{m+1}$. The space of shapes of all k-ads in general position is the projective shape space $P_0\Sigma_m^k$. Define a projective frame in $\mathbb{R}P^m$ to be an ordered system of $m+2$ points in general position. Let $I = i_1 < \ldots < i_{m+2}$ be an ordered subset of $\{1, \ldots, k\}$. A manifold structure on $P_I\Sigma_m^k$, the open dense subset of $P_0\Sigma_m^k$, of k-ads for which $(y_{i1}, \ldots, y_{i_{m+2}})$ is a projective frame in $\mathbb{R}P^m$, was derived in [36] as follows. The standard frame is defined to be $([e_1], \ldots, [e_{m+1}], [e_1 + e_2 + \ldots + e_{m+1}])$, where $e_j \in \mathbb{R}^{m+1}$ has 1 in the j-th coordinate and zeros elsewhere. Given two projective frames $(p_1, \ldots, p_{m+2})$ and $(q_1, \ldots, q_{m+2})$, there exists a unique $\alpha \in PGL(m)$ such that $\alpha(p_j) = q_j$ $(j = 1, \ldots, k)$. By ordering the points in a k-ad such that the first $m+2$ points are in general position, one may bring this ordered set, say, $(p_1, \ldots, p_{m+2})$, to the standard form by a unique $\alpha \in PGL(m)$. Then the ordered set of remaining $k-m-2$ points is transformed to a point in $(\mathbb{R}P^m)^{k-m-2}$. This provides a diffeomorphism between $P_I\Sigma_m^k$ and the product of $k-m-2$ copies of the real projective space $\mathbb{R}P^m$.

We will return to these manifolds again in Section 3.6. Now we turn to nonparametric inference on general manifolds.

3.3. Fréchet Means on Metric Spaces

Let (M, ρ) be a metric space, ρ being the distance, and let $f \geq 0$ be a given continuous increasing function on $[0, \infty)$. For a given probability measure

Q on (the Borel sigma field of) M, define the *Fréchet function* of Q as

$$F(p) = \int_M f(\rho(p, x))Q(dx), \quad p \in M. \tag{3.1}$$

Definition 3.2. Suppose $F(p) < \infty$ for some $p \in M$. Then the set of all p for which $F(p)$ is the minimum value of F on M is called the *Fréchet Mean set* of Q, denoted by C_Q. If this set is a singleton, say $\{\mu_F\}$, then μ_F is called the *Fréchet Mean* of Q. If $X_1, X_2, \ldots, X_n$ are independent and identically distributed (iid) M-valued random variables defined on some probability space $(\Omega, \mathcal{F}, P)$ with common distribution Q, and $Q_n \doteq \frac{1}{n}\sum_{j=1}^{n} \delta_{X_j}$ is the corresponding empirical distribution, then the Fréchet mean set of Q_n is called the *sample Fréchet mean set*, denoted by C_{Q_n}. If this set is a singleton, say $\{\mu_{F_n}\}$, then μ_{F_n} is called the *sample Fréchet mean.*

Proposition 3.1 proves the consistency of the sample Fréchet mean as an estimator of the Fréchet mean of Q.

Proposition 3.1. *Let M be a compact metric space. Consider the Fréchet function F of a probability measure given by* (3.1). *Given any $\epsilon > 0$, there exists an integer-valued random variable $N = N(\omega, \epsilon)$ and a P-null set $A(\omega, \epsilon)$ such that*

$$C_{Q_n} \subset C_Q^\epsilon \equiv \{p \in M : \rho(p, C_Q) < \epsilon\}, \; \forall n \geq N \tag{3.2}$$

outside of $A(\omega, \epsilon)$. In particular, if $C_Q = \{\mu_F\}$, then every measurable selection, μ_{F_n} from C_{Q_n} is a strongly consistent estimator of μ_F.

Proof. For simplicity of notation, we write $C = C_Q$, $C_n = C_{Q_n}$, $\mu = \mu_F$ and $\mu_n = \mu_{F_n}$. Choose $\epsilon > 0$ arbitrarily. If $C^\epsilon = M$, then (3.2) holds with $N = 1$. If $D = M \setminus C^\epsilon$ is nonempty, write

$$\begin{aligned} l &= \min\{F(p) : p \in M\} = F(q) \; \forall q \in C, \\ l + \delta(\epsilon) &= \min\{F(p) : p \in D\}, \; \delta(\epsilon) > 0. \end{aligned} \tag{3.3}$$

It is enough to show that

$$\max\{|F_n(p) - F(p)| : p \in M\} \longrightarrow 0 \text{ a.s., as } n \to \infty. \tag{3.4}$$

For if (3.4) holds, then there exists $N \geq 1$ such that, outside a P-null set $A(\omega, \epsilon)$,

$$\min\{F_n(p) : p \in C\} \leq l + \frac{\delta(\epsilon)}{3},$$

$$\min\{F_n(p) : p \in D\} \geq l + \frac{\delta(\epsilon)}{2}, \ \forall n \geq N. \tag{3.5}$$

Clearly (3.5) implies (3.2).

To prove (3.4), choose and fix $\epsilon' > 0$, however small. Note that $\forall p, p'$, $x \in M$,

$$|\rho(p, x) - \rho(p', x)| \leq \rho(p, p').$$

Hence

$$\begin{aligned} |F(p) - F(p')| &\leq \max\{|f(\rho(p, x)) - f(\rho(p', x))| : x \in M\} \\ &\leq \max\{|f(u) - f(u')| : |u - u'| \leq \rho(p, p')\}, \\ |F_n(p) - F_n(p')| &\leq \max\{|f(u) - f(u')| : |u - u'| \leq \rho(p, p')\}. \end{aligned} \tag{3.6}$$

Since f is uniformly continuous on $[0, R]$ where R is the diameter of M, so are F and F_n on M, and there exists $\delta(\epsilon') > 0$ such that

$$|F(p) - F(p')| \leq \frac{\epsilon'}{4}, \ |F_n(p) - F_n(p')| \leq \frac{\epsilon'}{4} \tag{3.7}$$

if $\rho(p, p') < \delta(\epsilon')$. Let $\{q_1, \ldots, q_k\}$ be a $\delta(\epsilon')$–net of M, i.e., $\forall \ p \in M$ there exists $q(p) \in \{q_1, \ldots, q_k\}$ such that $\rho(p, q(p)) < \delta(\epsilon')$. By the strong law of large numbers, there exists an integer-valued random variable $N(\omega, \epsilon')$ such that outside of a P-null set $A(\omega, \epsilon')$, one has

$$|F_n(q_i) - F(q_i)| \leq \frac{\epsilon'}{4} \ \forall i = 1, 2, \ldots, k; \text{ if } n \geq N(\omega, \epsilon'). \tag{3.8}$$

From (3.7) and (3.8) we get

$$\begin{aligned} |F(p) - F_n(p)| &\leq |F(p) - F(q(p))| + |F(q(p)) - F_n(q(p))| \\ &\quad + |F_n(q(p)) - F_n(p)| \\ &\leq \frac{3\epsilon'}{4} < \epsilon', \ \forall p \in M, \end{aligned}$$

if $n \geq N(\omega, \epsilon')$ outside of $A(\omega, \epsilon')$. This proves (3.4). □

Remark 3.1. Under an additional assumption guaranteeing the existence of a minimizer of F, Proposition 3.1 can be extended to all metric spaces whose closed and bounded subsets are all compact. We will consider such an extension elsewhere, thereby generalizing Theorem 2.3 in [10]. For statistical analysis on shape spaces which are compact manifolds, Proposition 3.1 suffices.

Remark 3.2. One can show that the reverse of (3.2) that is "$C_Q \subset C_{Q_n}^{\epsilon}$ $\forall$ $n \geq N(\omega, \epsilon)$" does not hold in general. See for example Remark 2.6 in [10].

Remark 3.3. In view of Proposition 3.1, if the Fréchet mean μ_F of Q exists as a unique minimizer of F, then every measurable selection of a sequence $\mu_{F_n} \in C_{Q_n}$ $(n \geq 1)$ converges to μ_F with probability one. In the rest of the paper it therefore suffices to define the sample Fréchet mean as a measurable selection from C_{Q_n} $(n \geq 1)$.

Next we consider the asymptotic distribution of μ_{F_n}. For Theorem 3.1, we assume M to be a differentiable manifold of dimension d. Let ρ be a distance metrizing the topology of M. The proof of the theorem is similar to that of Theorem 2.1 in [11]. Denote by D_r the partial derivative w.r.t. the rth coordinate $(r = 1, \ldots, d)$.

Theorem 3.1. *Suppose the following assumptions hold:*
A1 *Q has support in a single coordinate patch, (U, ϕ). [$\phi : U \longrightarrow \mathbb{R}^d$ smooth.] Let $Y_j = \phi(X_j)$, $j = 1, \ldots, n$.*
A2 *Fréchet mean μ_F of Q is unique.*
A3 *$\forall x, y \rightarrow h(x, y) = (\rho^{\phi})^2(x, y) = \rho^2(\phi^{-1}x, \phi^{-1}y)$ is twice continuously differentiable in a neighborhood of $\phi(\mu_F) = \mu$.*
A4 $\mathrm{E}\{\mathrm{D}_r h(Y, \mu)\}^2 < \infty$ $\forall r$.
A5 $\mathrm{E}\{\sup_{|u-v| \leq \epsilon} |\mathrm{D}_s \mathrm{D}_r h(Y, v) - \mathrm{D}_s \mathrm{D}_r h(Y, u)|\} \rightarrow 0$ *as* $\epsilon \rightarrow 0$ $\forall$ r, s.
A6 $\Lambda = ((\ \mathrm{E}\{\mathrm{D}_s \mathrm{D}_r h(Y, \mu)\}\ ))$ *is nonsingular.*
A7 $\Sigma = \mathrm{Cov}[\mathrm{grad}\ h(Y_1, \mu)]$ *is nonsingular.*
*Let $\mu_{F,n}$ be a measurable selection from the sample Frechet mean set. Then under the assumptions **A1-A7**,*

$$\sqrt{n}(\mu_n - \mu) \xrightarrow{\mathcal{L}} N(0, \Lambda^{-1}\Sigma(\Lambda')^{-1}). \tag{3.9}$$

3.4. Extrinsic Means on Manifolds

From now on, we assume that M is a Riemannian manifold of dimension d. Let G be a Lie group acting on M and let $J : M \rightarrow \mathbb{E}^N$ be a H-equivariant embedding of M into some euclidean space $\mathbb{E}^N$ of dimension N. For all our applications, H is compact. Then J induces the metric

$$\rho(x,y) = \|J(x) - J(y)\| \tag{4.1}$$

on M, where $\|.\|$ denotes Euclidean norm ($\|u\|^2 = \sum_{i=1}^N {u_i}^2 \ \forall u = (u_1, u_2, .., u_N)$). This is called the *extrinsic distance* on M.

For the Fréchet function F in (3.1), let $f(r) = r^2$ on $[0, \infty)$. This choice of the Fréchet function makes the Frechet mean computable in a number of important examples using Proposition 3.2. Assume $J(M) = \tilde{M}$ is a closed subset of $\mathbb{E}^N$.Then for every $u \in \mathbb{E}^N$ there exists a compact set of points in $\tilde{M}$ whose distance from u is the smallest among all points in $\tilde{M}$. We denote this set by

$$P_{\tilde{M}}u = \{x \in \tilde{M} : \|x - u\| \leq \|y - u\| \ \forall y \in \tilde{M}\}. \tag{4.2}$$

If this set is a singleton, u is said to be a *nonfocal point* of $\mathbb{E}^N$ (w.r.t. $\tilde{M}$), otherwise it is said to be a *focal point* of $\mathbb{E}^N$.

Definition 3.3. Let (M, ρ), J be as above. Let Q be a probability measure on M such that the Fréchet function

$$F(x) = \int \rho^2(x,y)Q(dy) \tag{4.3}$$

is finite. The Fréchet mean (set) of Q is called the *extrinsic mean (set)* of Q. If X_i, $i = 1, \ldots, n$ are iid observations from Q and $Q_n = \frac{1}{n}\sum_{i=1}^n \delta_{X_i}$, then the Fréchet mean(set) of Q_n is called the *extrinsic sample mean(set)*.

Let $\tilde{Q}$ and $\tilde{Q}_n$ be the images of Q and Q_n respectively in $\mathbb{E}^N$: $\tilde{Q} = Q \circ J^{-1}$, $\tilde{Q}_n = Q_n \circ J^{-1}$.

Proposition 3.2. *(a) If $\tilde{\mu} = \int_{\mathbb{E}^N} u\tilde{Q}(du)$ is the mean of $\tilde{Q}$, then the extrinsic mean set of Q is given by $J^{-1}(P_{\tilde{M}}\tilde{\mu})$. (b) If $\tilde{\mu}$ is a nonfocal point of $\mathbb{E}^N$ then the extrinsic mean of Q exists (as a unique minimizer of F).*

Proof. See Proposition 3.1, [10]. □

Corollary 3.1. *If $\tilde{\mu} = \int_{\mathbb{E}^N} u\tilde{Q}(du)$ is a nonfocal point of $\mathbb{E}^N$ then the extrinsic sample mean μ_n (any measurable selection from the extrinsic sample mean set) is a strongly consistent estimator of the extrinsic mean μ of Q.*

Proof. Follows from Proposition 3.1 for compact M. For the more general case, see [10]. □

3.4.1. *Asymptotic Distribution of the Extrinsic Sample Mean*

Although one can apply Theorem 3.1 here, we prefer a different, and more widely applicable approach, which does not require that the support of Q be contained in a coordinate patch. Let $\bar{Y} = \frac{1}{n}\sum_{j=1}^{n} Y_j$ be the (sample) mean of $Y_j = P(X_j)$. In a neighborhood of a nonfocal point such as $\tilde{\mu}$, $P(.)$ is smooth. Hence it can be shown that

$$\sqrt{n}[P(\bar{Y}) - P(\tilde{\mu})] = \sqrt{n}(d_{\tilde{\mu}}P)(\bar{Y} - \tilde{\mu}) + o_P(1) \tag{4.4}$$

where $d_{\tilde{\mu}}P$ is the differential (map) of the projection $P(.)$, which takes vectors in the tangent space of $\mathbb{E}^N$ at $\tilde{\mu}$ to tangent vectors of $\tilde{M}$ at $P(\tilde{\mu})$. Let $f_1, f_2, \ldots, f_d$ be an orthonormal basis of $T_{P(\tilde{\mu})}J(M)$ and $e_1, e_2, \ldots, e_N$ be an orthonormal basis (frame) for $T\mathbb{E}^N \approx \mathbb{E}^N$. One has

$$\sqrt{n}(\bar{Y} - \tilde{\mu}) = \sum_{j=1}^{N} \langle \sqrt{n}(\bar{Y} - \tilde{\mu}), e_j \rangle e_j,$$

$$\begin{aligned} d_{\tilde{\mu}}P(\sqrt{n}(\bar{Y} - \tilde{\mu})) &= \sum_{j=1}^{N} \langle \sqrt{n}(\bar{Y} - \tilde{\mu}), e_j \rangle d_{\tilde{\mu}}P(e_j) \\ &= \sum_{j=1}^{N} \langle \sqrt{n}(\bar{Y} - \tilde{\mu}), e_j \rangle \sum_{r=1}^{d} \langle d_{\tilde{\mu}}P(e_j), f_r \rangle f_r \\ &= \sum_{r=1}^{d} [\sum_{j=1}^{N} \langle d_{\tilde{\mu}}P(e_j), f_r \rangle \langle \sqrt{n}(\bar{Y} - \tilde{\mu}), e_j \rangle] f_r. \end{aligned} \tag{4.5}$$

Hence $\sqrt{n}[P(\bar{Y}) - P(\tilde{\mu})]$ has an asymptotic Gaussian distribution on the tangent space of $J(M)$ at $P(\tilde{\mu})$, with mean vector zero and a dispersion matrix (w.r.t. the basis vector $\{f_r : 1 \leq r \leq d\}$)

$$\Sigma = A'VA$$

where

$$A \equiv A(\tilde{\mu}) = ((\langle d_{\tilde{\mu}}P(e_j), f_r \rangle))_{1 \leq j \leq N, 1 \leq r \leq d}$$

and V is the $N \times N$ covariance matrix of $\tilde{Q} = Q \circ J^{-1}$ (w.r.t. the basis $\{e_j : 1 \leq j \leq N\}$). In matrix notation,

$$\sqrt{n}\bar{T} \xrightarrow{\mathcal{L}} N(0, \Sigma) \text{ as } n \to \infty, \tag{4.6}$$

where

$$T_j(\tilde{\mu}) = A'[\langle (Y_j - \tilde{\mu}), e_1 \rangle \ldots \langle (Y_j - \tilde{\mu}), e_N \rangle]', \; j = 1, \ldots, n$$

and

$$\bar{T} \equiv \bar{T}(\tilde{\mu}) = \frac{1}{n}\sum_{j=1}^{n} T_j(\tilde{\mu}).$$

This implies, writing $\mathcal{X}_d^2$ for the chi square distribution with d degrees of freedom,

$$n\bar{T}'\Sigma^{-1}\bar{T} \xrightarrow{\mathcal{L}} \mathcal{X}_d^2, \text{ as } n \to \infty. \tag{4.7}$$

A confidence region for $P(\tilde{\mu})$ with asymptotic confidence level $1 - \alpha$ is then given by

$$\{P(\tilde{\mu}) : n\bar{T}'\hat{\Sigma}^{-1}\bar{T} \leq \mathcal{X}_d^2(1-\alpha)\} \tag{4.8}$$

where $\hat{\Sigma} \equiv \hat{\Sigma}(\tilde{\mu})$ is the sample covariance matrix of $\{T_j(\tilde{\mu})\}_{j=1}^n$. The corresponding bootstrapped confidence region is given by

$$\{P(\tilde{\mu}) : n\bar{T}'\hat{\Sigma}^{-1}\bar{T} \leq c^*_{(1-\alpha)}\} \tag{4.9}$$

where $c^*_{(1-\alpha)}$ is the upper $(1-\alpha)$-quantile of the bootstrapped values U^*, $U^* = n\bar{T}^{*\prime}\hat{\Sigma}^{*-1}\bar{T}^*$ and $\bar{T}^*$, $\hat{\Sigma}^*$ being the sample mean and covariance respectively of the bootstrap sample $\{T_j^*(\bar{Y})\}_{j=1}^n$.

3.5. Intrinsic Means on Manifolds

Let (M, g) be a complete connected Riemannian manifold with metric tensor g. Then the natural choice for the distance metric ρ in Section 3.3 is the geodesic distance d_g on M. Unless otherwise stated, we consider the function $f(r) = r^2$ in (3.1) throughout this section and later sections. However one may take more general f. For example one may consider $f(r) = r^a$, for suitable $a \geq 1$.

Let Q be a probability distribution on M with finite Fréchet function

$$F(p) = \int_M d_g^2(p, m)Q(dm). \tag{5.1}$$

Let $X_1, \ldots, X_n$ be an iid sample from Q.

Definition 3.4. The Fréchet mean set of Q under $\rho = d_g$ is called the *intrinsic mean set* of Q. The Fréchet mean set of the empirical distribution Q_n is called the *sample intrinsic mean set.*

Before proceeding further, let us define a few technical terms related to Riemannian manifolds which we will use extensively in this section. For details on Riemannian Manifolds, see [19], [24] or [34].

1 *Geodesic*: These are curves γ on the manifold with zero acceleration. They are locally length minimizing curves. For example, consider great circles on the sphere or straight lines in $\mathbb{R}^d$.
2 *Exponential map*: For $p \in M$, $v \in T_pM$, we define $\exp_p v = \gamma(1)$, where γ is a geodesic with $\gamma(0) = p$ and $\dot{\gamma}(0) = v$.
3 *Cut locus*: For a point $p \in M$, define the cut locus $C(p)$ of p as the set of points of the form $\gamma(t_0)$, where γ is a unit speed geodesic starting at p and t_0 is the supremum of all $t > 0$ such that γ is distance minimizing from p to $\gamma(t)$. For example, $C(p) = \{-p\}$ on the sphere.
4 *Sectional Curvature*: Recall the notion of Gaussian curvature of two dimensional surfaces. On a Riemannian manifold M, choose a pair of linearly independent vectors $u, v \in T_pM$. A two dimensional submanifold of M is swept out by the set of all geodesics starting at p and with initial velocities lying in the two-dimensional section π spanned by u, v. The Gaussian curvature of this submanifold is called the sectional curvature at p of the section π.
5 *Injectivity Radius*: Define the injectivity radius of M as

$$\text{inj}(M) = \inf\{d_g(p, C(p)) : p \in M\}.$$

For example the sphere of radius 1 has injectivity radius equal to π.

Also let $r_* = \min\{\text{inj}(M), \frac{\pi}{\sqrt{\overline{C}}}\}$, where $\overline{C}$ is the least upper bound of sectional curvatures of M if this upper bound is positive, and $\overline{C} = 0$ otherwise. The exponential map at p is injective on $\{v \in T_p(M) : |v| < r_*\}$. By $B(p, r)$ we will denote an open ball with center $p \in M$ and radius r, and $\bar{B}(p, r)$ will denote its closure.

In case Q has a unique intrinsic mean μ_I, it follows from Proposition 3.1 and Remark 3.1 that the sample intrinsic mean μ_{nI} (a measurable selection from the sample intrinsic mean set) is a consistent estimator of μ_I. Broad conditions for the existence of a unique intrinsic mean are not known. From results due to [26] and [33], it follows that if the support of Q is in a geodesic ball of radius $\frac{r_*}{4}$, i.e. $\text{supp}(Q) \subseteq B(p, \frac{r_*}{4})$, then Q has a unique intrinsic mean. This result has been substantially extended by [31] which shows that if $\text{supp}(Q) \subseteq B(p, \frac{r_*}{2})$, then there is a unique local minimum of the Fréchet function F in that ball. Then we redefine the (local) intrinsic mean of Q as that unique minimizer in the ball. In that case one can show that the (local) sample intrinsic mean is a consistent estimator of the intrinsic mean of Q. This is stated in Proposition 3.3.

Proposition 3.3. *Let Q have support in $B(p, \frac{r_*}{2})$ for some $p \in M$. Then (a) Q has a unique (local) intrinsic mean μ_I in $B(p, \frac{r_*}{2})$ and (b) the sample intrinsic mean μ_{nI} in $B(p, \frac{r_*}{2})$ is a strongly consistent estimator of μ_I.*

Proof. (a) Follows from [31].
(b) Since $\mathrm{supp}(Q)$ is compact, $\mathrm{supp}(Q) \subseteq \bar{B}(p, r)$ for some $r < \frac{r_*}{2}$. From Lemma 1, [33], it follows that $\mu_I \in B(p, r)$ and μ_I is the unique intrinsic mean of Q restricted to $\bar{B}(p, r)$. Now take the compact metric space in Proposition 3.1 to be $\bar{B}(p, r)$ and the result follows. □

For the asymptotic distribution of the sample intrinsic mean, we may use Theorem 3.1. For that we need to verify assumptions A1-A7. Theorem 3.2 gives sufficient conditions for that. In the statement of the theorem, the usual partial order $A \geq B$ between $d \times d$ symmetric matrices A, B, means that $A - B$ is nonnegative definite.

Theorem 3.2. *Assume* $\mathrm{supp}(Q) \subseteq B(p, \frac{r_*}{2})$. *Let $\phi = \exp_{\mu_I}^{-1} : B(p, \frac{r_*}{2}) \longrightarrow T_{\mu_I}M (\approx \mathbb{R}^d)$. Then the map $y \mapsto h(x, y) = d_g^2(\phi^{-1}x, \phi^{-1}y)$ is twice continuously differentiable in a neighborhood of* 0 *and in terms of normal coordinates with respect to a chosen orthonormal basis for $T_{\mu_I}M$,*

$$\mathrm{D}_r h(x, 0) = -2x^r, \quad 1 \leq r \leq d, \tag{5.2}$$

$$[\mathrm{D}_r \mathrm{D}_s h(x, 0)] \geq \left[2\left\{\left(\frac{1 - f(|x|)}{|x|^2}\right) x^r x^s + f(|x|)\delta_{rs}\right\}\right]_{1 \leq r,s \leq d}. \tag{5.3}$$

Here $x = (x^1, \ldots, x^d)'$, $|x| = \sqrt{(x^1)^2 + (x^2)^2 + \ldots (x^d)^2}$ and

$$f(y) = \begin{cases} 1 \text{ if } \overline{C} = 0, \\ \sqrt{\overline{C}} y \frac{\cos(\sqrt{\overline{C}} y)}{\sin(\sqrt{\overline{C}} y)} \text{ if } \overline{C} > 0, \\ \sqrt{-\overline{C}} y \frac{\cosh(\sqrt{-\overline{C}} y)}{\sinh(\sqrt{-\overline{C}} y)} \text{ if } \overline{C} < 0. \end{cases} \tag{5.4}$$

There is equality in (5.3) *when M has constant sectional curvature $\overline{C}$, and in this case Λ has the expression:*

$$\Lambda_{rs} = 2\mathrm{E}\left\{\left(\frac{1 - f(|\tilde{X}_1|)}{|\tilde{X}_1|^2}\right) \tilde{X}_1^r \tilde{X}_1^s + f(|\tilde{X}_1|)\delta_{rs}\right\}, \quad 1 \leq r, s \leq d. \tag{5.5}$$

Λ is positive definite if $\mathrm{supp}(Q) \in B(\mu_I, \frac{r_*}{2})$.

Proof. See Theorem 2.2, [8]. □

From Theorem 3.2 it follows that $\Sigma = 4\text{Cov}(Y_1)$ where $Y_j = \phi(X_j)$, $j = 1, \ldots, n$ are the normal coordinates of the sample $X_1, \ldots, X_n$ from Q. It is nonsingular if $Q \circ \phi^{-1}$ has support in no smaller dimensional subspace of $\mathbb{R}^d$. That holds if for example Q has a density with respect to the volume measure on M.

3.6. Applications

In this section we apply the results of the earlier sections to some important manifolds. We start with the unit sphere S^d in $\mathbb{R}^{d+1}$.

3.6.1. S^d

Consider the space of all directions in $\mathbb{R}^{d+1}$ which can be identified with the unit sphere

$$S^d = \{x \in \mathbb{R}^{d+1} : \ \|x\| = 1\}.$$

Statistics on S^2, often called *directional statistics*, have been among the earliest and most widely used statistics on manifolds. (See, e.g., [42], [23], [35]). Among important applications, we cite paleomagnetism, where one may detect and/or study the shifting of magnetic poles on earth over geological times. Another application is the estimation of the direction of a signal.

3.6.1.1. *Extrinsic Mean on* S^d

The inclusion map $i : S^d \to \mathbb{R}^{d+1}$, $i(x) = x$ provides a natural embedding for S^d into $\mathbb{R}^{d+1}$. The extrinsic mean set of a probability distribution Q on S^d is then the set $P_{S^d}\tilde{\mu}$ on S^d closest to $\tilde{\mu} = \int_{\mathbb{R}^{d+1}} x\tilde{Q}(dx)$, where $\tilde{Q}$ is Q regarded as a probability measure on $\mathbb{R}^{d+1}$. Note that $\tilde{\mu}$ is non-focal iff $\tilde{\mu} \neq 0$ and then Q has a unique extrinsic mean $\mu = \frac{\tilde{\mu}}{\|\tilde{\mu}\|}$.

3.6.1.2. *Intrinsic Mean on* S^d

At each $p \in S^d$, endow the tangent space $T_pS^d = \{v \in \mathbb{R}^{d+1} : v.p = 0\}$ with the metric tensor $g_p : T_p \times T_p \to \mathbb{R}$ as the restriction of the scalar product at p of the tangent space of $\mathbb{R}^{d+1}$: $g_p(v_1, v_2) = v_1.v_2$. The geodesics are the big circles,

$$\gamma_{p,v}(t) = (\cos t|v|)p + (\sin t|v|)\frac{v}{|v|}. \tag{6.1}$$

The exponential map, $exp_p : T_pS^d \to S^d$ is

$$exp_p(v) = \cos(|v|)p + \sin(|v|)\frac{v}{|v|}, \tag{6.2}$$

and the geodesic distance is

$$d_g(p,q) = \arccos(p.q) \in [0, \pi]. \tag{6.3}$$

This space has constant sectional curvature 1 and injectivity radius π. Hence if Q has support in an open ball of radius $\frac{\pi}{2}$, then it has a unique intrinsic mean in that ball.

3.6.2. $\mathbb{R}P^d$

Consider the real projective space $\mathbb{R}P^d$ of all lines through the origin in $\mathbb{R}^{d+1}$. The elements of $\mathbb{R}P^d$ may be represented as $[u] = \{-u, u\}$ $(u \in S^d)$.

3.6.2.1. *Extrinsic Mean on* $\mathbb{R}P^d$

$\mathbb{R}P^d$ can be embedded into the space of $k \times k$ real symmetric matrices $S(k, \mathbb{R})$, $k = d+1$ via the *Veronese-Whitney embedding* $J : \mathbb{R}P^d \to S(k, \mathbb{R})$ which is given by

$$J([u]) = uu' = ((u_iu_j))_{1 \le i,j \le k} \ (u = (u_1, .., u_k)' \in S^d). \tag{6.4}$$

As a linear subspace of $\mathbb{R}^{k^2}$, $S(k, \mathbb{R})$ has the Euclidean distance

$$\|A - B\|^2 \equiv \sum_{1 \le i,j \le k} (a_{ij} - b_{ij})^2 = \text{Trace}(A - B)(A - B)'. \tag{6.5}$$

This endows $\mathbb{R}P^d$ with the extrinsic distance ρ given by

$$\rho^2([u], [v]) = \|uu' - vv'\|^2 = 2(1 - (u'v)^2). \tag{6.6}$$

Let Q be a probability distribution on $\mathbb{R}P^d$ and let $\tilde{\mu}$ be the mean of $\tilde{Q} = Q \circ J^{-1}$ considered as a probability measure on $S(k, \mathbb{R})$. Then $\tilde{\mu} \in S^+(k, \mathbb{R})$-the space of $k \times k$ real symmetric nonnegative definite matrices, and the projection of $\tilde{\mu}$ into $J(\mathbb{R}P^d)$ is given by the set of all uu' where u is a unit eigenvector of $\tilde{\mu}$ corresponding to the largest eigenvalue. Hence the projection is unique, i.e. $\tilde{\mu}$ is nonfocal iff its largest eigenvalue is simple, i.e., if the eigenspace corresponding to the largest eigenvalue is one dimensional. In that case the extrinsic mean of Q is $[u]$, u being a unit eigenvector in the eigenspace of the largest eigenvalue.

3.6.2.2. *Intrinsic Mean on* $\mathbb{R}P^d$

$\mathbb{R}P^d$ is a complete Riemannian manifold with geodesic distance

$$d_g([p],[q]) = \arccos(|p.q|) \in [0, \frac{\pi}{2}]. \tag{6.7}$$

It has constant sectional curvature 4 and injectivity radius $\frac{\pi}{2}$. Hence if the support of Q is contained in an open geodesic ball of radius $\frac{\pi}{4}$, it has a unique intrinsic mean in that ball.

3.6.3. Σ_m^k

Consider a set of k points in $\mathbb{R}^m$, not all points being the same. Such a set is called a k-ad or a configuration of k landmarks. We will denote a k-ad by the $m \times k$ matrix, $x = [x_1 \dots x_k]$ where x_i, $i = 1, \dots, k$ are the k landmarks from the object of interest. Assume $k > m$. The *direct similarity shape* of the k-ad is what remains after we remove the effects of translation, rotation and scaling. To remove translation, we subtract the mean $\bar{x} = \frac{1}{k}\sum_{i=1}^k x_i$ from each landmark to get the centered k-ad $w = [x_1 - \bar{x} \dots x_k - \bar{x}]$. We remove the effect of scaling by dividing w by its euclidean norm to get

$$u = [\frac{x_1 - \bar{x}}{\|w\|} \dots \frac{x_k - \bar{x}}{\|w\|}] = [u_1 u_2 \dots u_k]. \tag{6.8}$$

This u is called the *preshape* of the k-ad x and it lies in the unit sphere S_m^k in the hyperplane

$$H_m^k = \{u \in \mathbb{R}^{km} : \sum_{j=1}^k u_j = 0\}. \tag{6.9}$$

Thus the preshape space S_m^k may be identified with the sphere S^{km-m-1}. Then the shape of the k-ad x is the orbit of z under left multiplication by $m \times m$ rotation matrices. In other words $\Sigma_m^k = S^{km-m-1}/SO(m)$. The cases of importance are $m = 2, 3$. Next we turn to the case $m = 2$.

3.6.4. Σ_2^k

As pointed out in Sections 3.2.2 and 3.6.3, $\Sigma_2^k = S^{2k-3}/SO(2)$. For a simpler representation, we denote a k-ad in the plane by a set of k complex numbers. The preshape of this complex k-vector x is $z = \frac{x - \bar{x}}{\|x - \bar{x}\|}$, $x = (x_1, \dots, x_k) \in \mathbb{C}^k$, $\bar{x} = \frac{1}{k}\sum_{i=1}^k x_i$. z lies in the complex sphere

$$S_2^k = \{z \in \mathbb{C}^k : \sum_{j=1}^k |z_j|^2 = 1, \sum_{j=1}^k z_j = 0\} \tag{6.10}$$

which may be identified with the real sphere of dimension $2k-3$. Then the shape of x can be represented as the orbit

$$\sigma(x) = \sigma(z) = \{e^{i\theta}z\colon\ -\pi < \theta \leq \pi\} \tag{6.11}$$

and

$$\Sigma_2^k = \{\sigma(z)\colon z \in S_2^k\}. \tag{6.12}$$

Thus Σ_2^k has the structure of the complex projective space $\mathbb{C}P^{k-2}$ of all complex lines through the origin in $\mathbb{C}^{k-1}$, an important and well studied manifold in differential geometry (See [24], pp. 63-65, 97-100, [8]).

3.6.4.1. ***Extrinsic Mean on*** Σ_2^k

Σ_2^k can be embedded into $S(k, \mathbb{C})$, the space of $k \times k$ complex Hermitian matrices, via the *Veronese-Whitney embedding*

$$J\colon \Sigma_2^k \to S(k, \mathbb{C}),\ J(\sigma(z)) = zz^*. \tag{6.13}$$

J is equivariant under the action of $SU(k)$, the group of $k \times k$ complex matrices Γ such that $\Gamma^*\Gamma = I$, $\det(\Gamma) = 1$. To see this, let $\Gamma \in SU(k)$. Then Γ defines a diffeormorphism,

$$\Gamma : \Sigma_2^k \to \Sigma_2^k,\ \Gamma(\sigma(z)) = \sigma(\Gamma(z)). \tag{6.14}$$

The map ϕ_Γ on $S(k, \mathbb{C})$ defined by

$$\phi_\Gamma(A) = \Gamma A \Gamma^* \tag{6.15}$$

preserves distances and has the property

$$(\phi_\Gamma)^{-1} = \phi_{\Gamma^{-1}},\ \phi_{\Gamma_1\Gamma_2} = \phi_{\Gamma_1} \circ \phi_{\Gamma_2}. \tag{6.16}$$

That is, (6.15) defines a group homomorphism from $SU(k)$ into a group of isometries of $S(k, \mathbb{C})$. Finally note that $J(\Gamma(\sigma(z))) = \phi_\Gamma(J(\sigma(z)))$. Informally, the symmetries $SU(k)$ of Σ_2^k are preserved by the embedding J.

$S(k, \mathbb{C})$ is a (real) vector space of dimension k^2. It has the Euclidean distance,

$$\|A - B\|^2 = \sum_{i,j} |a_{ij} - b_{ij}\|^2 = \mathrm{Trace}(A - B)^2. \tag{6.17}$$

Thus the extrinsic distance ρ on Σ_2^k induced from the Veronese-Whitney embedding is given by

$$\rho^2(\sigma(x), \sigma(y)) = \|uu^* - vv^*\|^2 = 2(1 - |u^*v|^2), \tag{6.18}$$

where x and y are two k-ads, u and v are their preshapes respectively.

Let Q be a probability distribution on Σ_2^k and let $\tilde{\mu}$ be the mean of $\tilde{Q} = Q \circ J^{-1}$, regarded as a probability measure on $\mathbb{C}^{k^2}$. Then $\tilde{\mu} \in S_+(k, \mathbb{C})$: the space of $k \times k$ complex positive semidefinite matrices. Its projection into $J(\Sigma_2^k)$ is given by $P(\tilde{\mu}) = \{uu^*\}$ where u is a unit eigenvector of $\tilde{\mu}$ corresponding to its largest eigenvalue. The projection is unique, i.e. $\tilde{\mu}$ is nonfocal, and Q has a unique extrinsic mean μ_E, iff the eigenspace for the largest eigenvalue of $\tilde{\mu}$ is (complex) one dimensional, and then $\mu_E = \sigma(u)$, $u(\neq 0) \in$ eigenspace of the largest eigenvalue of $\tilde{\mu}$. Let $X_1, \ldots X_n$ be an iid sample from Q. If $\tilde{\mu}$ is nonfocal, the sample extrinsic mean μ_{nE} is a consistent estimator of μ_E and $J(\mu_{nE})$ has an asymptotic Gaussian distribution on the tangent space $T_{P(\tilde{\mu})} J(\Sigma_2^k)$ (see Section 3.4),

$$\sqrt{n}(J(\mu_{nE}) - J(\mu_E)) = \sqrt{n} d_{\tilde{\mu}} P(\overline{\tilde{X}} - \tilde{\mu}) + o_P(1) \xrightarrow{\mathcal{L}} N(0, \Sigma). \quad (6.19)$$

Here $\tilde{X}_j = J(X_j)$, $j = 1, \ldots, n$. In (6.19), $d_{\tilde{\mu}} P(\overline{\tilde{X}} - \tilde{\mu})$ has coordinates

$$\bar{T}(\tilde{\mu}) = (\sqrt{2}\mathrm{Re}(U_a^* \bar{\tilde{X}} U_k), \sqrt{2}\mathrm{Im}(U_a^* \bar{\tilde{X}} U_k))_{a=2}^{k-1} \quad (6.20)$$

with respect to the basis

$$\{(\lambda_k - \lambda_a)^{-1} U v_k^a U^*, (\lambda_k - \lambda_a)^{-1} U w_k^a U^*\}_{a=2}^{k-1} \quad (6.21)$$

for $T_{P(\tilde{\mu})} J(\Sigma_2^k)$ (see Section 3.3, [7]). Here $U = [U_1 \ldots U_k] \in SO(k)$ is such that $U^* \tilde{\mu} U = D \equiv \mathrm{Diag}(\lambda_1, \ldots, \lambda_k)$, $\lambda_1 \leq \ldots \leq \lambda_{k-1} < \lambda_k$ being the eigenvalues of $\tilde{\mu}$. $\{v_b^a : 1 \leq a \leq b \leq k\}$ and $\{w_b^a : 1 \leq a < b \leq k\}$ is the canonical orthonormal basis frame for $S(k, \mathbb{C})$, defined as

$$v_b^a = \begin{cases} \frac{1}{\sqrt{2}}(e_a e_b^t + e_b e_a^t), & a < b \\ e_a e_a^t, & a = b \end{cases}$$

$$w_b^a = \frac{i}{\sqrt{2}}(e_a e_b^t - e_b e_a^t), \; a < b \quad (6.22)$$

where $\{e_a : 1 \leq a \leq k\}$ is the standard canonical basis for $\mathbb{R}^k$.

Given two independent samples $X_1, \ldots X_n$ iid Q_1 and $Y_1, \ldots Y_m$ iid Q_2 on Σ_2^k, we may like to test if $Q_1 = Q_2$ by comparing their extrinsic mean shapes. Let μ_{iE} denote the extrinsic mean of Q_i and let μ_i be the mean of $Q_i \circ J^{-1}$ $i = 1, 2$. Then $\mu_{iE} = J^{-1}P(\mu_i)$, and we wish to test $H_0 : P(\mu_1) = P(\mu_2)$. Let $\tilde{X}_j = J(X_j)$, $j = 1, \ldots, n$ and $\tilde{Y}_j = J(Y_j)$, $j = 1, \ldots, m$. Let T_j, S_j denote the asymptotic coordinates for $\tilde{X}_j, \tilde{Y}_j$ respectively in $T_{P(\hat{\mu})}J(\Sigma_2^k)$ as defined in (6.20). Here $\hat{\mu} = \frac{n\bar{\tilde{X}} + m\bar{\tilde{Y}}}{m+n}$ is the pooled sample mean. We use the two sample test statistic

$$T_{nm} = (\bar{T} - \bar{S})' \left(\frac{1}{n}\hat{\Sigma}_1 + \frac{1}{m}\hat{\Sigma}_2\right)^{-1} (\bar{T} - \bar{S}). \tag{6.23}$$

Here $\hat{\Sigma}_1, \hat{\Sigma}_2$ denote the sample covariances of T_j, S_j respectively. Under H_0, $T_{nm} \xrightarrow{\mathcal{L}} \mathcal{X}_{2k-4}^2$ (see Section 3.4, [7]). Hence given level α, we reject H_0 if $T_{nm} > \mathcal{X}_{2k-4}^2(1 - \alpha)$.

3.6.4.2. ***Intrinsic Mean on*** Σ_2^k

Identified with $\mathbb{C}P^{k-2}$, Σ_2^k is a complete connected Riemannian manifold. It has all sectional curvatures bounded between 1 and 4 and injectivity radius of $\frac{\pi}{2}$ (see [24], pp. 97-100, 134). Hence if $supp(Q) \in B(p, \frac{\pi}{4})$, $p \in \Sigma_2^k$, it has a unique intrinsic mean μ_I in the ball.

Let $X_1, \ldots X_n$ be iid Q and let μ_{nI} denote the sample intrinsic mean. Under the hypothesis of Theorem 3.2,

$$\sqrt{n}(\phi(\mu_{nI}) - \phi(\mu_I)) \xrightarrow{\mathcal{L}} N(0, \Lambda^{-1}\Sigma\Lambda^{-1}). \tag{6.24}$$

However Theorem 3.2 does not provide an analytic computation of Λ, since Σ_2^k does not have constant sectional curvature. Proposition 3.4 below gives the precise expression for Λ. It also relaxes the support condition required for Λ to be positive definite.

Proposition 3.4. *With respect to normal coordinates,* $\phi : B(p, \frac{\pi}{4}) \to \mathbb{C}^{k-2} (\approx \mathbb{R}^{2k-4})$, Λ *as defined in Theorem 3.1 has the following expression:*

$$\Lambda = \begin{bmatrix} \Lambda_{11} & \Lambda_{12} \\ \Lambda_{12}' & \Lambda_{22} \end{bmatrix} \tag{6.25}$$

where for $1 \leq r, s \leq k-2$,

$$(\Lambda_{11})_{rs} = 2\mathrm{E}\big[\ d_1 \cot(d_1)\delta_{rs} - \frac{\{1-d_1\cot(d_1)\}}{d_1^2}(\mathrm{Re}\tilde{X}_{1,r})(\mathrm{Re}\tilde{X}_{1,s})$$

$$+ \frac{\tan(d_1)}{d_1}(\mathrm{Im}\tilde{X}_{1,r})(\mathrm{Im}\tilde{X}_{1,s})\ \big],$$

$$(\Lambda_{22})_{rs} = 2\mathrm{E}\big[\ d_1 \cot(d_1)\delta_{rs} - \frac{\{1-d_1\cot(d_1)\}}{d_1^2}(\mathrm{Im}\tilde{X}_{1,r})(\mathrm{Im}\tilde{X}_{1,s})$$

$$+ \frac{\tan(d_1)}{d_1}(\mathrm{Re}\tilde{X}_{1,r})(\mathrm{Re}\tilde{X}_{1,s})\ \big],$$

$$(\Lambda_{12})_{rs} = -2\mathrm{E}\big[\ \frac{\{1-d_1 cot(d_1)\}}{d_1^2}(\mathrm{Re}\tilde{X}_{1,r})(\mathrm{Im}\tilde{X}_{1,s})$$

$$+ \frac{\tan(d_1)}{d_1}(\mathrm{Im}\tilde{X}_{1,r})(\mathrm{Re}\tilde{X}_{1,s})\ \big]$$

where $d_1 = d_g(X_1, \mu_I)$ *and* $\tilde{X}_j \equiv (\tilde{X}_{j,1}, \ldots, \tilde{X}_{j,k-2}) = \phi(X_j)$, $j = 1, \ldots, n$. Λ *is positive definite if* $\mathrm{supp}(Q) \in B(\mu_I, 0.37\pi)$.

Proof. See Theorem 3.1, [8]. □

Note that with respect to a chosen orthonormal basis $\{v_1, \ldots, v_{k-2}\}$ for $T_{\mu_I}\Sigma_2^k$, ϕ has the expression

$$\phi(m) = (\tilde{m}_1, \ldots, \tilde{m}_{k-2})'$$

where

$$\tilde{m}_j = \frac{r}{\sin r} e^{i\theta} \bar{v}_j{}' z,\ r = d_g(m, \mu_I) = \arccos(|z_0'\bar{z}|),\ e^{i\theta} = \frac{z_0'\bar{z}}{|z_0'\bar{z}|}. \tag{6.26}$$

Here z, z_0 are the preshapes of m, μ_I respectively (see Section 3, [8]).

Given two independent samples $X_1, \ldots X_n$ iid Q_1 and $Y_1, \ldots Y_m$ iid Q_2, one may test if Q_1 and Q_2 have the same intrinsic mean μ_I. The test statistic used is

$$T_{nm} = (n+m)(\hat{\phi}(\mu_{nI}) - \hat{\phi}(\mu_{mI}))'\hat{\Sigma}^{-1}(\hat{\phi}(\mu_{nI}) - \hat{\phi}(\mu_{mI})). \tag{6.27}$$

Here μ_{nI} and μ_{mI} are the sample intrinsic means for the X and Y samples respectively and $\hat{\mu}$ is the pooled sample intrinsic mean. Then $\hat{\phi} = \exp_{\hat{\mu}}^{-1}$ gives normal coordinates on the tangent space at $\hat{\mu}$, and $\hat{\Sigma} = (m+n)\left(\frac{1}{n}\hat{\Lambda}_1^{-1}\hat{\Sigma}_1\hat{\Lambda}_1^{-1} + \frac{1}{m}\hat{\Lambda}_2^{-1}\hat{\Sigma}_2\hat{\Lambda}_2^{-1}\right)$, where (Λ_1, Σ_1) and (Λ_2, Σ_2)

are the parameters in the asymptotic distribution of $\sqrt{n}(\phi(\mu_{nI})-\phi(\mu_I))$ and $\sqrt{m}(\phi(\mu_{mI} - \phi(\mu_I))$ respectively, as defined in Theorem 3.1, and $(\hat{\Lambda}_1, \hat{\Sigma}_1)$ and $(\hat{\Lambda}_2, \hat{\Sigma}_2)$ are consistent sample estimates. Assuming H_0 to be true, $T_{nm} \xrightarrow{\mathcal{L}} \mathcal{X}^2_{2k-4}$ (see Section 4.1, [7]). Hence we reject H_0 at asymptotic level $1-\alpha$ if $T_{nm} > \mathcal{X}^2_{2k-4}(1-\alpha)$.

3.6.5. $R\Sigma^k_m$

For $m > 2$, the direct similarity shape space Σ^k_m fails to be a manifold. That is because the action of $SO(m)$ is not in general free (see, e.g., [30] and [39]). To avoid that one may consider the shape of only those k-ads whose preshapes have rank at least $m-1$. This subset is a manifold but not complete (in its geodesic distance). Alternatively one may also remove the effect of reflection and redefine shape of a k-ad x as

$$\sigma(x) = \sigma(z) = \{Az \colon A \in O(m)\} \tag{6.28}$$

where z is the preshape. Then $R\Sigma^k_m$ is the space of all such shapes where rank of z is m. In other words

$$R\Sigma^k_m = \{\sigma(z) \colon z \in S^k_m,\ \mathrm{rank}(z) = m\}. \tag{6.29}$$

This is a manifold. It has been shown that the map

$$J : R\Sigma^k_m \to S(k, \mathbb{R}),\ J(\sigma(z)) = z'z \tag{6.30}$$

is an embedding of the reflection shape space into $S(k, \mathbb{R})$ (see [2], [1], and [20]) and is H-equivariant where $H = O(k)$ acts on the right: $A\sigma(z) \doteq \sigma(zA')$, $A \in O(k)$.

Let Q be a probability distribution on $R\Sigma^k_m$ and let $\tilde{\mu}$ be the mean of $Q \circ J^{-1}$ regarded as a probability measure on $S(k, \mathbb{R})$. Then $\tilde{\mu}$ is positive semi-definite with rank atleast m. Let $\tilde{\mu} = UDU'$ be the singular value decomposition of $\tilde{\mu}$, where $D = \mathrm{Diag}(\lambda_1, \ldots, \lambda_k)$ consists of ordered eigen values $\lambda_1 \geq \ldots \geq \lambda_m \geq \ldots \geq \lambda_k \geq 0$ of $\tilde{\mu}$, and $U = [U_1 \ldots U_k]$ is a matrix in $SO(k)$ whose columns are the corresponding orthonormal eigen vectors. It has been shown in [6] that the extrinsic mean reflection shape set of Q has the following expression:

$$\left\{\mu \in R\Sigma^k_m \colon J(\mu) = \sum_{j=1}^{m}(\lambda_j - \bar{\lambda} + \frac{1}{m})U_j U_j'\right\} \tag{6.31}$$

where $\bar{\lambda} = \frac{1}{m}\sum_{j=1}^m \lambda_j$. The set in (6.31) is a singleton, and hence Q has a unique mean reflection shape μ iff $\lambda_m > \lambda_{m+1}$. Then $\mu = \sigma(u)$ where

$$u = [\sqrt{(\lambda_1 - \bar{\lambda} + \frac{1}{m})}U_1, \ldots, \sqrt{(\lambda_m - \bar{\lambda} + \frac{1}{m})}U_m]'. \tag{6.32}$$

3.6.6. $A\Sigma_m^k$

Let z be a centered k-ad in $H(m,k)$, and let $\sigma(z)$ denote its affine shape, as defined in Section 3.2.4. Consider the map

$$J : A\Sigma_m^k \to S(k,\mathbb{R}),\ J(\sigma(z)) \equiv P = FF' \tag{6.33}$$

where $F = [f_1 f_2 \dots f_m]$ is an orthonormal basis for the row space of z. This is an embedding of $A\Sigma_m^k$ into $S(k,\mathbb{R})$ with the image

$$J(A\Sigma_m^k) = \{A \in S(k,\mathbb{R})\colon A^2 = A,\ \text{Trace}(A) = m,\ A\mathbf{1} = 0\}. \tag{6.34}$$

It is equivariant under the action of $O(k)$ (see [18]).

Proposition 3.5. *Let Q be a probability distribution on $A\Sigma_m^k$ and let $\tilde{\mu}$ be the mean of $Q \circ J^{-1}$ in $S(k,\mathbb{R})$. The projection of $\tilde{\mu}$ into $J(A\Sigma_m^k)$ is given by*

$$P(\tilde{\mu}) = \{\sum_{j=1}^{m} U_j U_j'\} \tag{6.35}$$

where $U = [U_1 \dots U_k] \in SO(k)$ is such that $U'\tilde{\mu}U = D \equiv \text{Diag}(\lambda_1, \dots, \lambda_k)$, $\lambda_1 \geq \dots \geq \lambda_m \geq \dots \geq \lambda_k$. $\tilde{\mu}$ is nonfocal and Q has a unique extrinsic mean μ_E iff $\lambda_m > \lambda_{m+1}$. Then $\mu_E = \sigma(F')$ where $F = [U_1 \dots U_m]$.

Proof. See [41]. □

3.6.7. $P_0\Sigma_m^k$

Consider the diffeomorphism between $P_I\Sigma_m^k$ and $(\mathbb{R}P^m)^{k-m-2}$ as defined in Section 3.2.5. Using that one can embedd $P_I\Sigma_m^k$ into $S(m+1,\mathbb{R})^{k-m-2}$ via the Veronese Whitney embedding of Section 3.6.2 and perform extrinsic analysis in a dense open subset of $P_0\Sigma_m^k$.

3.7. Examples

3.7.1. *Example 1: Gorilla Skulls*

To test the difference in the shapes of skulls of male and female gorillas, eight landmarks were chosen on the midline plane of the skulls of 29 male and 30 female gorillas. The data can be found in [21], pp. 317-318. Thus we have two iid samples in Σ_2^k, $k = 8$. The sample extrinsic mean shapes

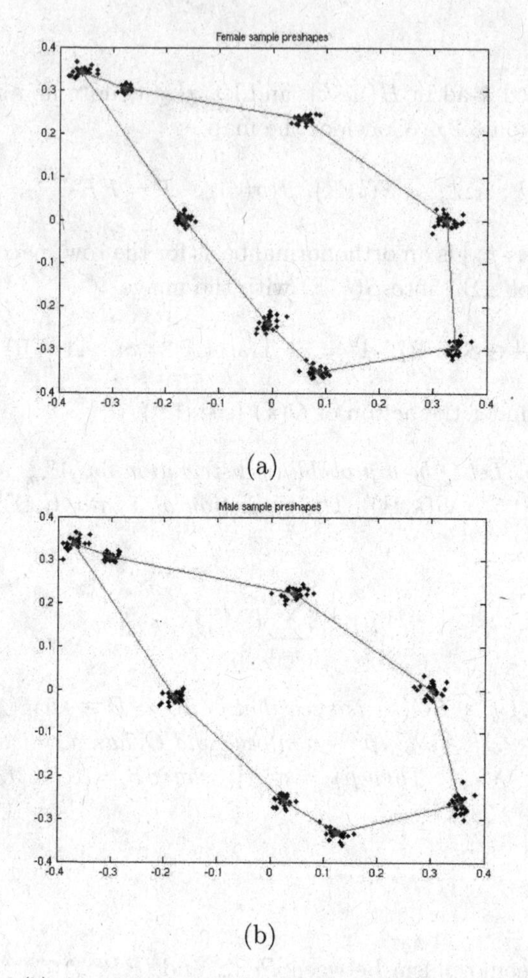

Fig. 3.1. (a) and (b) show 8 landmarks from skulls of 30 female and 29 male gorillas, respectively, along with the mean shapes. * correspond to the mean shapes' landmarks.

for the female and male samples are denoted by $\hat{\mu}_{1E}$ and $\hat{\mu}_{2E}$ where

$$\begin{aligned}\hat{\mu}_{1E} = \sigma[&-0.3586 + 0.3425i, 0.3421 - 0.2943i, 0.0851 - 0.3519i,\\ &-0.0085 - 0.2388i, -0.1675 + 0.0021i, -0.2766 + 0.3050i,\\ &0.0587 + 0.2353i, 0.3253],\\ \hat{\mu}_{2E} = \sigma[&-0.3692 + 0.3386i, 0.3548 - 0.2641i, 0.1246 - 0.3320i,\\ &0.0245 - 0.2562i, -0.1792 - 0.0179i, -0.3016 + 0.3072i,\\ &0.0438 + 0.2245i, 0.3022].\end{aligned}$$

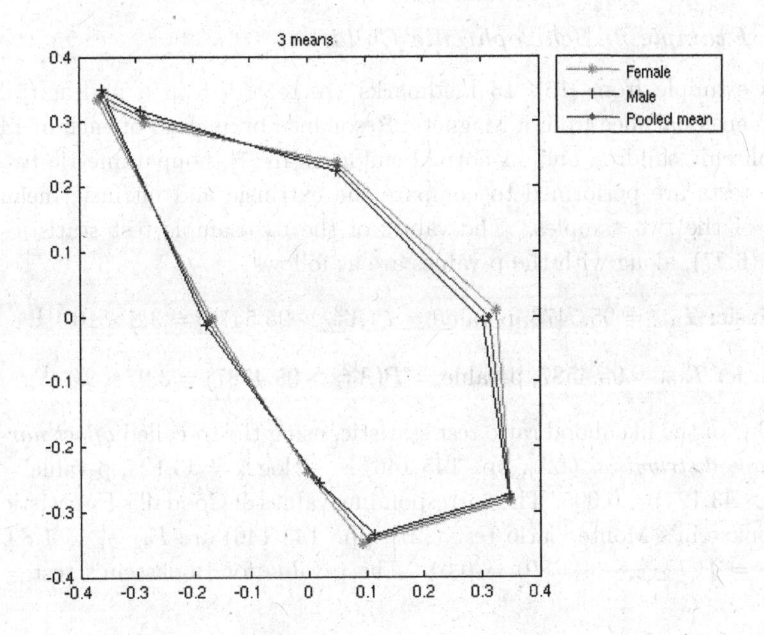

Fig. 3.2. The sample extrinsic means for the 2 groups along with the pooled sample mean, corresponding to Figure 3.1.

The corresponding intrinsic mean shapes are denoted by $\hat{\mu}_{1I}$ and $\hat{\mu}_{2I}$. They are very close to the extrinsic means ($d_g(\hat{\mu}_{1E}, \hat{\mu}_{1I}) = 5.5395 \times 10^{-7}$, $d_g(\hat{\mu}_{2E}, \hat{\mu}_{2I}) = 1.9609 \times 10^{-6}$). Figure 3.1 shows the preshapes of the sample k-ads along with that of the extrinsic mean. The sample preshapes have been rotated appropriately so as to minimize the Euclidean distance from the mean preshape. Figure 3.2 shows the preshapes of the extrinsic means for the two samples along with that of the pooled sample extrinsic mean. In [7], nonparametric two sample tests are performed to compare the mean shapes. The statistics (6.23) and (6.27) yield the following values:

$$\text{Extrinsic: } T_{nm} = 392.6, \text{ p-value} = P(\mathcal{X}_{12}^2 > 392.6) < 10^{-16}.$$

$$\text{Intrinsic: } T_{nm} = 391.63, \text{ p-value} = P(\mathcal{X}_{12}^2 > 391.63) < 10^{-16}.$$

A parametric F-test ([21], pp. 154) yields $F = 26.47$, p-value $= P(F_{12,46} > 26.47) = 0.0001$. A parametric (Normal) model for Bookstein coordinates leads to the Hotelling's T^2 test ([21], pp. 170-172) yields the p-value 0.0001.

3.7.2. *Example 2: Schizophrenic Children*

In this example from [15], 13 landmarks are recorded on a midsagittal two-dimensional slice from a Magnetic Resonance brain scan of each of 14 schizophrenic children and 14 normal children. In [7], nonparametric two sample tests are performed to compare the extrinsic and intrinsic mean shapes of the two samples. The values of the two-sample test statistics (6.23), (6.27), along with the p-values are as follows.

$$\text{Extrinsic: } T_{nm} = 95.5476, \text{ p-value} = P(\mathcal{X}^2_{22} > 95.5476) = 3.8 \times 10^{-11}.$$

$$\text{Intrinsic: } T_{nm} = 95.4587, \text{ p-value} = P(\mathcal{X}^2_{22} > 95.4587) = 3.97 \times 10^{-11}.$$

The value of the likelihood ratio test statistic, using the so-called *offset normal shape distribution* ([21], pp. 145-146) is $-2\log\Lambda = 43.124$, p-value $= P(\mathcal{X}^2_{22} > 43.124) = 0.005$. The corresponding values of Goodall's F-statistic and Bookstein's Monte Carlo test ([21], pp. 145-146) are $F_{22,572} = 1.89$, p-value $= P(F_{22,572} > 1.89) = 0.01$. The p-value for Bookstein's test $=$ 0.04.

3.7.3. *Example 3: Glaucoma Detection*

To detect any shape change due to Glaucoma, 3D images of the Optic Nerve Head (ONH) of both eyes of 12 rhesus monkeys were collected. One of the eyes was treated while the other was left untreated. 5 landmarks were recorded on each eye and their reflection shape was considered in $R\Sigma_3^k$, $k = 5$. For details on landmark registration, see [17]. The landmark coordinates can be found in [11]. Figure 3.3 shows the preshapes of the sample k-ads along with that of the mean shapes. The sample points have been rotated and (or) reflected so as to minimize their Euclidean distance from the mean preshapes. Figure 3.4 shows the preshapes of the mean shapes for the two eyes along with that of the pooled sample mean shape. In [1], 4 landmarks are selected and the sample mean shapes of the two eyes are compared. Five local coordinates are used in the neighborhood of the mean to compute Bonferroni type Bootstrap Confidence Intervals for the difference between the local reflection similarity shape coordinates of the paired glaucomatous versus control eye (see Section 6.1, [1] for details). It is found that the means are different at 1% level of significance.

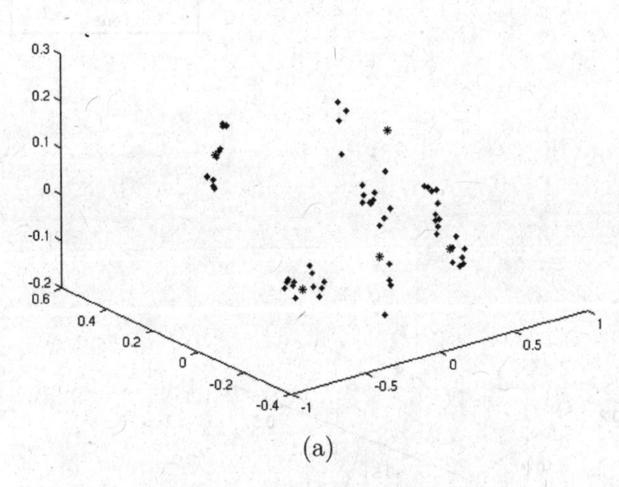

(a)

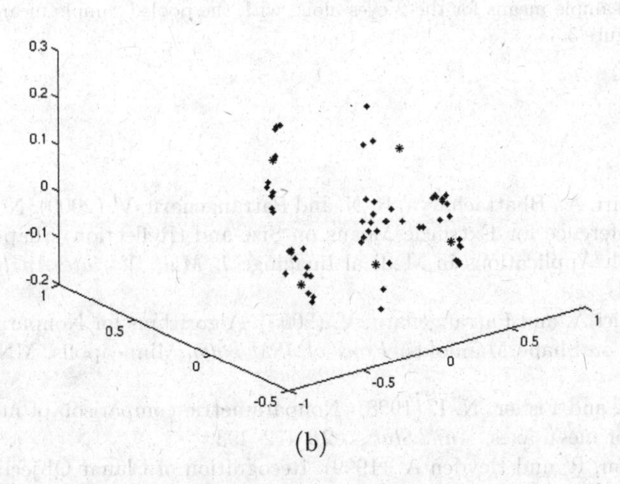

(b)

Fig. 3.3. (a) and (b) show 5 landmarks from treated and untreated eyes of 12 monkeys, respectively, along with the mean shapes. * correspond to the mean shapes' landmarks.

Acknowledgment

The authors thank the referee for his suggestions.

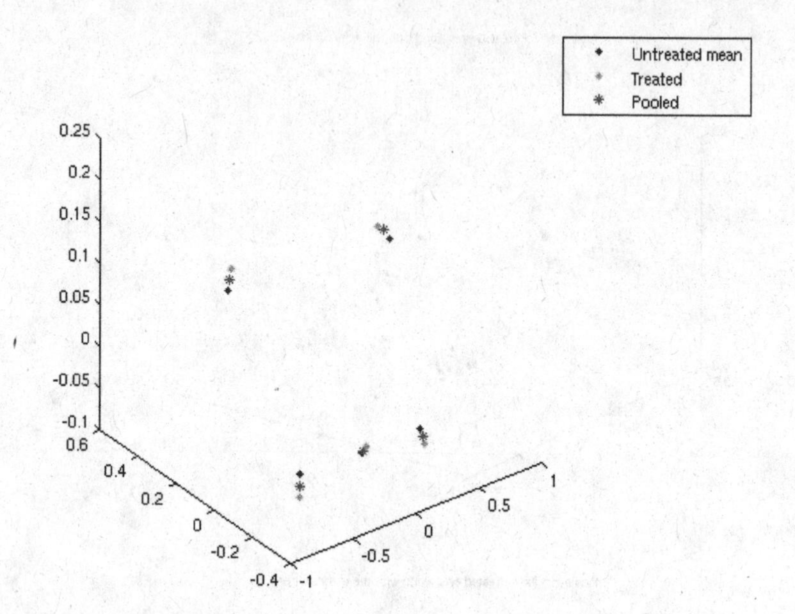

Fig. 3.4. The sample means for the 2 eyes along with the pooled sample mean, corresponding to Figure 3.3.

References

[1] Bandulasiri, A., Bhattacharya, R. N. and Patrangenaru, V. (2009). Nonparametric Inference for Extrinsic Means on Size-and-(Reflection)-Shape Manifolds with Applications in Medical Imaging. *J. Mult. Variate Analysis*. In Press.

[2] Bandulasiri A. and Patrangenaru, V. (2005). Algorithms for Nonparametric Inference on Shape Manifolds. *Proc. of JSM 2005*, Minneapolis, MN 1617–1622.

[3] Beran, R. and Fisher, N. I. (1998). Nonparametric comparison of mean directions or mean axes. *Ann. Statist.* **26** 472–493.

[4] Berthilsson, R. and Heyden A. (1999). Recognition of Planar Objects using the Density of Affine Shape. *Computer Vision and Image Understanding.* **76** 135–145.

[5] Berthilsson, R. and Astrom, K. (1999). Extension of affine shape. *J. Math. Imaging Vision.* **11** 119–136.

[6] Bhattacharya, A. (2009). Statistical Analysis on Manifolds: A Nonparametric Approach for Inference on Shape Spaces. *Sankhya Series* **A**. In Press.

[7] Bhattacharya, A. and Bhattacharya, R. (2008a). Nonparametric Statistics on Manifolds with Applications to Shape Spaces. *Pushing the Limits of Contemporary Statistics: Contributions in honor of J. K. Ghosh.* IMS Lecture Series.

[8] Bhattacharya, A. and Bhattacharya, R. (2008b). Statistics on Riemannian Manifolds: Asymptotic Distribution and Curvature. *Proc. Amer. Math. Soc.*

[9] Bhattacharya, R. N. and Patrangenaru, V. (2002). Nonparametric estimation of location and dispersion on Riemannian manifolds. *J. Statist. Plann. Inference.* **108** 23–35.

[10] Bhattacharya, R. N. and Patrangenaru, V. (2003). Large sample theory of intrinsic and extrinsic sample means on manifolds - I. *Ann. Statist.* **31** 1–29.

[11] Bhattacharya, R. and Patrangenaru, V. (2005). Large sample theory of intrinsic and extrinsic sample means on manifolds - II. *Ann. Statist.* **33** 1225–1259.

[12] Bigot, J. (2006). Landmark-based registration of curves via the continuous wavelet transform. *J. Comput. Graph. Statist.* **15** 542–564.

[13] Bookstein, F. (1978). *The Measurement of Biological Shape and Shape Change.* Lecture Notes in Biomathematics, **24**. Springer, Berlin.

[14] Bookstein, F. L. (1986). Size and shape spaces of landmark data (with discussion). *Statistical Science.* **1** 181–242.

[15] Bookstein, F. L. (1991). *Morphometric Tools for Landmark data: Geometry and Biology.* Cambridge Univ. Press.

[16] Boothby, W. (1986). *An Introduction to Differentiable Manifolds and Riemannian Geometry.* 2d ed. Academic Press, Orlando.

[17] Derado, G., Mardia, K. V., Patrangenaru, V. and Thompson, H. W. (2004). A Shape Based Glaucoma Index for Tomographic Images. *J. Appl. Statist.* **31** 1241–1248.

[18] Dimitrić, I. (1996). A note on equivariant embeddings of Grassmannians. *Publ. Inst. Math.* (Beograd) (N.S.) **59** 131–137.

[19] do Carmo, M. P. (1992). *Riemannian Geometry.* Birkhauser, Boston. English translation by F. Flaherty.

[20] Dryden, I. L., Kume, A., Le, H. and Wood, A. T. A. (2008). A multidimensional scaling approach to shape analysis. *Biometrika* **95** 779–798.

[21] Dryden, I. L. and Mardia, K. V. (1998). *Statistical Shape Analysis.* Wiley N.Y.

[22] Fisher, N. I., Lewis, T. and Embleton, B. J. J. (1987). *Statistical Analysis of Spherical Data.* Cambridge University Press, Cambridge.

[23] Fisher, N. I., Hall, P., Jing, B.-Y. and Wood, A. T. A. (1996). Improved pivotal methods for constructing confidence regions with directional data. *J. Amer. Statist. Assoc.* **91** 1062–1070.

[24] Gallot, S., Hulin, D. and Lafontaine, J. (1990). *Riemannian Geometry*, 2nd ed. Springer.

[25] Hendricks, H. and Landsman, Z. (1998). Mean location and sample mean location on manifolds: Asymptotics, tests, confidence regions. *J. Multivariate Anal.* **67** 227–243.

[26] Karchar, H. (1977). Riemannian center of mass & mollifier smoothing. *Comm. on Pure & Applied Math.* **30** 509–541.

[27] Kendall, D. G. (1977). The diffusion of shape. *Adv. Appl. Probab.* **9** 428–430.

[28] Kendall, D. G. (1984). Shape manifolds, Procrustean metrics, and complex projective spaces. *Bull. London Math. Soc.* **16** 81–121.

[29] Kendall, D. G. (1989) A survey of the statistical theory of shape. *Statist. Sci.* **4** 87–120.

[30] Kendall, D. G., Barden, D., Carne, T. K. and Le, H. (1999). *Shape & Shape Theory.* Wiley N.Y.

[31] Kendall, W. S. (1990). Probability, convexity, and harmonic maps with small image-I. Uniqueness and the fine existence. *Proc. London Math. Soc.* **61** 371–406.

[32] Kent, J. T. (1992). New directions in shape analysis. In *The Art of Statistical Science: A Tribute to G. S. Watson* (K. V. Mardia, ed.) 115–128. Wiley, New York.

[33] Le, H. (2001). Locating Frechet means with application to shape spaces. *Adv. Appl. Prob.* **33** 324–338.

[34] Lee, J. M. (1997). *Riemannian Manifolds: An Introduction to Curvature.* Springer, New York.

[35] Mardia, K. V. and Jupp, P. E. (1999). *Statistics of Directional Data.* Wiley, New York.

[36] Mardia, K. V. and Patrangenaru, V. (2005). Directions and projective shapes. *Ann. Statist.* **33** 1666–1699.

[37] Ramsay, J. O. and Silverman, B. W. (2005). *Functional Data Analysis.* Second ed. Springer, New York.

[38] Sepiashvili, D., Moura, J. M. F. and Ha, V. H. S. (2003). Affine-Permutation Symmetry: Invariance and Shape Space. *Proceedings of the 2003 Workshop on Statistical Signal Processing.* 293–296.

[39] Small, C. G. (1996). *The Statistical Theory of Shape.* Springer, New York.

[40] Sparr, G. (1992). Depth-computations from polihedral images. *Proc. 2nd European Conf. on Computer Vision*(ECCV-2) (G. Sandini, ed.). 378–386.

[41] Sugathadasa, M. S. (2006). *Affine and Projective Shape Analysis with Applications.* Ph.D. dissertation, Texas Tech University, Lubbock, TX.

[42] Watson, G. S. (1983). *Statistics on Spheres.* University of Arkansas Lecture Notes in the Mathematical Sciences, 6. Wiley, New York.

[43] Xia, M. and Liu, B. (2004). Image registration by "super-curves". *IEEE Trans. Image Process.* **13** 720–732.

Chapter 4

Reinforcement Learning — A Bridge Between Numerical Methods and Monte Carlo

Vivek S. Borkar

School of Technology and Computer Science,
Tata Institute of Fundamental Research,
Homi Bhabha Road, Mumbai 400005, India
*borkar@tifr.res.in**

This article advocates the viewpoint that reinforcement learning algorithms, primarily meant for approximate dynamic programming, can also be cast as a technique for estimating stationary averages and stationary distributions of Markov chains. In this role, they lie somewhere between standard deterministic numerical schemes and Markov chain Monte Carlo, and capture a trade-off between the advantages of either – lower per iterate computation than the former and lower variance than the latter. Issues arising from the 'curse of dimensionality' and convergence rate are also discussed.

4.1. Introduction

The genealogy of reinforcement learning goes back to mathematical psychology ([16], [17], [20], [21], [37]). The current excitement in the field, however, is spurred by its more recent application to dynamic programming, originating in the twin disciplines of machine learning and control engineering ([8], [34], [36]). A somewhat simplistic but nevertheless fairly accurate view of these schemes is that they replace a classical recursive scheme for dynamic programming by a stochastic approximation based incremental scheme, which exploits the averaging properties of stochastic approximation in order to do away with the conditional expectation operator intrinsic to the former. In particular, these schemes can be used for simulation based or online, possibly approximate, solution of dynamic programmes.

*Research supported in part by a J. C. Bose Fellowship from Dept. of Science and Technology, Govt. of India, and a grant from General Motors India Science Lab.

In this article, we advocate a somewhat different, albeit related, application for this paradigm, viz., to solve two classical problems for Markov chains: the problem of estimating the stationary expectation of a prescribed function of an ergodic Markov chain, and that of estimating stationary distribution of such a chain. The former problem is the linear, or 'policy evaluation' variant of the learning algorithm for average cost dynamic programming equation ([1], [26], [38], [39]). The latter in turn has been of great importance in queuing networks (see, e.g., [35]) and more recently, in the celebrated '*PageRank*' scheme for ranking of web sites ([27]). The scheme proposed here is a variant of a general scheme proposed recently in [5], [11] for estimating the Perron-Frobenius eigenvector of a nonnegative matrix. See [19] for an earlier effort in this direction.

Analogous ideas have also been proposed in [9] in a related context.

The article is organized as follows: As stochastic approximation theory forms the essential backdrop for this development, the next section is devoted to recapitulating the essentials thereof that are relevant here. Section 4.3 recalls the relevant results of [1], [26], [39] in the context of the problem of estimating the stationary expectation of a prescribed function of an ergodic Markov chain. A 'function approximation' variant of this, which addresses the curse of dimensionality by adding another layer of approximation, is described in Section 4.4. Section 4.5 first recalls the relevant developments of [5], [11] and then highlights their application to estimating stationary distributions of ergodic Markov chains. Section 4.6 discusses the issues that affect the speed of such algorithms and suggests some techniques for accelerating their convergence. These draw upon ideas from [2] and [13]. Section 4.7 concludes with brief pointers to some potential research directions.

4.2. Stochastic Approximation

The archetypical stochastic approximation algorithm is the d-dimensional recursion

$$x_{n+1} = x_n + a(n)[h(x_n, Y_n) + M_{n+1} + \varepsilon_n], \tag{4.1}$$

where, for $\mathcal{F}_n \stackrel{def}{=} \sigma(x_m, Y_m, M_m; m \leq n), n \geq 0$, the following hold:

- $(M_n, \mathcal{F}_n), n \geq 0$, is a *martingale difference sequence*, i.e., $\{M_n\}$ is an $\{\mathcal{F}_n\}$-adapted integrable process satisfying

$$E[M_{n+1}|\mathcal{F}_n] = 0 \;\; \forall n.$$

  Furthermore, it satisfies

$$E[\|M_{n+1}\|^2|\mathcal{F}_n] \leq K(1 + \|x_n\|^2) \;\; \forall n$$

  for some $K > 0$.
- $\{Y_n\}$ is a process taking values in a finite state space S, satisfying

$$P(Y_{n+1} = y|\mathcal{F}_n) = q_{x_n}(y|Y_n)$$

  for a continuously parametrized family of transition probability functions $x \to q_x(\cdot|\cdot)$ on S. For each fixed x, it is a Markov chain with transition matrix $q_x(\cdot|\cdot)$, assumed to be irreducible. In particular, it then has a unique stationary distribution ν_x.
- $\{\varepsilon_n\}$ is a random sequence satisfying $\varepsilon_n \to 0$ a.s.
- $\{a(n)\}$ are positive step sizes satisfying

$$\sum_n a(n) = \infty; \; \sum_n a(n)^2 < \infty. \tag{4.2}$$

- $h : \mathcal{R}^d \times S \to \mathcal{R}^d$ is Lipschitz in the first argument.

Variations and generalizations of these conditions are possible, but the above will suffice for our purposes. The 'o.d.e.' approach to the analysis of stochastic approximation, which goes back to [18], [28], views (4.1) as a noisy discretization of the ordinary differential equation (o.d.e. for short)

$$\dot{x}(t) = \bar{h}(x(t)). \tag{4.3}$$

Here $\bar{h}(x) \stackrel{def}{=} \sum_{i \in S} h(x,i)\nu_x(i)$ will be Lipschitz under our hypotheses. This ensures the well-posedness of (4.3). Recall that a set A is invariant for (4.3) if any trajectory $x(t), t \in \mathcal{R}$, of (4.3) which is in A at time $t = 0$ remains in A for all $t \in \mathcal{R}$. It is said to be internally chain transitive if in addition, for any $x, y \in A$, and $\epsilon, T > 0$, there exist $n \geq 2$ and $x = x_1, x_2, \ldots, x_n = y$ in A such that the trajectory of (4.3) initiated at x_i intersects the open ϵ-ball centered at x_{i+1} at some time $t_i \geq T$ for all $1 \leq i < n$. Suppose

$$\sup_n \|x_n\| < \infty \;\; \text{a.s.} \tag{4.4}$$

Then a result due to Benaim ([6]) states that almost surely, x_n converges to a compact connected internally chain transitive invariant set of (4.3). In case the only such sets are the equilibria of (4.3) (i.e., points where $\bar{h}$ vanishes), then x_n converges to the set of equilibria of (4.3) a.s. If these are finitely many, then a.s. x_n converges to some possibly sample path dependent equilibrium. In particular, when there is only one such, say x^*, $x_n \to x^*$ a.s. This is the situation of greatest interest in most algorithmic applications.

The 'stability' condition (4.4) usually needs to be separately established. There are several criteria for the purpose. One such, adapted from [12], is the following: Suppose the limit

$$h_\infty(x) \stackrel{def}{=} \lim_{a \uparrow \infty} \frac{\bar{h}(ax)}{a} \tag{4.5}$$

exists for all x, and the o.d.e.

$$\dot{x}(t) = h_\infty(x(t)) \tag{4.6}$$

has the origin as the globally asymptotically stable equilibrium. Then (4.4) holds. (Note that h_∞ will perforce be Lipschitz with the same Lipschitz constant as $\bar{h}$ and therefore the well-posedness of (4.6) is not an issue.) This is a slight variant of Theorem 2.1 of [12] and can be proved analogously.

Let $\Gamma(i)$ denote the $d \times d$ matrix with all entries zero except the (i, i)-th entry, which is one. A special case of the foregoing is the iteration

$$x_{n+1} = x_n + a(n)\Gamma(X_n)[h(x_n, Y_n) + M_{n+1} + \varepsilon_n],$$

where $\{X_n\}$ is an irreducible Markov chain on the index set $\{1, 2, \ldots, d\}$. This is a special case of 'asynchronous stochastic approximation' studied in [10]. It corresponds to the situation when only one component of the vector x_n, viz., the X_nth, is updated at time n for all n, in accordance with the above dynamics. The preceding theory then says that the o.d.e. limit will change to

$$\dot{x}(t) = \Lambda\bar{h}(x(t)).$$

Here Λ is a diagonal matrix whose ith diagonal element equals the stationary probability of i for the chain $\{X_n\}$. It thus captures the effect of different components being updated with different relative frequencies. While this change does not matter for some dynamics (in the sense that it does not alter their asymptotic behaviour), it is undesirable for some others. One way of getting around this ([10]) is to replace the step size $a(n)$ by $a(\nu(X_n, n))$, where $\nu(i, n) \stackrel{def}{=} \sum_{m=0}^{n} I\{X_m = i\}$ is the *local clock* at i: the number of times ith component got updated till time n. Under some additional hypotheses on $\{a(n)\}$, this ensures that Λ above gets replaced by $\frac{1}{d}$ times the identity matrix, whereby the above o.d.e. has the same qualitative behaviour as (4.3), but with a change of time-scale. See [10] for details.

4.3. Estimating Stationary Averages

The problem of interest here is the following: Given an irreducible Markov chain $\{X_n\}$ on a finite state space $\mathcal{S}$ ($= \{1, 2, \ldots, s\}$, say) with transition probability matrix $P = [[p(j|i)]]_{i,j \in \mathcal{S}}$, and the unique stationary distribution η, we want to estimate the stationary average $\beta \stackrel{def}{=} \sum_{i \in \mathcal{S}} f(i)\eta(i)$ for a prescribed $f : \mathcal{S} \to \mathcal{R}$. The standard Monte Carlo approach would be to use the sample average

$$\frac{1}{N} \sum_{m=1}^{N} f(X_m) \tag{4.7}$$

for large N as the estimate. This is justified by the strong law of large numbers for Markov chains, which states that (4.7) converges a.s. to β as $N \uparrow \infty$. Even in cases where η is known, there may be strong motivation for going for a Markov chain with stationary distribution η instead of i.i.d. random variables with law η: in typical applications, the latter are hard to simulate, but the Markov chain, which usually has a simple local transition rule, is not. This is the case, e.g., in many statistical physics applications.

The problem in most applications is that this convergence is very slow. This typically happens because $\mathcal{S}$ is very large and the chain makes only local moves, moving from any $i \in \mathcal{S}$ to one of its 'neighbors'. Thus the very aspect that makes it easy to simulate works against fast convergence. For example, one may have $\mathcal{S} = \{1, 2, \ldots, M\}^d$ for some $M, d >> 1$ and the chain may move from any point therein only to points which differ in at most one of the d components by at most 1. In addition the state space may be 'nearly decoupled' into 'almost invariant' subsets such that the transitions between them are rare, and thus the chain spends a long time in whichever of these sets it finds itself in. As a result, the chain is very slow in scanning the entire state space, leading to slow convergence of the above estimate. Also, the variance of this estimate can be quite high. There are many classical techniques for improving the performance of Monte Carlo, such as importance sampling, use of control variates, antithetic variates, stratified sampling, etc. See [3] for an extensive and up to date account of these.

An alternative linear algebraic view of the problem would be to look at the associated Poisson equation:

$$V(i) = f(i) - \beta' + \sum_j p(j|i)V(j), \ i \in \mathcal{S}. \tag{4.8}$$

This is an equation in the pair $(V(\cdot), \beta')$. Under our hypotheses, β' is uniquely characterized as $\beta' = \beta$ and $V(\cdot)$ is unique up to an additive scalar. As clear from (4.8), this is the best one can expect, since the equation is unaltered if one adds a scalar to all components of V. We shall denote by H the set of V such that (V, β) satisfies (4.8). Thus $H = \{V : V = V^* + c\mathbf{e}, c \in \mathcal{R}\}$ where (V^*, β) is any fixed solution to (4.8), and $\mathbf{e} \stackrel{def}{=}$ the vector of all 1's. The *relative value iteration* algorithm for solving (4.8) is given by the iteration

$$V_{n+1}(i) = f(i) - V_n(i_0) + \sum_j p(j|i)V_n(j), \ i \in \mathcal{S}, \tag{4.9}$$

where $i_0 \in \mathcal{S}$ is a fixed state.

Remarks: This is a special case of the more general relative value iteration algorithm for solving the (nonlinear) dynamic programming equation associated with the average cost control of a controlled Markov chain ([7], section 4.3). The convergence results for the latter specialized to the present case lead to the conclusions $V_n \to V$ and $V_n(i_0) \to \beta$, where (V, β) is the unique solution to (4.8) corresponding to $V(i_0) = \beta$. The choice of the 'offset' $V_n(i_0)$ above is not unique, one can replace it by $g(V_n)$ for any g satisfying $g(\mathbf{e}) = 1, g(x + c\mathbf{e}) = g(x) + c$ for all $c \in \mathcal{R}$ – see [1], p. 684. This choice will dictate which solution of (4.8) gets picked in the limit. Nevertheless, in all cases, $g(V_n) \to \beta$, which is the quantity of interest. In what follows, the analysis will have to be suitably modified in a few places for choices of g other than the specific one considered here.

To obtain a reinforcement learning variant of (4.9), one follows a standard schematic. Suppose that at time n, $X_n = i$. The first step is to replace the conditional expectation on the right hand side of (4.9) by an actual evaluation at the observed next state, i.e., by

$$f(i) - V_n(i_0) + V_n(X_{n+1}).$$

The second step is to make an incremental correction towards this quantity, i.e., to replace $V_n(i)$ by a convex combination of itself and the above, with a small weight $a(n)$ for the latter. The remaining components $V_n(j), j \neq i$, remain unaltered. Thus the update is

$$\begin{aligned}
&V_{n+1}(i) \\
&= (1 - a(n)I\{X_n = i\})V_n(i) + a(n)I\{X_n = i\}[f(i) - V_n(i_0) + V_n(X_{n+1})] \\
&= V_n(i) + a(n)I\{X_n = i\}[f(i) - V_n(i_0) + V_n(X_{n+1}) - V_n(i)]. \qquad (4.10)
\end{aligned}$$

This can be rewritten as

$$V_{n+1}(i) = V_n(i) + a(n)I\{X_n = i\}[T_i(V_n) - V_n(i_0) + M_{n+1}(i) - V_n(i)],$$

where $T(\cdot) = [T_1(\cdot), \ldots, T_s(\cdot)]^T$ is given by

$$T_k(x) \stackrel{def}{=} f(k) + \sum_j p(j|k)x_j$$

for $x = [x_1, \ldots, x_s]^T \in \mathcal{R}^s$, and for $n \geq 0$,

$$M_{n+1}(j) \stackrel{def}{=} f(j) + V_n(X_{n+1}) - T_j(V_n), \ n \geq 0, 1 \leq j \leq s,$$

is a martingale difference sequence w.r.t. $\mathcal{F}_n \stackrel{def}{=} \sigma(X_m, m \leq n)$. Let D denote the diagonal matrix whose ith diagonal element is $\eta(i)$ for $i \in \mathcal{S}$. Then the counterpart of (4.3) for this case is

$$\begin{aligned}
\dot{x}(t) &= D\,(T(x(t)) - x_{i_0}(t)\mathbf{e} - x(t)) \\
&= \tilde{T}(x(t)) - x_{i_0}(t)\eta - x(t) \qquad (4.11)
\end{aligned}$$

where $\tilde{T}(x) \stackrel{def}{=} (I - D)x + DT(x)$, $x \in \mathcal{R}^s$. We shall analyze (4.11) by relating it to a secondary o.d.e.

$$\dot{\tilde{x}}(t) = \tilde{T}(\tilde{x}(t)) - \beta\eta - \tilde{x}(t). \qquad (4.12)$$

It is easily verified that the map $\hat{T}$ defined by $\hat{T}(x) = \tilde{T}(x) - \beta\eta$ is non-expansive w.r.t. the max-norm : $\|x\|_\infty^\eta \stackrel{def}{=} \max_i |\frac{x_i}{\eta_i}|$, and has H as its set of fixed points. Thus by the results of [14], $\tilde{x}(t)$ converges to some point in H, depending on $\tilde{x}(0)$, as $t \uparrow \infty$. One can then mimic the arguments of Lemmas 3.1–Theorem 3.4 of [1] step for step to claim successively that:

- If $x(0) = \tilde{x}(0)$, then $x(t) - \tilde{x}(t)$ is of the form $r(t)\eta$ for some $r(t)$ that converges to a constant as $t \uparrow \infty$. (The only changes required in the arguments of [1] are that, (i) in the proof of Lemma 3.3 therein, use the weighted span seminorm $\|x\|_{\eta,s} \stackrel{def}{=} \max_i \left(\frac{x_i}{\eta_i}\right) - \min_i \left(\frac{x_i}{\eta_i}\right)$ instead of the span semi-norm $\|x\|_s \stackrel{def}{=} \max_i x_i - \min_i x_i$, and, (ii) $r(\cdot)$ satisfies a different convergent o.d.e., viz.,

$$\dot{r}(t) = -\eta(i_0)r(t) + (\beta - \tilde{x}_{i_0}(t)),$$

than the one used in [1][a].)
- $x(t)$ therefore converges to an equilibrium point of (4.11), which is seen to be unique and equal to the unique element V^* of H characterized by $V^*(i_0) = \beta$. In fact, V^* is the unique globally asymptotically stable equilibrium for (4.11).

Stochastic approximation theory then guarantees that $V_n \to V^*$ a.s. if we establish (4.4), i.e., that $\sup_n \|V_n\| < \infty$ a.s. For this purpose, note that the corresponding o.d.e. (4.6) is simply (4.11) with $f(\cdot) \equiv 0$, for which an analysis similar to that for (4.11) shows that the origin is the globally asymptotically stable equilibrium. The stability test of [12] mentioned above then ensures the desired result.

The scheme (4.10) combines the deterministic numerical method (4.9) with a Monte Carlo simulation and stochastic approximation to exploit the averaging effect of the latter. Note, however, that unlike pure Monte Carlo, it does *conditional averaging* instead of averaging. The determination of the desired stationary average from this is then a consequence of an algebraic relationship between the two specified by (4.8). The net gain is that the part stochastic, part algebraic scheme has lower variance than pure Monte Carlo because of the simpler conditional expectations involved, and therefore more graceful convergence. But it has higher per iterate computation. On the other hand, it has higher variance than the deterministic relative value iteration – the latter has zero variance! But then it has lower per iterate computation because it does only local updates and does away with the conditional expectation operation. (It is worth noting that in some applications, this is not even an option because the conditional probabilities are not explicitly available, only a simulation/experimental device

[a] Here $r(\cdot)\eta$ is seen to satisfy the o.d.e. $\dot{r}(t)\eta = (D(P-I) - \eta(i_0)I)r(t)\eta + (\beta - \tilde{x}_{i_0}(t))\eta$. Left-multiply both sides by $\mathbf{e}^T$.

that conforms to them is.) In this sense, the reinforcement learning algorithm (4.10) captures a trade-off between pure numerical methods and pure Monte Carlo.

In the form stated above, however, the scheme also inherits one major drawback each from numerical methods and Monte Carlo when s is very large. From the former, it inherits the notorious 'curse of dimensionality'. From the latter it inherits slow convergence due to slow approach to stationarity already mentioned earlier, a consequence of the local movement of the underlying Markov chain and possible occurrence of 'quasi-invariant' subsets of the state space. This motivates two important modifications of the basic scheme. The first is aimed at countering the curse of dimensionality and uses the notion of (linear) function approximation. That is, we approximate V^* by a linear combination of a moderate number of *basis functions* (or *features* in the parlance of artificial intelligence, these can also be thought of as *approximate sufficient statistics* – see [8]). These are kept fixed and their weights are learnt through a learning scheme akin to the above, but lower dimensional than the original. We outline this in the next section. The second problem, that of slow mixing, can be alleviated by using the devices of *conditional importance sampling* from [2] or *split sampling* from [13], either separately or together. This will be described in section 4.6.

Before doing so, however, we describe an important related development. Recall the notion of control variates ([3]). These are zero mean random variables $\{\xi_n\}$ incorporated into the Monte Carlo scheme such that if we evaluate

$$\frac{1}{N}\sum_{m=1}^{N}(f(X_m)+\xi_m) \tag{4.13}$$

instead of (4.7), it has lower variance but the same asymptotic limit as (4.7). Consider the choice

$$\xi_n = \sum_j p(j|X_n)V^*(j) - V^*(X_n),\ n \geq 0,$$

where (V^*, β) satisfy (4.8). Then $f(X_n) + \xi_n = \beta\ \forall n$ and the variance is in fact zero! The catch, of course, is that we do not know V^*. If, however, we have a good approximation thereof, these would serve as good control variates. This idea is pursued in [24], where the 'V^*' for a limiting deterministic problem, the so called fluid limit, is used for arriving at a good choice of control variates (see also [31], Chapter 11). The scheme described above could also be thought of as one that adaptively generates control variates.

4.4. Function Approximation

We now describe an approximation to the aforementioned scheme that tries to beat down the curse of dimensionality at the expense of an additional layer of approximation and the ensuing approximation error. Function approximation based reinforcement learning schemes for precisely this problem have been proposed and analyzed in [38] and [39]. The one sketched below, though similar in spirit, is new and appears here for the first time.

The essential core of the scheme is the approximation $V(i) \approx \sum_{j=1}^{M} r_j \phi_j(i)$, $i \in \mathcal{S}$, where M is significantly smaller than $s = |\mathcal{S}|$, $\phi_j : \mathcal{S} \to \mathcal{R}$ are basis functions prescribed a priori and $\{r_j\}$ are weights that need to be estimated. For a function $g : \mathcal{S} \to \mathcal{R}$, we shall use g to denote both the function itself and the vector $[g(1), \ldots, g(s)]^T$, depending on the context. For the specific problem under consideration, we impose the special requirement that $\phi_1 = f, \phi_2 = \mathbf{e}$. Let $\Phi \stackrel{def}{=} [[\varphi_{ij}]]_{1 \leq i \leq s, 1 \leq j \leq M}$, where $\varphi_{ij} \stackrel{def}{=} \phi_j(i)$ and $\phi(i) \stackrel{def}{=} [\varphi_{i1}, \ldots, \varphi_{iM}]^T$. Thus $f = \Phi u_1, \mathbf{e} = \Phi u_2$, where u_i is the unit vector in the ith direction. Consider the 'approximate Poisson equation'

$$\Phi r = P \Phi r - \tilde{\beta} \mathbf{e} + f.$$

This is obtained[b] by replacing V by Φr in (4.8). Left-multiplying this equation on both sides first by Φ^T and then by $(\Phi^T \Phi)^{-1}$, we get

$$r = (\Phi^T \Phi)^{-1} \Phi^T (P \Phi r - \tilde{\beta} \mathbf{e} + f).$$

Note that $(y \stackrel{def}{=} \Phi r, \tilde{\beta})$ then satisfies the 'projected Poisson equation'

$$y = \hat{\Pi} P y - \tilde{\beta} \mathbf{e} + f, \tag{4.14}$$

where $\hat{\Pi} \stackrel{def}{=} \Phi (\Phi^T \Phi)^{-1} \Phi^T$ is the projection onto the range of Φ. (In particular, $\hat{\Pi}$ leaves $\mathbf{e}, f$ invariant, a fact used here.) This is our initial approximation of the original problem, which will be modified further in what

[b] This is a purely formal substitution, as clearly this equation will not have a solution unless $V = \Phi r$ + a scalar multiple of $\mathbf{e}$.

follows. By analogy with the preceding section, this suggests the algorithm

$$r_{n+1} = r_n + a(n)[B_n^{-1}\phi(X_n)\left(\phi^T(X_{n+1})r_n - (\phi^T(i_0)r_n)\mathbf{e} + f(X_n)\right) - r_n], \tag{4.15}$$

where

$$B_n \stackrel{def}{=} \frac{1}{n+1}\sum_{m=0}^{n}\phi(X_m)\phi^T(X_m).$$

Also, i_0 is a fixed state in $\cup_k\{j : \phi_k(j) > 0\}$. Note that $\{B_n^{-1}\}$ can be computed iteratively by: $B_n^{-1} = (n+1)\tilde{B}_n^{-1}$, where

$$\tilde{B}_{n+1}^{-1} = \tilde{B}_n^{-1} - \frac{\tilde{B}_n^{-1}\phi(X_{n+1})\phi^T(X_{n+1})\tilde{B}_n^{-1}}{1 + \phi^T(X_{n+1})\tilde{B}_n^{-1}\phi(X_{n+1})}.$$

This follows from the Sherman-Morrison-Woodbury formula ([22], p. 50). If $\tilde{B}_n$ is not invertible in early iterations, one may add to it δ times the identity matrix for a small $\delta > 0$, or use pseudo-inverse in place of inverse. The error this causes will be asymptotically negligible because of the time averaging. By ergodicity of $\{X_n\}$, $B_n \to \Phi^T D\Phi$ a.s.

The limiting o.d.e. can be written by inspection as

$$\dot{r}_t = (\Phi^T D\Phi)^{-1}\Phi^T D(P\Phi r_t - \phi^T(i_0)r_t\mathbf{e} + f) - r_t$$

Let $\Pi \stackrel{def}{=} \Phi(\Phi^T D\Phi)^{-1}\Phi^T D$. Then $\mathbf{e}$ is invariant under Π, i.e., it is an eigenvector of Π corresponding to the eigenvalue 1. To see this, note that $\sqrt{D}\Pi(\sqrt{D})^{-1}$ is the projection operator onto $Range(\sqrt{D}\Phi)$, whence $\sqrt{D}\mathbf{e} = \sqrt{D}\phi_2$ is invariant under it. This is equivalent to the statement that $\mathbf{e}$ is invariant under Π. A similar argument shows that f is invariant under Π. Let $y(t) \stackrel{def}{=} \Phi r_t$. Then the above o.d.e. leads to

$$\dot{y}(t) = \Pi P y(t) - y_{i_0}(t)\mathbf{e} + f - y(t).$$

The following easily proven facts are from [38]:

1 Π is the projection to $Range(\Phi)$ w.r.t. the weighted norm $\|\cdot\|_D$ defined by $\|x\|_D \stackrel{def}{=} (\sum_i \eta(i)x_i^2)^{\frac{1}{2}}$.
2 P is nonexpansive w.r.t. $\|\cdot\|_D$: $\|Px\|_D \leq \|x\|_D$.

Note that P is nonexpansive w.r.t. the norm $\|x\|_D$, hence so is ΠP. Also, $\mathbf{e}$ is its unique eigenvector corresponding to eigenvalue 1 and the remaining eigenvalues are in the interior of the unit circle. Thus $\Pi P - \mathbf{e}u_{i_0}^T = \Pi(P - \mathbf{e}u_{i_0}^T)$ has eigenvalues strictly inside the unit disc of the complex plane. (Recall that u_j for each j denotes the unit vector in jth coordinate direction.) Finally, $\Pi(P - \mathbf{e}u_{i_0}^T) - I$ then has eigenvalues with strictly negative real parts. These considerations lead to:

- the map $y \to \Pi(P - \mathbf{e}u_{i_0}^T)y + f$ is a $\|\cdot\|_D$-contraction and has a unique fixed point y^*, and,
- the above o.d.e. is a stable linear system and converges to y^*.

Thus $r_t \to r^* \stackrel{def}{=} (\Phi^T\Phi)^{-1}\Phi^T y^*$. Just as in the preceding section, an identical analysis of the o.d.e. with f replaced by the zero vector leads to a.s. boundedness of the iterates, leading to $r_n \to r^*$ a.s. The quantity $\phi^T(i_0)r^*$ then serves as our estimate of β. Note that we have solved the approximate Poisson equation

$$y = \Pi P y - \tilde{\beta}\mathbf{e} + f$$

and not (4.14). This change is caused by the specific sampling scheme we used.

4.5. Estimating Stationary Distribution

Here we consider the problem of estimating the stationary distribution of an irreducible finite state Markov chain. It is the right Perron-Frobenius eigenvector of its transposed transition matrix, corresponding to the Perron-Frobenius eigenvalue 1. More generally, one can consider the problem of estimating the Perron-Frobenius eigenvector (say, $\hat{q}$) of an irreducible non-negative matrix $Q = [[q_{ij}]]$, corresponding to the Perron-Frobenius eigenvalue λ^*. Thus we have $Q\hat{q} = \lambda^*\hat{q}$. One standard numerical method for this is the 'power method':

$$q_{n+1} = \frac{Qq_n}{\|Qq_n\|}, \ n \geq 1,$$

beginning with some initial guess q_0 with strictly positive components. The normalization on the right hand side is somewhat flexible, e.g., one may use

$$q_{n+1} = \frac{Qq_n}{q_n(i_0)}, \ n \geq 1, \ i_0 \in \mathcal{S} \text{ fixed},$$

with $q_0(i_0) > 0$. The following scheme, a special case of the learning algorithm studied in [11], may be considered a stochastic approximation version of this. It can also be viewed as a multiplicative analog of (4.9). We shall consider the unique choice of $\hat{q}$ for which $\hat{q}(i_0) = \lambda^*$. Write $Q = LP$ where

L is a diagonal matrix with ith diagonal element $\ell(i) \stackrel{def}{=} \sum_j q_{ij}$ and P is an irreducible stochastic matrix. Then

$$q_{n+1}(j) = q_n(j) + a(\nu(j,n))I\{X_n = j\}\Big[\frac{\ell(X_n)q_n(X_{n+1})}{q_n(i_0)} - q_n(j)\Big],\ j \in \mathcal{S},$$

with q_0 in the positive orthant and $\{a(n)\}$ satisfying the additional conditions stipulated in [10]. Under suitable conditions on the step sizes (see the discussion at the end of section 4.2 above), the limiting o.d.e. is

$$\dot{q}_t = \frac{LPq_t}{q_t(i_0)} - q_t = \frac{Qq_t}{q_t(i_0)} - q_t. \tag{4.16}$$

This can be analyzed by first analyzing a 'secondary o.d.e.'

$$\dot{\tilde{q}}_t = \frac{Q\tilde{q}_t}{\lambda^*} - \tilde{q}_t. \tag{4.17}$$

Under our hypotheses, $Q/\lambda^* - I$ has a unique normalized eigenvector q^* corresponding to the eigenvalue zero, viz., the normalized Perron-Frobenius eigenvector of Q. Then aq^* is an equilibrium point for this linear system for all $a \in \mathcal{R}$. All other eigenvalues are in the left half of the complex plane. Thus $\tilde{q}_t \to a^*q^*$ for an $a^* \in \mathcal{R}$ that depends on the initial condition. In turn, one can show that for the same initial condition, $q_t = \psi(t)\tilde{q}_{\tau(t)}$, where

$$\tau(t) \stackrel{def}{=} \int_0^t \frac{\lambda^*}{q_{t'}(i_0)}dt',\ \psi(t) \stackrel{def}{=} exp(\int_0^{\tau(t)} \Big(1 - \frac{q_{t'}(\tau^{-1}(t'))}{\lambda^*}\Big)dt'.$$

This is verified by first checking that $\psi(\cdot)\tilde{q}_{\tau(\cdot)}$ does satisfy (4.16) and then invoking the uniqueness of the solution to the latter. Now argue as in Lemma 4.1 of [11] to conclude that for $q_0 \in$ the positive orthant, $q_t \to \hat{q}$, a scalar multiple of q^* corresponding to $\hat{q}(i_0) = \lambda^*$.

Letting

$$\bar{h}(q) = \frac{LPq}{q_{i_0}} - q,$$

one has $h_\infty(q) = -q$, where the l.h.s. is defined as in (4.5). By the stability test of [12], we then have $\sup_n \|q_n\| < \infty$ a.s. In view of the foregoing, this leads to $q_n \to \hat{q}$ a.s.

Once again, in view of the curse of dimensionality, we need to add another layer of approximation via the function approximation $q \approx \Phi r$, where

Φ, r are analogous to the preceding section. The function approximation variant of the above is

$$r_{n+1} = r_n + a(n)\Big[B_n^{-1}\frac{\ell(X_n)\phi(X_n)\phi^T(X_{n+1})}{(\phi^T(i_0)r_n)\vee\epsilon} - I\Big]r_n, \tag{4.18}$$

where B_n is as before and $\epsilon > 0$ is a small prescribed scalar introduced in order to avoid division by zero. We need the following additional restriction on the basis functions:

(†) $\phi_1, \ldots, \phi_M$ are orthogonal vectors in the positive cone of $\mathcal{R}^s$, and the submatrix of P corresponding to $\cup_k\{i : \phi_k(i) > 0\}$ is irreducible.

This scheme is a variant of the scheme proposed and analyzed in [5]. (It has one less explicit averaging operation. As a rule of thumb, any additional averaging makes the convergence more graceful at the expense of its speed.) Let $A \stackrel{def}{=} \Phi^T DQ\Phi, B \stackrel{def}{=} \Phi^T D\Phi$. The limiting o.d.e. is the same as that of [5], viz.,

$$\dot{r}(t) = \left(\frac{B^{-1}A}{(\phi^T(i_0)r(t))\vee\epsilon} - I\right)r(t). \tag{4.19}$$

For $\epsilon = 0$, this reduces to

$$\dot{r}(t) = \left(\frac{B^{-1}A}{\phi^T(i_0)r(t)} - I\right)r(t). \tag{4.20}$$

Let $W \stackrel{def}{=} \sqrt{D}\Phi$ and $\mathcal{M} \stackrel{def}{=} \sqrt{D}Q\sqrt{D}^{-1}$. Then $B = W^TW$. Let $\breve{\Pi} = W(W^TW)^{-1}W^T$. Setting $Y(t) \stackrel{def}{=} Wr(t)$, (4.20) may be rewritten as

$$\dot{Y}(t) = \left(\frac{\breve{\Pi}\mathcal{M}}{\phi^T(i_0)r(t)} - I\right)Y(t) = \left(\sqrt{\eta(i_0)}\frac{\breve{\Pi}\mathcal{M}}{Y_{i_0}(t)} - I\right)Y(t). \tag{4.21}$$

Under (†), $\breve{\Pi}\mathcal{M}$ can be shown to be a nonnegative matrix ([5]). Let γ^* denote the Perron-Frobenius eigenvalue of $\breve{\Pi}\mathcal{M}$. Consider the associated *secondary* o.d.e.

$$\dot{Y}(t) = \left(\frac{\breve{\Pi}\mathcal{M}}{\gamma^*} - I\right)Y(t). \tag{4.22}$$

Let z^* be the Perron-Frobenius eigenvector of $\check{\Pi}\mathcal{M}$ uniquely specified by the condition $z^*(i_0) = \sqrt{\eta(i_0)}\gamma^*$. The convergence of (4.22), and therefore that of (4.21), to z^* is established by arguments analogous to those for (4.16), (4.17) (see [5] for details). Now let $\epsilon > 0$. One can use Theorem 1, p. 339, of [25] to conclude that given any $\delta > 0$, there exists an $\epsilon^* > 0$ small enough such that for $\epsilon \in (0, \epsilon^*)$, the corresponding $Y(t)$ converges to the δ-neighborhood of z^*. But once in a small neighborhood of z^*, $Y_{i_0}(t) \approx \sqrt{\eta(i_0)}\gamma^* > 0$. Hence, if $\epsilon << \gamma^*$, (4.19) reduces to (4.20) in this neighborhood and therefore $Y(t) \to z^*$. In turn, one then has $r_n \to r^* = (W^T W)^{-1} W^T z^*$ a.s. by familiar arguments. Once again, we have solved an approximation to the original eigenvalue problem, viz., the eigenvalue problem for $\check{\Pi}\mathcal{M}$. Φr^*, resp. γ^* are then our approximations to $\hat{q}, \lambda^*$.

Recalling our original motivation of estimating stationary distribution of a Markov chain, note that Q in fact will be the transpose of its transition matrix and λ^* is a priori known to be one. Thus we may replace the denominator $q_n(i_0)$ in the second term on the right hand side of the algorithm by 1. This is a linear iteration for which the limiting o.d.e. is in fact the secondary o.d.e. (4.17) with $\lambda^* = 1$. This is not, however, possible for (4.21), as γ^* may not be 1.

We conclude this section by outlining a related problem. Consider the case when Q is an irreducible nonnegative and positive definite symmetric matrix. The problem once again is to find the Perron-Frobenius eigenvector of A. This can be handled exactly as above, except that the Perron-Frobenius eigenvalue need not be one. An important special case is when $Q = A^T A$ where A is the adjacency matrix of an irreducible graph. This corresponds to Kleinberg's *HITS* scheme for web page ranking, an alternative to the *PageRank* mentioned above ([27]). The Perron-Frobenius eigenvector then corresponds to the unique stationary distribution for a *vertex-reinforced random walk* on S with reinforcement matrix Q ([33]).

4.6. Acceleration Techniques

For reasons mentioned earlier, the convergence of these algorithms can be slow even after dimensionality reduction, purely due to the structure of the underlying Markov chain. We shall discuss two ideas that can be used to advantage for speeding things up. The first is that of *conditional importance sampling.* This was introduced in [2] in the context of rare event simulation and has also been subsequently used in [39]. Recall that importance sampling for evaluating a stationary average $\sum_{i \in S} f(i)\eta(i)$ amounts to replacing (4.7) by

$$\frac{1}{N}\sum_{m=1}^{N} f(\tilde{X}_m)\Lambda_m, \tag{4.23}$$

where:

- $\{\tilde{X}_m\}$ is another Markov chain on $\mathcal{S}$ with the same initial law, with its transition matrix $[[\tilde{p}(\cdot|\cdot)]]_{i,j\in\mathcal{S}}$ satisfying the condition:

$$p(j|i) > 0 \Longleftrightarrow \tilde{p}(j|i) > 0 \ \forall\ i, j, \tag{4.24}$$

and,

- $\Lambda_n \stackrel{def}{=} \Pi_{m=0}^{n-1} \frac{p(X_{m+1}|X_m)}{\tilde{p}(X_{m+1}|X_m)}$ is the likelihood ratio at time n.

This is the simplest version, other more sophisticated variations are possible ([3]). Clearly (4.23) is asymptotically unbiased, i.e., its mean approaches the desired mean $\sum_{i\in\mathcal{S}} f(i)\eta(i)$ as $N \uparrow \infty$. The idea is to choose $\{\tilde{X}_n\}$ to accelerate the convergence. This, however, can lead to inflating the variance, which needs to be carefully managed. See [3] for an account of this broad area.

To motivate conditional importance sampling, note that our algorithms are of the form

$$\begin{aligned} x_{n+1}(i) = x_n(i)& \\ &+ a(n)I\{X_n = i\}[F_{X_{n+1}}(x_n) + g_i(x_n) + M_{n+1} + \zeta_{n+1}(i)], \\ &\qquad 1 \le i \le d, \end{aligned} \tag{4.25}$$

where $d \geq 1, F = [F_1, \ldots, F_d], g = [g_1, \ldots, g_d] : \mathcal{R}^d \to \mathcal{R}^d$ are prescribed maps, $\{M_n\}$ is a martingale difference sequence as before, and $\zeta_n = [\zeta_n(1), \ldots, \zeta_n(d)], n \geq 1$, an asymptotically negligible 'error' sequence, i.e., $\zeta_n \to 0$ a.s. The o.d.e. limit is

$$\dot{x}(t) = D\left(PF(x(t)) + g(x(t))\right).$$

The idea here is to replace $\{X_n\}$ by $\{\tilde{X}_n\}$ as above and (4.25) by

$$\begin{aligned} x_{n+1}(i) = x_n(i) + a(n)I\{\tilde{X}_{n+1} = i\}\left(\frac{p(\tilde{X}_{n+1}|\tilde{X}_n)}{\tilde{p}(\tilde{X}_{n+1}|\tilde{X}_n)}\right)\Big[F_{\tilde{X}_{n+1}}(x_n) + g_i(x_n) \\ + M_{n+1} + \zeta_{n+1}(i)\Big], \qquad 1 \le i \le d. \end{aligned} \tag{4.26}$$

This will have the o.d.e. limit

$$\dot{x}(t) = \tilde{D}\left(PF(x(t)) + g(x(t))\right),$$

where $\tilde{D}$ is the diagonal matrix with ith diagonal element $= \tilde{\eta}(i) \stackrel{def}{=}$ the stationary probability of state i for $\{\tilde{X}_n\}$. The advantages are:

1 $\tilde{D}$ can then be tailored to be a more suitable matrix (e.g., a scalar multiple of identity) by an appropriate choice of $\tilde{p}(\cdot|\cdot)$. $\tilde{D} \neq D$ can, however, be a problem when the convergence of the algorithm (at least theoretically) critically depends on having D in place, see, e.g., section 4.4 above.

2 As already noted, a major reason for slow mixing is the occurrence of nearly invariant sets. For the reversible case, a neat theoretical basis for this behaviour is available in terms of the spectrum of transition matrices - see, e.g., [15]. To work around this, $\{\tilde{X}_n\}$ can be chosen to be rapidly mixing, i.e., the laws of $\tilde{X}_n$ converge to the stationary law $\tilde{\eta}$ much faster than the corresponding rate for convergence of laws of the original chain $\{X_n\}$ to η. This can be achieved by increasing the probability of links which connect the nearly invariant sets to each other, so that the chain moves across their boundary more often. In addition, one may also introduce new edges which enhance mixing. This means that we relax (4.24) to

$$p(j|i) > 0 \Longrightarrow \tilde{p}(j|i) > 0 \ \forall \ i, j.$$

See, e.g., [23], which discusses such a scenario in the context of MCMC. In particular, the new edges can be chosen to alleviate the problems caused by local movement of the chain, by introducing the so called 'long jumps'. There will, however, be tradeoffs involved. For example, too many such transitions, which do not contribute anything to the learning process explicitly (because the correction term on r.h.s. of (4.26) is then zero), will in fact slow it down. Furthermore, there is often significantly higher computational cost associated with simulating these 'long jumps'.

3 An important difference between this and the traditional importance sampling is that this involves only a one step likelihood ratio for single transitions, not a full likelihood ratio accumulated over time from a regeneration time or otherwise. This tremendously reduces the problem of high variance, at the cost of higher per iterate computation.

This scheme, however, does require that the transition probabilities are known in an explicit analytic form, or are easy to compute or estimate. This need not always be the case. Another scheme which does away with this need is the split sampling proposed in [13], which does so at the expense of essentially doubling the simulation budget. Here we generate at each time n *two* $\mathcal{S}$-valued random variables, Y_n and Z_n, such that $\{Y_n\}$ is an irreducible Markov chain on $\mathcal{S}$ with transition matrix (say) $[[q(\cdot|\cdot)]]$ and stationary distribution γ, and Z_n is conditionally independent of $\{Z_m, Y_m; m < n\}$, given Y_n, with conditional law $P(Z_n = j|Y_n = i) = p(j|i) \ \forall \ i, j$. Then the scheme is

$$x_{n+1}(i) = x_n(i) + a(n)I\{Y_n = i\}[F_{Z_n}(x) + g_i(x) + M_{n+1} + \zeta_{n+1}(i)], \ 1 \leq i \leq d. \tag{4.27}$$

The limiting o.d.e. is

$$\dot{x}(t) = \hat{D}\left(PF(x(t)) + g(x(t))\right),$$

where $\hat{D}$ is a diagonal matrix whose ith diagonal element is $\gamma(i)$. In addition to the advantage of not requiring the explicit knowledge of $p(\cdot|\cdot)$ (though we do require a simulator that can simulate the transitions governed by $p(\cdot|\cdot)$), this also decouples the issues of mixing and that of obtaining the correct transition probabilities on the right hand side of the limiting o.d.e. We can choose $\{Y_n\}$, e.g., to be a rapidly mixing Markov chain with uniform stationary distribution (see, e.g., [32]), or even i.i.d. uniform random variables when they are easy to generate (unfortunately this is not always an easy task when the dimension is very high, as computer scientists well know). This scheme does away with the need for conditional importance sampling for speed-up, though one could also consider a combined scheme that combines both, i.e., a scheme which sets $P(Z_n = j|Y_n = i) = \tilde{p}(j|i) \ \forall \ i, j$ in addition to the above and replaces (4.27) by

$$x_{n+1}(i) = x_n(i) + a(n)\left(\frac{p(Z_n|Y_n)}{\tilde{p}(Z_n|Y_n)}\right)I\{Y_n = i\}[F_{Z_n}(x_n) + g_i(x_n) + M_{n+1} + \zeta_{n+1}(i)], \quad 1 \leq i \leq d. \tag{4.28}$$

4.7. Future Directions

This line of research is quite young and naturally has ample research opportunities. There are the usual issues that go with stochastic algorithms, such as convergence rates, sample complexity, etc. In addition, one important theme not addressed in this article is that of obtaining good error bounds for function approximations. While some of the references cited above do have something to say about this, it is still a rather open area. Its difficulty is further compounded by the fact that the approximation error will depend on the choice of basis functions and there are no clear guidelines for this choice, though some beginnings have been made. These include clustering techniques based on graph clustering ([29], [30]) and random projections ([4]). Finally, the theoretical convergence proof of some schemes crucially depends on the states being sampled according to the stationary distribution of the Markov chain, which is inconvenient for the acceleration techniques mentioned above. There is a need for making the schemes robust vis-a-vis the sampling strategy.

References

[1] Abounadi, J., Bertsekas, D. P. and Borkar, V. S. (2001). Learning algorithms for Markov decision processes with average cost. *SIAM J. Control and Optimization.* **40** 681–698.

[2] Ahamed, T. P. I., Borkar, V. S. and Juneja, S. (2006). Adaptive importance sampling technique for Markov chains using stochastic approximation. *Operations Research.* **54** 489–504.

[3] Asmussen, S. and Glynn, P. W. (2007). *Stochastic Simulation: Algorithms and Analysis.* Springer Verlag, New York.

[4] Barman, K. and Borkar, V. S. (2008). A note on linear function approximation using random projections. *Systems and Control Letters.* **57** 784–786.

[5] Basu, A., Bhattacharya, T. and Borkar, V. S. (2008). A learning algorithm for risk-sensitive cost. *Mathematics of Operations Research.* **33** 880–898.

[6] Benaim, M. (1999). Dynamics of stochastic approximation algorithms. In *Le Séminaire de Probabilités* (J. Azéma, M. Emery, M. Ledoux and M. Yor, eds.), Springer Lecture Notes in Mathematics No. 1709, Springer Verlag, Berlin–Heidelberg, 1–68.

[7] Bertsekas, D. P. (2007). *Dynamic Programming and Optimal Control, Vol. 2* (3rd edition). Athena Scientific, Belmont, Mass.

[8] Bertsekas, D. P. and Tsitsiklis, J. N. (1996). *Neurodynamic Programming.* Athena Scientific, Belmont, Mass.

[9] Bertsekas, D. P. and Yu, H. (2007). Solution of large systems of equations using approximate dynamic programming methods. Report LIDS–2754, Lab. for Information and Decision Systems, M.I.T.

[10] Borkar, V. S. (1998). Asynchronous stochastic approximation. *SIAM J. Control and Optim.* **36** 840–851 (Correction note in *ibid.* **38** 662–663).

[11] Borkar, V. S. (2002). Q-learning for risk-sensitive control. *Math. Op. Research.* **27** 294–311.

[12] Borkar, V. S. and Meyn, S. P. (2000). The O.D.E. method for convergence of stochastic approximation and reinforcement learning. *SIAM J. Control and*

Optimization. **38** 447–469.

[13] Borkar, V. S., Pinto, J. and Prabhu, T. (2009). A new reinforcement learning algorithm for optimal stopping. **19** 91–113.

[14] Borkar, V. S. and Soumyanath, K. (1997). A new analog parallel scheme for fixed point computation, part I: Theory. *IEEE Trans. Circuits and Systems I: Fund. Theory Appl.* **44** 351–355.

[15] Bovier, A., Eckhoff, M., Gayrard, V. and Klein, M. (2002). Metastability and low lying spectra in reversible Markov chains. *Comm. in Math. Physics.* **228** 219–255.

[16] Bush, R. R. and Mosteller, F. (1951). A mathematical model of simple learning. *Psychological Review.* **58** 313–323.

[17] Bush, R. R. and Mosteller, F. (1951). A model for stimulus generalization and discrimination. *Psychological Review.* **58** 413–423.

[18] Derevitskii, D. P. and Fradkov, A. L. (1974). Two models for analyzing the dynamics of adaptation algorithms. *Automation and Remote Control.* **35** 59–67.

[19] Desai, P. Y. and Glynn, P. W. (2001). A Markov chain perspective on adaptive Monte Carlo algorithms. *Proc. of the 2001 Winter Simulation Conference* (B. A. Peters, J. S. Smith, D. J. Medeiros, and M. W. Rohrer, eds.), 379–384.

[20] Estes, K. W. (1950). Towards a statistical theory of learning. *Psychological Review.* **57** 94–107.

[21] Estes, K. W. and Burke, C. J. (1953). A theory of stimulus variability in learning. *Psychological Review.* **60** 276–286.

[22] Golub, G. H. and Van Loan, C. F. (1996). *Matrix Computations* (3rd edition). Johns Hopkins Uni. Press, Baltimore, MD.

[23] Guan, Y. and Krone, S. M. (2007). Small-world MCMC and convergence to multi-modal distributions: from slow mixing to fast mixing. *Annals of Probability* . **17** 284–304.

[24] Henderson, S., Meyn, S. P. and Tadić, V. B. (2003). Performance evaluation and policy selection in multiclass networks. *Discrete Event Dynamic Systems.* **13** 149–189.

[25] Hirsch, M. W. (1989). Convergent activation dynamics in continuous time networks. *Neural Networks.* **2** 331–349.

[26] Konda, V. R. and Borkar, V. S. (2000). Actor-critic type learning algorithms for Markov decision processes. *SIAM J. Control and Optimization.* **38** 94–123.

[27] Langville, A. N. and Meyer, C. D. (2006). *Google's PageRank and Beyond: The Science of Search Engine Rankings.* Princeton Uni. Press, Princeton, NJ.

[28] Ljung, L. (1977). Analysis of recursive stochastic algorithms. *IEEE Trans. on Auto. Control.* **22** 551–575.

[29] Mahadevan, M. and Maggioni, M. (2007). Proto-value functions: a Laplacian framework for learning representation and control in Markov decision processes. *Journal of Machine Learning Research.* **8** 2169–2231.

[30] Mathew, V. (2007). *Automated spatio-temporal abstraction in reinforcement*

learning. M.S. Thesis, Dept. of Computer Science and Engg., Indian Inst. of Technology, Madras.

[31] Meyn, S. P. (2007). *Control Techniques for Complex Networks.* Cambridge Uni. Press, Cambridge, U.K.

[32] Montenegro, R. and Tetali, P. (2006). *Mathematical Aspects of Mixing Times in Markov Chains.* Foundations and Trends in Theoretical Computer Science, Vol. 1:3, NOW Publishers, Boston–Delft.

[33] Pemantle, R. (1992). Vertex-reinforced random walk. *Probability Theory and Related Fields.* **92** 117–136.

[34] Powell, W. B. (2007). *Approximate Dynamic Programming: Solving the Curses of Dimensionality.* John Wiley, Hoboken, NJ.

[35] Serfozo, R. (1999). *Introduction to Stochastic Networks.* Springer Verlag, NY.

[36] Sutton, R. S. and Barto, A. G. (1998). *Reinforcement Learning.* MIT Press, Cambridge, MA.

[37] Thorndike, E. L. (1898). Animal intelligence: an experimental study of the associative processes in animals. *Psychological Review, Monograph Supplement.* **2** No. 8.

[38] Tsitsiklis, J. N. and Van Roy, B. (1997). Average cost temporal-difference learning. *Automatica.* **35** 1799–1808.

[39] Yu, H. and Bertsekas, D. P. (2006). Convergence results based on some temporal difference methods based on least squares. Report LIDS–2697, Lab. for Information and Decision Systems, M.I.T.

Chapter 5

Factors, Roots and Embeddings of Measures on Lie Groups

S. G. Dani
School of Mathematics,
Tata Institute of Fundamental Research,
Homi Bhabha Road, Colaba, Mumbai 400 005, India
dani@math.tifr.res.in

5.1. Introduction

Let G be a locally compact second countable group. We denote by $P(G)$ the space of all probability measures on G equipped with the weak$*$-topology with respect to bounded continuous functions, viz. a sequence $\{\mu_i\}$ in $P(G)$ converges to μ in $P(G)$, and we write $\mu_i \to \mu$, if $\int \varphi d\mu_i \to \varphi d\mu$ for all bounded continuous functions φ. On $P(G)$ we have the convolution product $*$ of measures making it a topological semigroup; namely $(\lambda, \mu) \mapsto \lambda * \mu$ is a continuous map of $P(G) \times P(G)$ into $P(G)$.

For $g \in G$ let δ_g denote the point mass concentrated at g, namely the probability measure such that $\delta_g(E) = 1$ for a Borel subset E of G if and only if $g \in E$. It can be seen that $\{\delta_g \mid g \in G\}$ is a closed subset of $P(G)$ and the map $g \mapsto \delta_g$ gives an embedding of G as a closed subset of $P(G)$, which is also a homomorphism of the semigroup G into the semigroup $P(G)$.

Notation 5.1. In the sequel we suppress the notation $*$ (as is common in the area) and write the product $\lambda * \mu$ of $\lambda, \mu \in P(G)$ as $\lambda\mu$, and similarly for $n \geq 2$ the n-fold product of $\mu \in P(G)$ by μ^n. Also, for any $g \in G$ and $\mu \in P(G)$ we shall write $g\mu$ for $\delta_g * \mu$ and similarly $\mu * \delta_g$ by μg. In view of the observations above this change in notation is unambiguous. For $\lambda \in P(G)$ we denote by $\operatorname{supp}\lambda$ the support of λ, namely the smallest closed subset with measure 1. For any closed subgroup H of G we shall also denote by $P(H)$ the subsemigroup of $P(G)$ consisting of all $\lambda \in P(G)$ such that $\operatorname{supp}\lambda$ is contained in H.

With regard to the definitions and the discussion in the sequel it would be convenient to bear in mind the following connection between probability measures and "random walks" on G. To each $\mu \in P(G)$ there corresponds a random walk on G with μ as its *transition probability*, namely the Markov process such that for any $a \in G$ and any measurable subset E of G the probability of starting at $a \in G$ and landing in aE (in one step) equals $\mu(E)$.

Definition 5.1. Let $\mu \in P(G)$.

i) A probability measure $\lambda \in P(G)$ is called a nth *root* of μ if $\lambda^n = \mu$.

ii) A probability measure $\sigma \in P(G)$ is called a *factor* of μ if there exists $\rho \in P(G)$ such that $\rho\sigma = \sigma\rho = \mu$.

It may be noted that a factor in the above sense is a "two sided factor" in the semigroup structure of $P(G)$. We will not have an occasion to consider one-sided factors in the usual sense, and the term factor will consistently mean a two-sided factor.

Remark 5.1. i) Every root of μ is a factor of μ. On the other hand in general factors form a much larger class of measures, even for point measures.

ii) Given $g \in G$, $\mu \in P(G)$ is a factor of δ_g only if $\mu = \delta_h$ for some $h \in G$ which commutes with g; if furthermore μ is a root of δ_g then the element h is a root of g in G.

Remark 5.2. If λ is a nth root of μ, $n \geq 2$, then the random walk corresponding to μ is the n-step iterate of the random walk corresponding to the nth root λ. Similarly factorisation of μ corresponds to factorisation of the corresponding random walks.

The main aim of this article is to discuss results about the sets of roots and factors of probability measures. Much of the study of these was inspired by the so called *embedding problem*, which we will now recall.

Definition 5.2. A probability measure μ is said to be *infinitely divisible* if it has nth roots for all natural numbers n.

In the (algebraic) study of semigroups an element is said to be "divisible" if it has roots of all orders, and the term "infinitely" as above is redundant, but in probability theory it has been a tradition, since the early years of classical probability to use the phrase "infinitely divisible".

Definition 5.3. A family $\{\mu_t\}_{t>0}$ of probability measures on G is called a *one-parameter convolution semigroup* if $\mu_{s+t} = \mu_s\mu_t$ for all $s, t > 0$, and it is said to be a *continuous one-parameter convolution semigroup* if the map $t \mapsto \mu_t$, $t > 0$, is continuous.

A probability measure μ is said to be *embeddable* if there exists a continuous one-parameter convolution semigroup $\{\mu_t\}$ such that $\mu = \mu_1$.

Every embeddable measure is infinitely divisible, since given $\mu = \mu_1$ in a one-parameter convolution semigroup $\{\mu_t\}_{t>0}$, for any $n \geq 2$, $\mu_{1/n}$ is a n th root of μ.

There is a rich analytic theory for embeddable measures obtaining in particular a Lévy-Khintchine representation theorem for these measures. Such a theory was developed by G.A. Hunt in the case of connected Lie groups, and has been extended to locally compact groups by Heyer, Hazod and Siebert (see [9]).

In the light of Hunt's theory, K.R. Parthasarathy ([13]) raised the question whether one can embed a given infinitely divisible probability measure in a one-parameter convolution semigroup, in particular to obtain a Lévy-Khintchine representation for it; this would of course involve some extra condition on G, since infinite divisibility does not always imply embeddability; e.g. if G is the group of rational numbers with the discrete topology then δ_1 is infinitely divisible, but it cannot be embeddable. Indeed, for the classical groups $\mathbb{R}^d$, $d \geq 1$, every infinitely divisible probability measure is embeddable. A locally compact group G is said to have the *embedding property* if every infinitely divisible probability measure on G is embeddable. It was shown in [13] that compact groups have the embedding property; an analogous result was also proved for measures on symmetric spaces of non-compact type, but we shall not be concerned with it here. Parthasarathy's work inspired a folklore conjecture that every connected Lie group has the embedding property. This conjecture is not yet fully settled, though it is now known to be true for a large class of Lie groups. We will discuss the details in this respect in the last section.

We will conclude this section by recalling a result which illustrates how the study of the set of roots plays a role in the embedding problem.

Definition 5.4. A probability measure μ is said to be *strongly root compact* if the set $\{\lambda^k \mid \lambda^n = \mu \text{ for some } n \in \mathbb{N}, 1 \leq k \leq n\}$, has compact closure in $P(G)$.

Theorem 5.1. *Let G be a Lie group. If $\mu \in P(G)$ is infinitely divisible and strongly root compact then it is embeddable.*

Such a result is known, in place of Lie groups, also for a larger class of locally compact groups; see [12], Corollary 3.7.

In view of Theorem 5.1 to prove that a Lie group G has the embedding property it suffices to show the following: given $\mu \in P(G)$ infinitely divisible there exists a closed subgroup H of G and a root ν of μ such that $\operatorname{supp} \nu$ is contained in H and, viewed as a probability measure on H, it is infinitely divisible and strongly root compact. Proving existence of such H and ν has been one of the strategies for proving the embedding theorem.

5.2. Some Basic Properties of Factors and Roots

Let G be a locally compact second countable group and $\mu \in P(G)$. We begin by introducing some notation associated with μ. We denote by $G(\mu)$ the smallest closed subgroup containing $\operatorname{supp} \mu$, or equivalently the smallest closed subgroup whose complement has measure 0. Let $N(\mu)$ denote the normaliser of $G(\mu)$ in G, namely

$$N(\mu) = \{g \in G \mid gxg^{-1} \in G(\mu) \text{ for all } x \in G(\mu)\}.$$

Then $N(\mu)$ is a closed subgroup of G.

The following is an interesting simple lemma.

Lemma 5.1. *Let $\mu \in P(G)$ and λ be a factor of μ. Then $\operatorname{supp} \lambda$ is contained in $N(\mu)$.*

Proof. Let $\nu \in P(G)$ be such that $\mu = \lambda\nu = \nu\lambda$. Then we have $\operatorname{supp} \mu = \overline{(\operatorname{supp} \lambda)(\operatorname{supp} \nu)} = \overline{(\operatorname{supp} \nu)(\operatorname{supp} \lambda)}$. Let $g \in \operatorname{supp} \lambda$ and consider any $x \in (\operatorname{supp} \nu)(\operatorname{supp} \lambda)$, say $x = yz$ with $y \in \operatorname{supp} \nu$ and $z \in \operatorname{supp} \lambda$. Then $gxg^{-1} = gyzg^{-1}$. Picking any $w \in \operatorname{supp} \nu$ we can therefore write gxg^{-1} as $(gy)(zw)(gw)^{-1}$. As gy, zw and gw are contained in $(\operatorname{supp} \lambda)(\operatorname{supp} \nu) \subseteq \operatorname{supp} \mu$ this shows that $gxg^{-1} \in G(\mu)$. As this holds for all $x \in (\operatorname{supp} \nu)(\operatorname{supp} \lambda)$ and the latter set is dense in $\operatorname{supp} \mu$ it follows that gxg^{-1} is contained in $G(\mu)$ for all $x \in \operatorname{supp} \mu$, and in turn for all $x \in G(\mu)$. Hence $\operatorname{supp} \lambda$ is contained in $N(\mu)$. □

Note that $G(\mu)$ is a closed normal subgroup of $N(\mu)$ and we may form the quotient group $N(\mu)/G(\mu)$. Let $p : N(\mu) \to N(\mu)/G(\mu)$ be the quotient homomorphism. Consider the image $p(\mu)$ of μ in $N(\mu)/G(\mu)$. It is the point mass at the identity element in $N(\mu)/G(\mu)$. Let λ be any factor of μ. Then $p(\lambda)$ is a factor of $p(\mu)$, and since the latter is a point mass so is $p(\lambda)$. Hence there exists $g \in N(\mu)$ such that $\operatorname{supp}\lambda$ is contained in $gG(\mu) = G(\mu)g$. If furthermore λ is a root of μ, say $\lambda^n = \mu$, then $p(\lambda)^n = p(\mu)$ and in this case g as above is such that $p(g)^n$ is the identity, so $g^n \in G(\mu)$. These observations may be summed up as follows:

Lemma 5.2. *If λ is a factor of μ then there exist $g \in N(\mu)$ and a $\sigma \in P(G(\mu))$ such that $\lambda = \sigma g$. If furthermore $\lambda^n = \mu$ for some $n \in \mathbb{N}$ then $g^n \in G(\mu)$.*

Let λ be a root of μ, say $\lambda^n = \mu$ with $n \in \mathbb{N}$. Let $g \in N(\mu)$ and $\sigma \in P(G(\mu))$, as obtained above, such that $g^n \in G(\mu)$ and $\lambda = \sigma g$. Then we have

$$\mu = \lambda^n = (\sigma g)^n = \sigma(g\sigma g^{-1})(g^2\sigma g^{-2})\cdots(g^{(n-1)}\sigma g^{-(n-1)})g^n.$$

Let $\Theta_g : G(\mu) \to G(\mu)$ be the automorphism defined by $\Theta_g(x) = gxg^{-1}$ for all $x \in G(\mu)$; note that $\alpha \mapsto \Theta_g(\alpha)$, $\alpha \in P(G(\mu))$, defines a homomorphism of the $P(G(\mu))$. From the identity we see that

Lemma 5.3. *λ is a nth root of μ if and only if it is of the form σg with $\sigma \in P(G(\mu))$ and $g \in N(\mu)$ such that $\sigma\Theta_g(\sigma)\cdots\Theta_g^{n-1}(\sigma)g^n = \mu$.*

The point about this characterisation is that the relation in the conclusion is entirely within $G(\mu)$ on which the measure μ lives. The measure σ may be viewed as an "affine nth root" of μ in $G(\mu)$, depending on the automorphism Θ_g and the translating element g^n from $G(\mu)$. It is more convenient when the translating element g^n is the identity element. The notion of affine nth root in this sense is studied in [7] (the results there have some consequences to the embedding problem, which however are beyond the scope of the present article). In general it may not be possible to choose the element g (in its $G(\mu)$ coset in $N(\mu)$) to be such that g^n is the identity element. However there are many natural situations in which this is possible.

We denote by $F(\mu)$ the set of all factors of μ in $P(G)$. The next result is about sequences in $F(\mu)$, and in particular shows that $F(\mu)$ is a closed set. It may be noted here that for a given $n \in \mathbb{N}$ the set of all nth roots can be readily seen to be a closed set. On the other hand the set of *all* roots of μ is in general not closed, as can be seen, for example, from the fact that the roots of unity form a dense subset of the circle group.

Proposition 5.1. *Let $\{\lambda_i\}$ be a sequence in $F(\mu)$, and let $\{\nu_i\}$ in $P(G)$ be such that $\lambda_i\nu_i = \nu_i\lambda_i = \mu$ for all i. Then there exists a sequence $\{x_i\}$ in $N(\mu)$ such that the following holds: the sequences $\{x_i\lambda_i\}$, $\{\lambda_i x_i\}$, $\{x_i^{-1}\nu_i\}$ and $\{\nu_i x_i^{-1}\}$ have compact closures in $P(G)$; in turn the sequences $\{x_i\mu x_i^{-1}\}$ and $\{x_i^{-1}\mu x_i\}$ are contained in a compact subset of $P(G)$.*

The first part of the assertion may be seen from the proof of Proposition 1.2 in [2]; (the statement of the Proposition there is not in this form, but the proof includes this assertion along the way); see also [1], Proposition 4.2 for another presentation of the proof. The assertion about $\{x_i\mu x_i^{-1}\}$ and $\{x_i^{-1}\mu x_i\}$ follows immediately from the first statement, and the relation between the measures: indeed, $x_i\mu x_i^{-1} = (x_i\lambda_i)(\nu_i x_i^{-1})$, and $x_i^{-1}\mu x_i = (x_i^{-1}\nu_i)(\nu_i x_i^{-1})$, yields the desired conclusion.

Corollary 5.1. *$F(\mu)$ is a closed subset of $P(G)$.*

Proof. Let $\{\lambda_i\}$ be a sequence in $F(\mu)$ converging to $\lambda \in P(G)$, and $\{\nu_i\}$ be such that $\lambda_i\nu_i = \nu_i\lambda_i = \mu$ for all i. Let $\{x_i\}$ be a sequence in $N(\mu)$ as in Proposition 5.1. Since $\{\lambda_i\}$ and $\{\lambda_i x_i\}$ have compact closures, it follows that $\{x_i\}$ has compact closure in $N(\mu)$. In turn, together with the fact that $\{\nu_i x_i^{-1}\}$ has compact closure this implies that $\{\nu_i\}$ has compact closure. Passing to a subsequence we may assume that it converges, to say $\nu \in P(G)$. Then $\lambda\nu = \nu\lambda = \mu$, and hence $\lambda \in F(\mu)$, which shows that $F(\mu)$ is closed. □

5.3. Factor Sets

Let G be a locally compact group and $\mu \in P(G)$. In this section we will discuss the factor set of μ, under certain conditions on G. As before we denote the set of factors of μ by $F(\mu)$. Let

$$Z(\mu) = \{g \in G \mid gx = xg \text{ for all } x \in \operatorname{supp}\mu\},$$

the centraliser of $\operatorname{supp}\mu$ (or equivalently of $G(\mu)$) in G. Also let

$$T(\mu) = \{g \in G \mid g\mu = \mu g\}.$$

Then $Z(\mu)$ and $T(\mu)$ are closed subgroups and $Z(\mu)$ is contained in $T(\mu)$. We note that if λ is a factor of μ then for any $x \in T(\mu)$, $x\lambda$ is also a factor of μ; if ν is such that $\mu = \lambda\nu = \nu\lambda$ then $(x\lambda)(\nu x^{-1}) = x\mu x^{-1} = \mu = \nu\lambda = (\nu x^{-1})(x\lambda)$.

Thus $T(\mu)$ (and also $Z(\mu)$) act on the space $F(\mu)$ (viewed with the subspace topology from $P(G)$). The key question is how large are the quotient spaces $F(\mu)/T(\mu)$, $F(\mu)/Z(\mu)$, and specifically whether there exist compact sets of representatives for the actions.

In the light of Proposition 5.1 the question is related to understanding sequences $\{x_i\}$ in G such that $\{x_i \mu x_i^{-1}\}$ and $\{x_i^{-1} \mu x_i\}$ are contained in compact sets. Consider the action of G on $P(G)$, with the action of $g \in G$, which we shall denote by Φ_g, given by $\lambda \mapsto g\lambda g^{-1}$. Then $\{\Phi_{x_i}(\mu)\}$ and $\{\Phi_{x_i^{-1}}(\mu)\}$ are contained in compact sets, and we want to know whether this implies that $\{x_i Z(\mu)\}$ is relatively compact in $G/Z(\mu)$, or $\{x_i T(\mu)\}$ is relatively compact in $G/T(\mu)$. A general scheme for studying asymptotic behaviour of measures under the action of a sequence of automorphisms of the group is discussed in [1], where a variety of applications are indicated, including to the study of the factor sets of measures. Questions involving orbit behaviour typically have better accessibility in the framework algebraic groups, and in the present instance also the known results are based on techniques from the area. We shall now briefly recall the set up, in a relatively simpler form, and then the results.

Let G be a subgroup of $GL(d, \mathbb{R})$, $d \geq 2$. Then G is said to be *algebraic* if there exists a polynomial $P(x_{ij})$ in d^2 variables x_{ij}, $i, j = 1, \ldots, d$, such that $G = \{(g_{ij}) \mid P(g_{ij}) = 0\}$; (normally, over a general field, one takes a set of polynomials, but over the field of real numbers one polynomial suffices). Also, it is said to be *almost algebraic* if it is an open subgroup of an algebraic subgroup. Almost algebraic subgroups form a rich class of Lie groups. To that end we may mention that given a connected Lie subgroup G of $SL(d, \mathbb{R})$ the smallest almost algebraic subgroup $\tilde{G}$ containing G is such that G is a normal subgroup of $\tilde{G}$, $\tilde{G}/G$ is isomorphic to $\mathbb{R}^k$ for some k, and $[\tilde{G}, \tilde{G}] = [G, G]$; in particular if G_1 and G_2 are two connected Lie subgroups of $GL(d, \mathbb{R})$ whose commutator subgroups are different then the corresponding almost algebraic subgroups are distinct. In the sequel we will suppress the inclusion of the groups G in $GL(d, \mathbb{R})$ as above, and think of them independently as "almost algebraic groups", the $GL(d, \mathbb{R})$ being in the background.

A connected Lie group is said to be *W-algebraic* if i) AdG is an almost algebraic subgroup of $GL(\mathfrak{G})$, where $\mathfrak{G}$ is the Lie algebra of G and ii) for any compact subgroup C contained in the centre of G and $x \in G$, $\{g \in C \mid g = xyx^{-1}y^{-1}$ for some $y \in G\}$ is finite. The class of W-algebraic groups includes all almost algebraic connected Lie groups, all connected semisimple Lie groups and also, more generally, all connected Lie groups whose nilradical is simply connected.

The following result was proved in [6].

Theorem 5.2. *Let G be a W-algebraic group and $\mu \in P(G)$. Then $F(\mu)/Z(\mu)$ is compact. In particular, $F(\mu)/T(\mu)$ and $T(\mu)/Z(\mu)$ are compact.*

For the case of almost algebraic groups (which in particular are W-algebraic) a proof of the theorem may be found in [4]. The question was also studied earlier in [3] and the same conclusion was upheld under a condition termed as "weakly algebraic", but the method there is more involved.

There are examples to show that if either of the conditions in Theorem 5.2 do not hold then the conclusion $F(\mu)/Z(\mu)$ is compact need not hold; see [3]. Let us only recall here the example pertaining to the second condition (in a slightly modified form than in [3]).

Example 5.1. Let H be the Heisenberg group consisting of 3×3 upper triangular unipotent matrices. Let Z be the (one-dimensional) centre of H and D be a nonzero cyclic subgroup of Z. Let $G = H/D$. Then $T = Z/D$ is a compact subgroup forming the center of G, and G/T is topologically isomorphic to $\mathbb{R}^2$. On G we can have a probability measure μ which is invariant under the action of T by translations, and such that $G(\mu) = G$. Then for any $g \in G$ the T-invariant probability measure supported on gT is a factor of μ. On the other hand, since $G(\mu) = G$, $Z(\mu) = T$. It follows that $F(\mu)/Z(\mu)$ cannot be compact.

In all the known examples where $F(\mu)/Z(\mu)$ is not compact for a probability measure μ on connected Lie group G, the construction involves in fact that $T(\mu)/Z(\mu)$ is non-compact. It is not known whether there exist a connected Lie group G and a $\mu \in P(G)$ such that $F(\mu)/T(\mu)$ is non-compact.

Remark 5.3. Conditions under which $\{x_i \mu x_i^{-1}\}$ can be relatively compact for a probability measure μ and a sequence $\{x_i\}$ in the group are not well understood. When this holds for a μ for a sequence $\{x_i\}$ not contained in a compact subset, μ is said to be *collapsible*. Some partial results were obtained on this question in [8], and were applied in the study of decay of concentration functions of convolution powers of probability measures.

5.4. Compactness

Let G be a locally compact second countable group, and let $\mu \in P(G)$. The set $R(\mu) = \{\lambda^k \mid \lambda^n = \mu \text{ for some } n \in \mathbb{N}, 1 \leq k \leq n\}$ is called the *root set* of μ. Recall that μ is said to be strongly root compact if the root set is relatively compact and, by Theorem 5.1, when this holds infinitely divisibility of μ implies embeddability. Clearly $R(\mu)$ is contained in the factor set $F(\mu)$, so μ is strongly root compact when $F(\mu)$ is compact. In the light of the results of the previous section we have the following.

Corollary 5.2. *Let G be a W-algebraic group and let μ be such that $Z(\mu)$ is compact. Then μ is strongly root compact. In particular, if G is an almost algebraic group with compact center then any $\mu \in P(G)$ such that $G(\mu) = G$ (namely such that* $\operatorname{supp}\mu$ *is not contained in any proper closed subgroup), is strongly root compact.*

The following proposition enables extending the class of measures for which strong root compactness holds (see [11], Proposition 8).

Proposition 5.2. *Let G and H be locally compact second countable groups and suppose there is a continuous surjective homomorphism $\psi : H \to G$ such that the kernel of ψ is a compactly generated subgroup contained in the center of H. Let $\nu \in P(H)$ and X be a subset of $R(\nu)$ such that $\psi(X)$ is relatively compact in $P(G)$. Then X is a relatively compact subset of $P(G)$.*

Corollary 5.3. *Let G be a connected Lie group and suppose that there exists a closed subgroup Z contained in the center such that G/Z is topologically isomorphic to a W-algebraic group. Let $\eta : G \to G/Z$ be the quotient homomorphism. If $\mu \in P(G)$ is such that $\eta(\mu)$ is strongly root compact then μ is strongly root compact.*

We note in this respect that every closed subgroup contained in the center of a connected Lie subgroup is compactly generated (see [10]), so Proposition 5.2 applies.

By an inductive argument using the above corollary one can prove the following.

Corollary 5.4. *i) If G is a connected nilpotent Lie group then every $\mu \in P(G)$ is strongly root compact.*

ii) If G is an almost algebraic group and $\mu \in P(G)$ is such that $\operatorname{supp}\mu$ *is not contained in any proper almost algebraic subgroup of G then μ is strongly root compact.*

It may be mentioned that there is also a group theoretic condition called Böge strong root compactness which implies strong root compactness of every measure on the group. Various groups including compact groups, connected nilpotent Lie groups and also connected solvable groups G for which all eigenvalues of $\mathrm{Ad}g$, $g \in G$, are real (in this case G is said to have real roots) are known to satisfy the Böge strong root compactness condition. The reader is referred to [9], Chapter III for details; see also [12].

5.5. Roots

Let G be a locally compact group and let $\mu \in P(G)$. We now discuss the set of roots of μ. For $n \in \mathbb{N}$ we denote by $R_n(\mu)$ the set of nth roots of μ, namely $\lambda \in P(G)$ such that $\lambda^n = \mu$.

Let $Z(\mu)$ and $T(\mu)$ be the subgroups as before. We note that if $\lambda \in R_n(\mu)$ and $g \in T(\mu)$ then $g\lambda g^{-1} \in R_n(\mu)$, since $(g\lambda g^{-1})^n = g\lambda^n g^{-1} = g\mu g^{-1} = \mu$; (note that a translate of a root by an element of $T(\mu)$ or $Z(\mu)$ need not be a root). Thus we have an action of $T(\mu)$ on each $R_n(\mu)$, with the action of $g \in T(\mu)$ given by $\lambda \mapsto g\lambda g^{-1}$ for all $\lambda \in R_n(\mu)$. The object in this case will be to understand the quotient space of $R_n(\mu)$ under this action.

Let us first discuss a special case. Let $G = SL(d, \mathbb{C})$ and $\mu = \delta_I$, where I is the identity matrix. Then for $n \in \mathbb{N}$, $R_n(\mu)$ consists of $\{\delta_x \mid x \in G, x^n = I\}$. Every x in this is diagonalisable, viz. has the form gdg^{-1}, for some $g \in G$ and $d = \mathrm{diag}\,(\sigma_1, \ldots, \sigma_d)$, with each σ_i a nth root of unity; there are only finitely many of these diagonal matrices. Since $\mu = \delta_I$, we have $Z(\mu) = G$, so the diagonal matrices as above form a set of representatives for the quotient space of $R_n(\mu)$ under the action of $T(\mu)$ defined above. In particular the quotient space is finite. Analogous assertion holds for any point mass over an algebraic group.

The following theorem provides a generalisation of this picture in the special case, to more general probability measures, over a class of Lie groups G. In the general case the quotient is shown to be a compact set (in place of being finite in the special case). The condition that we need on G is described in the following Proposition (see [5], Proposition 2.5).

Proposition 5.3. *Let G be a connected Lie group. Then the following conditions are equivalent.*

i) there exists a representation $\rho : G \to GL(d, \mathbb{R})$ for some $d \in \mathbb{N}$ such that the kernel of ρ is discrete.

ii) if R is the radical of G then $[R, R]$ is a closed simply connected nilpotent Lie subgroup.

A connected Lie group satisfying either of the equivalent conditions is said to be of *class* $\mathcal{C}$. Groups of class $\mathcal{C}$ include all linear groups, all simply connected Lie groups (by Ado's theorem), and all semisimple Lie groups (through the adjoint representation), and thus form a large class.

Remark 5.4. We note however that not all connected Lie groups are of class $\mathcal{C}$. An example of this may be given as follows. Let H and D be an in Example 5.1 and let $N = H/D$. Then N is not of class $\mathcal{C}$; in fact it can be shown that for any finite-dimensional representation of N, the one-dimensional center $T = Z/D$ of N is contained in the kernel of the representation. An example of a non-nilpotent Lie group which is not of class $\mathcal{C}$ may be constructed from this as follows. Note that N/T is isomorphic to $\mathbb{R}^2$. The group of rotations of $\mathbb{R}^2$ extends, over the quotient homomorphism of N onto N/T, to a group of automorphisms of N, say C. Let $G = C \cdot N$, the semidirect product. Then G is not of class $\mathcal{C}$. Similarly $SL(2, \mathbb{R})$, viewed as a group of automorphisms of $\mathbb{R}^2$ extends to a group of automorphisms of N and the corresponding semidirect product is a non-solvable Lie group that is not of class $\mathcal{C}$.

For any $\lambda \in P(G)$ we denote by $Z^0(\lambda)$ the connected component of the identity in $Z(\lambda)$.

Theorem 5.3. *Let G be a connected Lie group of class $\mathcal{C}$ and let $\mu \in P(G)$. Let $n \in \mathbb{N}$ and $\{\lambda_i\}$ be a sequence in $R_n(\mu)$. Then there exists a sequence $\{z_i\}$ in $Z^0(\mu)$ such that $\{z_i \lambda_i z_i^{-1}\}$ is relatively compact. Moreover, the sequence $\{z_i\}$ has also the property that if $m \in \mathbb{N}$ and $\{\nu_i\}$ is a sequence in $R_{mn}(\mu)$ such that $\nu_i^m = \lambda_i$ and $Z^0(\nu_i) = Z^0(\lambda_i)$ for all i, then $\{z_i \nu_i z_i^{-1}\}$ is relatively compact.*

Let $n \in \mathbb{N}$ and let $\sim$ denote the equivalence relation on $R_n(\mu)$ defined by $\lambda \sim \lambda'$, for $\lambda, \lambda' \in R_n(\mu)$ if there exists a $g \in Z^0(\mu)$ such that $\lambda' = g\lambda g^{-1}$. Then the first assertion in the theorem shows in particular that the quotient space $R_n(\mu)/\sim$ is compact. The point in second statement is that the same z_i's work for n th as well as mn th roots, and this can be seen to be useful in taking "inverse limits". Such limits appear in the proof of the embedding theorem in [5].

We will now sketch a part of the proof of the theorem, recalling some key ingredients, which could be of independent interest. For simplicity we shall assume that G an almost algebraic group. Let $n \in \mathbb{N}$ and a sequence $\{\lambda_i\}$ in $R_n(\mu)$ be given. By Theorem 5.2 there exists a sequence $\{x_i\}$ in $Z(\mu)$ such that $\{\lambda_i x_i\}$ is relatively compact. Thus we have a sequence of *translates* by elements of $Z(\mu)$ forming a relatively compact set. Our aim would be to find a sequence of conjugates contained in a compact set. To achieve this we proceed as follows. As $\{\lambda_i x_i\}$ is relatively compact, so is $\{(\lambda_i x_i)^n\}$. Recall that each λ_i can be written as $\sigma_i y_i$, with $\sigma_i \in P(G(\mu))$ and $y_i \in N(\mu)$. Then for any i, $(\lambda_i x_i)^n = (\sigma_i y_i x_i)^n = (\sigma_i y_i x_i)(\sigma_i y_i x_i)\cdots(\sigma_i y_i x_i)$, and a straightforward computation using that $x \in Z(\mu)$ shows that $(\sigma_i y_i x_i)^n$ equals $(\sigma_i y_i)^n(\alpha_i(x_i)\alpha_i^2(x_i)\cdots\alpha_i^n(x_i))$, where α_i is the automorphism of $Z(\mu)$ defined by $\alpha_i(z) = y_i z y_i^{-1}$ for all $z \in Z(\mu)$. Therefore $(\lambda_i x_i)^n = \lambda_i^n(\alpha_i(x_i)\alpha_i^2(x_i)\cdots\alpha_i^n(x_i)) = \mu(\alpha_i(x_i)\alpha_i^2(x_i)\cdots\alpha_i^n(x_i))$. As $\{\lambda_i x_i\}$ has compact closure in $P(G)$ it follows that $\{(\alpha_i(x_i)\alpha_i^2(x_i)\cdots\alpha_i^n(x_i))\}$ is relatively compact in $Z(\mu)$.

We now recall an interesting property of nilpotent Lie groups, called "affine root rigidity".

Theorem 5.4. *Let N be a connected nilpotent Lie group. Let $n \in \mathbb{N}$ and $\{\alpha_i\}$ be a sequence of automorphisms of N such that $\alpha_i^n = I$, the identity automorphism of N, for all $i \in \mathbb{N}$. Let $\{x_i\}$ be a sequence in N such that $\{(\alpha_i(x_i)\alpha_i^2(x_i)\cdots\alpha_i^n(x_i))\}$ is relatively compact. Then there exists a sequence $\{\xi_i\}$ in N such that $\{\xi_i^{-1}x_i\}$ is relatively compact and $\alpha_i(\xi_i)\alpha_i^2(\xi_i)\cdots\alpha_i^n(\xi_i) = e$, the identity element of N.*

While a priori the subgroup $Z(\mu)$ need not be nilpotent it turns out that using some structure theory of almost algebraic groups one can reduce to the case when the sequence $\{x_i\}$ as above is contained in the nilradical of $Z^0(\mu)$, so this theorem can be applied to the above context. Following the computation backwards one can now see that $\{\lambda_i\xi_i\}$ is relatively compact and moreover $(\lambda_i\xi_i)^n = \mu$, namely the translates $\lambda_i\xi_i$ are roots of μ.

This brings us to another interesting fact, again involving nilpotent Lie groups:

Theorem 5.5. *Let G be a locally compact second countable group and let $\mu \in P(G)$. Let N be a subgroup of $Z(\mu)$ which is isomorphic to a simply connected nilpotent Lie group, and normalized by $N(\mu)$. Let $n \in \mathbb{N}$, $\lambda \in P(G)$, and $z \in N$ be such that $(\lambda\xi)^n = \lambda^n = \mu$. Then there exists $\zeta \in N$ such that $\lambda\xi = \zeta\rho\zeta^{-1}$.*

This shows that the translates we had are also conjugates by another sequence from the same subgroup. This proves the first assertion in Theorem 5.3. The second part involves keeping track of how the conjugations operate when we go to higher roots, using again certain properties of nilpotent Lie groups.

5.6. One-Parameter Semigroups

In this section we discuss one-parameter semigroups and embeddings. One may think of "one-parameter semigroups" parametrised either by positive reals, or just by positive rationals (parametrisation by other subsemigroups of positive reals may also be considered, but we shall not go into that here). Via the study of the factor sets of measures the following was proved in [3].

Theorem 5.6. *Let G be a connected Lie group. Then any homomorphism φ of the semigroup $\mathbb{Q}^+$ (positive rationals, under addition) into $P(G)$ extends to a homomorphism of $\mathbb{R}^+$ (positive reals) to $P(G)$.*

This reduces the embedding problem for infinitely divisible measures to finding a rational embedding. (Actually our proof of the embedding result in [5] does not make serious use of this, but it would help to see the problem in perspective). It may also be noted that the task of finding a rational embedding of $\mu \in P(G)$ is equivalent to finding a sequence λ_k in $P(G)$ such that $\lambda_k^{k!} = \mu$ and $\lambda_k^k = \lambda_{k-1}$ for all k; this produces a homomorphism from $\mathbb{Q}^+$ to $P(G)$ given by $\frac{p}{q} \mapsto \lambda_q^{p(q-1)!}$. While by infinite divisibility μ admits $k!$ th roots for all k, we need to have them matching as above; an arbitrarily picked $k!$ th root may a priori not have any nontrivial roots at all.

Let us now come to the embedding problem, i.e. embedding a given infinitely divisible probability measure μ, on a Lie group, in a continuous one-parameter semigroup. Recall that by Theorem 5.1 if μ is strongly root compact then μ is embeddable, and in particular the conclusion holds for the strongly root compact measures as noted in § 4. In particular it was known by the 1970's that all nilpotent Lie groups and solvable Lie groups with real roots have the embedding property. The reader is referred to [9] for details. In [4] it was shown that all connected Lie groups of class $\mathcal{C}$ (see § 5) have the embedding property. A simpler and more transparent proof of the result was given in [5] using Theorem 5.3.

Theorem 5.7. *Let G be a Lie group of class $\mathcal{C}$. Let $\mu \in P(G)$ be infinitely divisible and let $r : \mathbb{N} \to P(G)$ be a map such that $r(m)^m = \mu$ for all $m \in \mathbb{N}$. Then there exist sequences $\{m_i\}$ in $\mathbb{N}$ and $\{z_i\}$ in $Z^0(\mu)$, and $n \in \mathbb{N}$ such that n divides m_i for all i and the sequence $\{z_i r(m_i)^{m_i/n} z_i^{-1}\}$ (consisting of n th roots of μ) converges to a n th root ν of μ which is infinitely divisible and strongly root compact on the subgroup $Z(Z^0(\nu))$, the centraliser of $Z^0(\nu)$ in G. Hence μ is embeddable.*

An overall idea of the proof is as follows. A subset M of $\mathbb{N}$ is said to be infinitely divisible if for every $k \in \mathbb{N}$ there exists $m \in M$ such that k divides m. Let $M = \{m_i\}$ be an infinitely divisible set and $n \in \mathbb{N}$. By infinite divisibility of μ we can find a sequence ρ_i in $P(G)$ such that for all i, ρ_i is a $m_i n$th root of μ. Then $\rho_i^{m_i}$ is a nth root of μ for all i. By Theorem 5.3 there exists a sequence $\{z_i\}$ in $Z^0(\mu)$ such that $\{z_i \rho_i^{m_i} z_i^{-1}\}$ is relatively compact. Note that any limit point of the sequence is a nth root of μ which is infinitely divisible in G. We need the limit to be such that it is simultaneously infinitely divisible and strongly root compact in a suitable subgroup of H. For this we need to pick the set M as above and n suitably, which involves in particular analysing how $Z^0(\lambda)$ changes over the roots λ, of higher and higher order. For full details the reader is referred to the proof in original ([5]).

We conclude with some miscellaneous comments concerning the status of the embedding problem.

i) For a general connected Lie group G, not necessarily of class $\mathcal{C}$, we get a "weak embedding theorem":

Theorem 5.8. *Let G be a connected Lie group and $\mu \in P(G)$ be infinitely divisible. Let T the maximal compact subgroup of $\overline{[R, R]}$, where R is the solvable radical of G. Let $p : G \to G/T$ the quotient homomorphism, and let $M = p^{-1}(Z^0(p(\mu)))$. Then there exists a sequence $\{\zeta_i\}$ in M such that $\zeta_i \mu \zeta_i^{-1}$ converges to an embeddable measure.*

This can be deduced from Theorem 5.7, using the fact that G/T as above is of class $\mathcal{C}$ (see [5]).

ii) The embedding problem has been studied also for various other classes of groups: discrete subgroups of $GL(d, \mathbb{R})$ (Dani and McCrudden), Finitely generated subgroups of $GL(n, \mathbb{A})$, where $\mathbb{A}$ is the field of algebraic numbers (Dani and Riddhi Shah), p-adic groups (Riddhi Shah, McCrudden - Walker); see [12] for some details and references.

References

[1] Dani, S. G. (2006). Asymptotic behaviour of measures under automorphisms. *Probability Measures on Groups: Recent Directions and Trends*, pp. 149–178, Tata Inst. Fund. Res., Mumbai, and Narosa Publishing House, Delhi; international distribution by the American Mathematical Society.

[2] Dani, S. G. and McCrudden, M. (1988). Factors, roots and embeddability of measures on Lie groups. *Math. Zeits.* **199** 369–385.

[3] Dani, S. G. and McCrudden, M. (1988). On the factor sets of measures and local tightness of convolution semigroups over Lie groups. *J. Theor. Probability* **1** 357–370.
[4] Dani, S. G. and McCrudden, M. (1992). Embeddability of infinitely divisible distributions on linear Lie groups. *Invent. Math.* **110** 237–261.
[5] Dani, S. G. and McCrudden, M. (2007). Convolution roots and embeddings of probability measures on Lie groups. *Adv. Math.* **209** 198–211.
[6] Dani, S. G. and Raja, C. R. E. (1998). Asymptotics of measures under group automorphisms and an application to factor sets. *Lie Groups and Ergodic Theory* (Mumbai, 1996), 59–73, Tata Inst. Fund. Res. Stud. Math., 14, Tata Inst. Fund. Res., Bombay, and Narosa Publishing House, Delhi.
[7] Dani, S. G. and Schmidt, K. (2002). Affinely infinitely divisible distributions and the embedding problem. *Math. Res. Lett.* **9** 607–620.
[8] Dani, S. G. and Shah, R. (1997). Collapsible probability measures and concentration functions on Lie groups. *Math. Proc. Cambridge Philos. Soc.* **122** 105–113.
[9] Heyer, H. (1977). *Probability Measures on Locally Compact Groups.* Springer Verlag.
[10] Hochschild, G. (1965). *The Structure of Lie Groups.* Holden-Day.
[11] McCrudden, M. (1981). Factors and roots of large measures on connected Lie groups. *Math. Z.* **177** 315–322.
[12] McCrudden, M. (2006). The embedding problem for probabilities on locally compact groups. *Probability Measures on Groups: Recent Directions and Trends*, pp. 331–363, Tata Inst. Fund. Res., Mumbai, and Narosa Publishing House, Delhi; international distribution by the American Mathematical Society.
[13] Parthasarathy, K. R. (1967). On the imbedding of an infinitely divisible distribution in a one-parameter convolution semigroup. *Theor. Probab. Appl.* **12** 373–380.
[14] Parthasarathy, K. R. (1967). *Probability Measures on Metric Spaces.* Academic Press.

Chapter 6

Higher Criticism in the Context of Unknown Distribution, Non-independence and Classification

Aurore Delaigle[1] and Peter Hall[2]

[1] *Department of Mathematics,*
University of Bristol,
Bristol, BS8 4JS, UK

[2] *Department of Mathematics and Statistics,*
University of Melbourne,
*VIC, 3010, Australia**

Higher criticism has been proposed as a tool for highly multiple hypothesis testing or signal detection, initially in cases where the distribution of a test statistic (or the noise in a signal) is known and the component tests are statistically independent. In this paper we explore the extent to which the assumptions of known distribution and independence can be relaxed, and we consider too the application of higher criticism to classification. It is shown that effective distribution approximations can be achieved by using a threshold approach; that is, by disregarding data components unless their significance level exceeds a sufficiently high value. This method exploits the good relative accuracy of approximations to light-tailed distributions. In particular, it can be effective when the true distribution is founded on something like a Studentised mean, or on an average of related type, which is commonly the case in practice. The issue of dependence among vector components is also shown not to be a serious difficulty in many circumstances.

6.1. Introduction

Donoho and Jin (cf. [8]) developed higher-criticism methods for hypothesis testing and signal detection. Their methods are founded on the assumption that the test statistics are independent and, under the null hypothesis, have a known normal distribution. However, in some applications of higher

*The first author is affiliated to both the universities, whereas the second author is affiliated to the second.

criticism, for example to more elaborate hypothesis testing problems and to classification, these assumptions may not be tenable. For example, we may have to estimate the distributions from data, by pooling information from components that have "neighbouring" indices, and the assumption of independence may be violated.

Taken together, these difficulties place obstacles in the way of using higher-criticism methods for a variety of applications, even though those techniques have potential performance advantages. We describe the effects that distribution approximation and data dependence can have on results, and suggest ways of alleviating problems caused by distribution approximation. We show too that thresholding, where only deviations above a particular value are considered, can produce distinct performance gains. Thresholding permits the experimenter to exploit the greater relative accuracy of distribution approximations in the tails of a distribution, compared with accuracy towards the distribution's centre, and thereby to reduce the tendency of approximation errors to accumulate. Our theoretical arguments take sample size to be fixed and the number of dimensions, p, to be arbitrarily large.

Thresholding makes it possible to use rather crude distribution approximations. In particular, it permits the approximations to be based on relatively small sample sizes, either through pooling data from a small number of nearby indices, or by using normal approximations based on averages of relatively small datasets. Without thresholding, the distribution approximations used to construct higher-criticism signal detectors and classifiers would have to be virtually root-p consistent.

We shall provide theoretical underpinning for these ideas, and explore them numerically; and we shall demonstrate that higher criticism can accommodate significant amounts of local dependence, without being seriously impaired. We shall further show that, under quite general conditions, the higher-criticism statistic can be decomposed into two parts, of which one is stochastic and of smaller order than p^ϵ for any positive ϵ, and the other is purely deterministic and admits a simple, explicit formula. This simplicity enables the effectiveness of higher criticism to be explored quite generally, for distributions where the distribution tails are heavy, and also for distributions that have relatively light tails, perhaps through being convolutions of heavy-tailed distributions. These comments apply to applications to both signal detection and classification.

In the contexts of independence and signal detection, [8] used an approach alternative to that discussed above. They employed delicate, empirical-process methods to develop a careful approximation, on the $\sqrt{\log \log p}$ scale, to the null distribution of the higher-criticism statistic. It is unclear from their work whether the delicacy of the log-log approximation is essential, or whether significant latitude is available for computing critical points. We shall show that quite crude bounds can in fact be used, in both the dependent and independent cases. Indeed, any critical point on a scale that is of smaller order than p^ϵ, for each $\epsilon > 0$, is appropriate.

Higher-criticism methods for signal detection have their roots in unpublished work of Tukey; see [8] for discussion. Optimal, but more tightly specialised, methods for signal detection were developed by [17–19] and [20], broadly in the context of techniques for multiple comparison (see e.g. the methods of Bonferroni, [30] and [26]), for simultaneous hypothesis testing (e.g. [9] and [23]) and for moderating false-discovery rates (e.g. [2], [28] and [1]). Model-based approaches to the analysis of high-dimensional microarray data include those of [29], [16], [12] and [11]. Related work on higher criticism includes that of [25] and [4]. Higher-criticism classification has been discussed by [15], although this work assumed that test statistic distributions are known exactly. Applications of higher criticism to signal detection in astronomy include those of [22], [5, 6], [7] and [21]. [14] discussed properties of higher criticism under long-range dependence.

Our main theoretical results are as follows. Theorem 6.1, in section 6.3.1, gives conditions under which the higher-criticism statistic, based on a general approximation to the unknown test distributions, can be expressed in terms of its "ideal" form where the distributions are known, plus a negligible remainder. This result requires no assumptions about independence. Theorem 6.2, in section 6.3.2, gives conditions on the strength of dependence under which the higher-criticism statistic can be expressed as a purely deterministic quantity plus a negligible remainder. Theorem 6.3, in section 6.3.3, describes properties of the deterministic "main term" in the previous result. Discussion in sections 6.3.3 and 6.4 draws these three results together, and shows that they lead to a variety of properties of signal detectors and classifiers based on higher criticism. These properties are explored numerically in section 6.5.

6.2. Methodology

6.2.1. *Higher-criticism signal detection*

Assume we observe independent random variables $Z_1, \ldots, Z_p$, where each Z_j is normally distributed with mean μ_j and unit variance. We wish to test, or at least to assess the validity of, the null hypothesis H_0 that each μ_j equals ν_j, a known quantity, versus the alternative hypothesis that one or more of the μ_j are different from ν_j. If each ν_j equals zero then this context models signal detection problems where the null hypothesis states that the signal is comprised entirely of white noise, and the alternative hypothesis indicates that a nondegenerate signal is present.

A higher-criticism approach to signal detection and hypothesis testing, a two-sided version of a suggestion by [8], can be based on the statistic

$$\mathrm{hc} = \inf_{u\,:\,\psi(u)>C} \psi(u)^{-1/2} \sum_{j=1}^{p} \{I(|Z_j - \nu_j| \le u) - \Psi(u)\}, \qquad (6.1)$$

where $\Psi(u) = 2\,\Phi(u) - 1$ is the distribution function of $|Z_j - \nu_j|$ under H_0, Φ is the standard normal distribution function, $\psi(u) = p\,\Psi(u)\,\{1 - \Psi(u)\}$ equals the variance of $\sum_j \{I(|Z_j - \nu_j| \le u) - \Psi(u)\}$ under H_0, and C is a positive constant.

The statistic at (6.1) provides a way of assessing the statistical significance of p tests of significance. In particular, H_0 is rejected if hc takes too large a negative value. This test enjoys optimality properties, in that it is able to detect the presence of nonzero values of μ_j up to levels of sparsity and amplitude that are so high and so low, respectively, that no test can distinguish between the null and alternative hypotheses ([8]).

6.2.2. *Generalising and adapting to an unknown null distribution*

When employed in the context of hypothesis testing (where the ν_js are not necessarily equal to zero), higher-criticism could be used in more general settings, where the centered Z_js are not identically distributed. Further, instead of assuming that the ν_js are prespecified, they could be taken equal to the jth component of the empirical mean of a set of n_W identically distributed random p-vectors $W_1, \ldots, W_{n_W}$, where $W_i = (W_{i1}, \ldots, W_{ip})$ has the distribution of $(Z_1, \ldots, Z_p)$ under the null hypothesis H_0. Here, H_0 asserts the equality of the mean components of the vector Z, and of the vectors $W_1, \ldots, W_{n_W}$, whose distribution is known except for the mean which is estimated by its empirical counterpart. There, we could redefine hc by replacing, in (6.1), ν_j by $\bar{W}_{.j} = n_W^{-1} \sum_j W_{ij}$ and Ψ by the distribution Ψ_{Wj}, say, of $|Z_j - \bar{W}_{.j}|$ under the null hypothesis. This gives, in place of hc at (6.1),

$$\mathrm{hc}_W = \inf_{u \in \mathcal{U}_W} \psi_W(u)^{-1/2} \sum_{j=1}^{p} \{I(|Z_j - \bar{W}_{.j}| \leq u) - \Psi_{Wj}(u)\}, \tag{6.2}$$

where

$$\psi_W(u) = \sum_{j=1}^{p} \Psi_{Wj}(u) \{1 - \Psi_{Wj}(u)\} \tag{6.3}$$

and, given $C, t > 0$, $\mathcal{U}_W = \mathcal{U}_W(C, t)$ is the set of u for which $u \geq t$ and $\psi_W(u) \geq C$. Here t denotes a threshold, and the fact that, in the definition of $\mathcal{U}_W$, we confine attention to $u > t$, means that we restrict ourselves to values of u for which distribution approximations are relatively accurate; see section 6.4.2 for details.

Further, in practical applications it is often unrealistic to argue that Ψ (respectively, Ψ_{Wj}), is known exactly, and we should replace Ψ in (6.1) and in ψ (respectively, Ψ_{Wj} in (6.2) and in ψ_W) by an estimator $\widehat{\Psi}$ of Ψ (respectively, $\widehat{\Psi}_{Wj}$ of Ψ_{Wj}). This leads to an empirical approximation, $\widehat{\mathrm{hc}} = \widehat{\mathrm{hc}}(C, t)$, to hc:

$$\widehat{\mathrm{hc}} = \inf_{u \in \widehat{\mathcal{U}}} \widehat{\psi}(u)^{-1/2} \sum_{j=1}^{p} \{I(|Z_j - \nu_j| \leq u) - \widehat{\Psi}(u)\},$$

where $\widehat{\psi}(u) = p\,\widehat{\Psi}(u)\{1 - \widehat{\Psi}(u)\}$ and $\widehat{\mathcal{U}} = \widehat{\mathcal{U}}(C, t)$ is the set of u for which $u \geq t$ and $\widehat{\psi}(u) \geq C$, and to an empirical approximation $\widehat{\mathrm{hc}}_W = \widehat{\mathrm{hc}}_W(C, t)$, to hc_W:

$$\widehat{\mathrm{hc}}_W = \inf_{u \in \widehat{\mathcal{U}}_W} \widehat{\psi}_W(u)^{-1/2} \sum_{j=1}^{p} \{I(|Z_j - \bar{W}_{.j}| \leq u) - \widehat{\Psi}_{Wj}(u)\}, \tag{6.4}$$

where

$$\widehat{\psi}_W(u) = \sum_{j=1}^{p} \widehat{\Psi}_{Wj}(u) \{1 - \widehat{\Psi}_{Wj}(u)\} \tag{6.5}$$

and $\widehat{\mathcal{U}}_W = \widehat{\mathcal{U}}_W(C, t)$ is the set of u for which $u \geq t$ and $\widehat{\psi}_W(u) \geq C$. Here t denotes a threshold, and the fact that, in the definition of $\mathcal{U}_W$, we confine attention to $u > t$, means that we restrict ourselves to values of u for which distribution approximations are relatively accurate; see section 6.4.2 for details.

Estimators of Ψ_{Wj} are, broadly speaking, of two types: either they depend strongly on the data, or they depend on the data only through the way these have been collected, for instance through sample size. In the first case, $\widehat{\Psi}_{Wj}$ might, for example, be computed directly from the data, for example by pooling vector components for nearby values of the index j. (In genomic examples, "nearby" does not necessarily mean close in terms of position on the chromosome; it is often more effectively defined in other ways, for example in the sense of gene pathways.)

Examples of the second type come from an important class of problems where the variables W_{ij} are obtained by averaging other data. For example, they can represent Studentised means, $W_{ij} = N_{Wj}^{1/2}\,\bar{U}_{Wij}/S_{Wij}$, or Student t statistics for two-sample tests, $W_{ij} = N_{Wj}^{1/2}\,(\bar{U}_{Wij,1} - \bar{U}_{Wij,2})/(S^2_{Wij,1} + S^2_{Wij,2})^{1/2}$, where, for $i = 1, \ldots, n_W$ and $j = 1, \ldots, p$, $\bar{U}_{Wij,k}$ and $S^2_{Wij,k}$, $k = 1, 2$ denote respectively the empirical mean and empirical variance of N_{Wj} independent and identically distributed data; or statistics computed in a related way. See e.g. [24], [27], [13] and [10]. In such cases, if Z_j and $W_{1j}, \ldots, W_{nj}$ were identically distributed, $Z_j - \bar{W}_{\cdot j}$ would be approximately normally distributed with variance $\tau_W = 1 + n_W^{-1}$, and $\widehat{\Psi}_{Wj}$ would be the distribution function of the normal $\mathrm{N}(0, \tau_W)$ distribution, not depending on the index j, and depending on the data only through the number n_W of observations. See section 6.3.3 for theoretical properties for this type of data.

Formula (6.5), giving an empirical approximation to the variance of the series on the right-hand side of (6.4), might seem to suggest that, despite the increased generality we are capturing by using empirical approximations to the distribution functions Ψ_{Wj}, we are continuing to assume that the vector components $W_{i1}, \ldots, W_{ip}$ are independent. However, independence is not essential. By choosing the threshold, t, introduced earlier in this section, to be sufficiently large, the independence assumption can be removed while retaining the validity of the variance approximation at (6.5). See section 6.4.2.

6.2.3. *Classifiers based on higher criticism*

The generality of the higher-criticism approximation in section 6.2.2 leads directly to higher-criticism methods for classification. To define the classification problem, assume we have data, in the form of independent random samples of p-dimensional vectors $\mathcal{X} = \{X_1, \ldots, X_{n_X}\}$ from population Π_X, and $\mathcal{Y} = \{Y_1, \ldots, Y_{n_Y}\}$ from population Π_Y, and a new, independent observation, Z, from either Π_X or Π_Y. (In our theoretical work the sample sizes n_X and n_Y will be kept fixed as p increases.) We wish to assign Z to one of the populations. In the conventional case where p is small relative to sample size, many different techniques have been developed for solving this problem. However, in the setting in which we are interested, where p is large and the sample size is small, these methods can be ineffective, and better classification algorithms can be obtained by using methods particularly adapted to the detection of sparse signals.

Let X_{ij}, Y_{ij} and Z_j denote the jth components of X_i, Y_i and Z, respectively. Assume that $\mu_{Xj} = E(X_{ij})$ and $\mu_{Yj} = E(Y_{ij})$ do not depend on i, that the distributions of the components are absolutely continuous, and that the distributions of the vectors $(X_{i_1 1} - \mu_{X1}, \ldots, X_{i_1 p} - \mu_{Xp})$, $(Y_{i_2 1} - \mu_{Y1}, \ldots, Y_{i_2 p} - \mu_{Yp})$ and $(Z_1 - E(Z_1), \ldots, Z_p - E(Z_p))$ are identical to one another, for $1 \le i_1 \le n_X$ and $1 \le i_2 \le n_Y$.

In particular, for each i_1, i_2 and j the distributions of $X_{i_1 j}$ and $Y_{i_2 j}$ differ only in location. This assumption serves to motivate methodology, and is a convenient platform for theoretical arguments. Of course, many other settings can be addressed, but they are arguably best treated using their intrinsic features. Instances of particular interest include those where each component distribution is similar to a Studentised mean. A particular representation of this type, involving only location changes, will be discussed extensively in section 6.3.3. Other variants, where non-zero location also entails changes in shape, can be treated using similar arguments, provided the shape-changes can be parametrised.

With W denoting X or Y we shall write V_{Wj} for a random variable having the distribution of $Z_j - \bar{W}_{\cdot j}$, given that Z is drawn from Π_W. If $n_X = n_Y$ then the distribution of V_{Wj} depends only on j, not on choice of W. Let $\bar{X}_{\cdot j} = {n_X}^{-1} \sum_i X_{ij}$ and define $\bar{Y}_{\cdot j}$ analogously. Let Ψ_{Wj} be the distribution function of $|V_{Wj}|$, and put $\Delta_{Wj}(u) = I(|Z_j - \bar{W}_{\cdot j}| \le u) - \Psi_{Wj}(u)$.

If Z is from Π_W then, for each j, $|Z_j - \bar{W}_{\cdot j}|$ has distribution function Ψ_{Wj}, and so, for each fixed u, $\Delta_{Wj}(u)$ has expected value zero. On the other hand, since the distributions of X_{ij} and Y_{ij} may differ in location, then, if Z is not from Π_W, $P(|Z_j - \bar{W}_{\cdot j}| \le u)$ may take a lesser value than it does when Z is from Π_W, with the result that the expected value of $\Delta_{Wj}(u)$ can be strictly negative. Provided an estimator of Ψ_{Wj} is available for $W = X$ and $W = Y$, this property can be used to motivate a classifier.

In particular, defining $\widehat{\mathrm{hc}}_X$ and $\widehat{\mathrm{hc}}_Y$ as at (6.4), we should classify Z as coming from Π_X if $\widehat{\mathrm{hc}}_X \geq \widehat{\mathrm{hc}}_Y$, and as coming from Π_Y otherwise.

6.3. Theoretical Properties

6.3.1. *Effectiveness of approximation to* hc_W *by* $\widehat{\mathrm{hc}}_W$

We start by studying the effectiveness of the approximation by $\widehat{\mathrm{hc}}_W$ to hc_W, where hc_W and $\widehat{\mathrm{hc}}_W$ are defined as at (6.2) and (6.4), respectively (arguments similar to those given here can be used to demonstrate the effectiveness of the approximation by $\widehat{\mathrm{hc}}$ to hc). To embed the case of hypothesis testing in that of classification, we express the problem of hypothesis testing as one where the vector Z comes from a population Π_Z, equal to either Π_X or Π_W, and where the data vectors $W_1, \ldots, W_{nW}$ come from Π_W, with $\Pi_W = \Pi_Z$ under H_0, $\Pi_W = \Pi_T$ otherwise, and (Π_Z, Π_T) denoting one of (Π_X, Π_Y) or (Π_Y, Π_X). We assume throughout section 6.3 that each $\widehat{\Psi}_{Wj}$ is, with probability 1, a continuous distribution function satisfying $\widehat{\Psi}_{Wj}(x) \to 0$ as $x \downarrow 0$ and $\widehat{\Psi}_{Wj}(x) \to 1$ as $x \uparrow \infty$. We also make the following additional assumptions.

Condition A

(A1) The threshold, $t = t(p)$, varies with p in such a manner that: For each $C > 0$ and for $W = X$ and Y, $\sup_{u \in \mathcal{U}_W(C,t)} \psi_W(u)^{-1/2} \sum_j |\widehat{\Psi}_{Wj}(u) - \Psi_{Wj}(u)| = o_p(1)$.

(A2) For a constant $u_0 > 0$, for each of the two choices of W, and for all sufficiently large p, ψ_W is nonincreasing and strictly positive on $[u_0, \infty)$; and the probability that $\widehat{\psi}_W$ is nonincreasing on $[u_0, \infty)$ converges to 1 as $p \to \infty$.

The reasonableness of (A1) is taken up in section 6.4.2, below. The first part of (A2) says merely that ψ_W inherits the monotonicity properties of its component parts, $\Psi_{Wj}(1 - \Psi_{Wj})$. Indeed, if Ψ_{Wj} is the distribution function of a distribution that has unbounded support, then $\Psi_{Wj}(1 - \Psi_{Wj})$ is nonincreasing and strictly positive on $[u_0, \infty)$ for some $u_0 > 0$, and (A2) asks that the same be true of $\psi_W = \sum_j \Psi_{Wj}(1 - \Psi_{Wj})$. This is of course trivial if the distributions Ψ_{Wj} are identical. The second part of (A2) states that the same is true of the estimator, $\widehat{\psi}_W$, of ψ_W, which condition is satisfied if, for example, the observations represent Studentised means.

The next theorem shows that, under sufficient conditions, $\widehat{\mathrm{hc}}_W$ is an effective approximation to hc_W. Note that we make no assumptions about independence of vector components, or about the population from which Z comes. In particular, the theorem is valid for data drawn under both the null and alternative hypotheses.

Theorem 6.1. *Let* $0 < C_1 < C_2 < C_3 < \infty$ *and* $0 < C < C_3$, *and assume that* $\psi_W(t) \geq C_3$ *for all sufficiently large* p. *If* (A1) *and* (A2) *hold then, with* $W = X$ *or* Y,

$$\widehat{\mathrm{hc}}_W(C,t) = \{1 + o_p(1)\} \inf_{u \in \widehat{\mathcal{U}}_W(C,t)} \psi_W(u)^{-1/2} \times \sum_{j=1}^{p} \{I(|Z_j - \bar{W}_{\cdot j}| \leq u) - \Psi_{Wj}(u)\} + o_p(1), \quad (6.6)$$

$$\begin{aligned} \mathrm{hc}_W(C_1,t) + o_p(1) &\leq \{1 + o_p(1)\}\, \widehat{\mathrm{hc}}_W(C_2,t) + o_p(1) \\ &\leq \{1 + o_p(1)\}\, \mathrm{hc}_W(C_3,t) + o_p(1). \quad (6.7) \end{aligned}$$

We shall see in the next section that, in many cases of interest, when Z is not drawn from Π_W, the higher-criticism statistic hc_W tends, with high probability, to be negative and does not converge to zero as $p \to \infty$. Our results in the next section also imply that, when Z comes from Π_W, the last, added remainders $o_p(1)$ on the far right-hand sides of (6.6) and (6.7) are of smaller order than the earlier quantities on the right. Together, these properties justify approximating hc_W by $\widehat{\mathrm{hc}}_W$.

6.3.2. *Removing the assumption of independence*

We now study the properties of higher-criticism statistics in cases where the components are not independent. To illustrate the type of dependence that we have in mind, let us consider the case where Z is drawn from Π_W, and the variables $V_j = Z_j - \bar{W}_{\cdot j}$ form a mixing sequence with exponentially rapidly decreasing mixing coefficients. The case where the mixing coefficients decrease only polynomially fast, as functions of p, can also be treated; see Remark 6.3.

To give a specific example, note that the cases of moving-average processes or autoregressions, of arbitrary (including infinite) order, fall naturally into the setting of exponentially fast mixing. Indeed, assume for simplicity that the variables V_j form a stochastic process, not necessarily stationary, that is representable as

$$V_j = \sum_{k=1}^{\infty} \alpha_{jk}\, \xi_{j-k},$$

where the α_{jk}'s are constants satisfying $|\alpha_{jk}| \le \text{const.}\, \rho^k$ for all j and k, $0 < \rho < 1$, and the disturbances ξ_j are independent with zero means and uniformly bounded variances. Given $c \ge 1$, let ℓ denote the integer part of $c \log p$, and put

$$V_j' = \sum_{k=1}^{\ell} \alpha_{jk}\, \xi_{j-k} \,.$$

Then, by Markov's inequality,

$$P(|V_j - V_j'| > u) \le u^{-2} E\biggl(\sum_{k=\ell+1}^{\infty} \alpha_{jk}\, \xi_{j-k} \biggr)^2 = O(u^{-2} \rho^{2\ell}) \,,$$

uniformly in $u > 0$, $c \ge 1$ and integers j. By taking $u = p^{-C}$ for $C > 0$ arbitrarily large, and then choosing $c \ge \frac{3}{2}\, C \,|\log \rho|^{-1}$, we deduce that the approximants V_j' have the following two properties: (a) $P(|V_j - V_j'| \le p^{-C}) = 1 - O(p^{-C})$, uniformly in $1 \le j \le p$; and (b) for each r in the range $2 \le r \le p$, and each sequence $1 \le j_1 < \ldots < j_r \le p$ satisfying $j_{k+1} - j_k \ge c \log p + 1$ for $1 \le k \le r-1$, the variables V_{j_k}', for $1 \le k \le r$, are stochastically independent.

The regularity condition (B1), below, captures this behaviour in greater generality. There, we let V_{Wj}, for $1 \le j \le p$, have the joint distribution of the respective values of $Z_j - \bar{W}_{\cdot j}$ when Z is drawn from Π_W. At (B2), we also impose (a) a uniform Hölder smoothness condition on the respective distribution functions χ_{Wj} of V_{Wj}, (b) a symmetry condition on χ_{Wj}, and (c) a restriction which prevents the upper tail of Ψ_{Wj}, for each j and W, from being pathologically heavy.

Condition B

(B1) For each $C, \epsilon > 0$, and each of the two choices of W, there exists a sequence of random variables V_{Wj}', for $1 \le j \le p$, with the properties: (a) $P(|V_{Wj} - V_{Wj}'| \le p^{-C}) = 1 - O(p^{-C})$, uniformly in $1 \le j \le p$; and (b) for all sufficiently large p, for each r in the range $2 \le r \le p$, and each sequence $1 \le j_1 < \ldots < j_r \le p$ satisfying $j_{k+1} - j_k \ge p^{\epsilon}$ for $1 \le k \le r-1$, the variables V_{Wj_k}', for $1 \le k \le r$, are stochastically independent.

(B2) (a) For each of the two choices of W there exist constants $C_1, C_2 > 0$, the former small and the latter large, such that $|\chi_{Wj}(u_1) - \chi_{Wj}(u_2)| \le C_2\, |u_1 - u_2|^{C_1}$, uniformly in $u_1, u_2 > 0$, $1 \le j \le p < \infty$ and $W = X$ or Y; (b) the function $G_{Wj}(u, v) = P(|V_{Wj} + v| \le u)$ is nonincreasing in $|v|$ for each fixed u, each choice of W and each j; and (c) $\max_{1 \le j \le p} \{1 - \Psi_{Wj}(u)\} = O(u^{-\epsilon})$, for $W = X, Y$ and for some $\epsilon > 0$.

Part (b) of (B2) holds if each distribution of V_{Wj} is symmetric and unimodal.

As explained in the previous section, in both the hypothesis testing and classification problems we can consider that $W = X$ or Y, indicating the population from which we draw the sample against which we check the new data value Z. Let $\mu_Z = \mu_X$ if Z is from Π_X, and $\mu_Z = \mu_Y$ otherwise, and define $\nu_{WZj} = \mu_{Zj} - \mu_{Wj}$,

$$\overline{\text{hc}}_{WZ}(C, t) = \sup_{u \in \mathcal{U}_W(C,t)} \psi_W(u)^{-1/2} \times \sum_{j=1}^{p} \left\{ P(|V_{Wj}| \le u) - P(|V_{Wj} + \nu_{WZj}| \le u) \right\}. \quad (6.8)$$

In view of (B2)(b), the quantity within braces in the definition of $\overline{\text{hc}}_{WZ}$ is nonnegative, and so $\overline{\text{hc}}_{WZ} \ge 0$. Theorem 6.2 below describes the extent to which the statistic hc_W, a random variable, can be approximated by the deterministic quantity $\overline{\text{hc}}_{WZ}$.

Theorem 6.2. *Let $C > 0$ be fixed, and take the threshold, $t = t(p)$, to satisfy $t \ge 0$ and $\psi_W(t) \ge C$, thus ensuring that $\mathcal{U}_W(C, t)$ is nonempty. Let* hc_W *and* $\overline{\text{hc}}_{WZ}$ *denote* $\text{hc}_W(C, t)$ *and* $\overline{\text{hc}}_{WZ}(C, t)$, *respectively. If* (B1) *and* (B2) *hold then for each* $\epsilon > 0$,

$$\text{hc}_W = -\{1 + o_p(1)\}\, \overline{\text{hc}}_{WZ} + O_p(p^{\epsilon}). \quad (6.9)$$

An attractive feature of (6.9) is that it separates the "stochastic" and "deterministic" effects of the higher-criticism statistic hc_W. The stochastic effects go into the term $O_p(p^\epsilon)$. The deterministic effects are represented by $\overline{\text{hc}}_{WZ}$. When the data value Z is from the same population Π_W as the dataset with which it is compared, each $\nu_{WZj} = 0$ and so, by (6.8), $\overline{\text{hc}}_{WZ} = 0$. Property (6.9) therefore implies that, when Z is from Π_W, $\text{hc}_W = O_p(p^\epsilon)$ for each $\epsilon > 0$. In other cases, where Z is drawn from a population different from that from which come the data with which Z is compared, $\overline{\text{hc}}_{WZ}$ is generally nonzero. In such instances the properties of hc_W can be computed by relatively straightforward, deterministic calculations based on $\overline{\text{hc}}_{WZ}$. In particular, when $W \ne Z$, if $\overline{\text{hc}}_{WZ}$ is of order larger than p^ϵ for some $\epsilon > 0$ (see (6.13) below), then it follows directly that the probability of rejecting the null hypothesis, in the hypothesis testing problem, or of correct classification, in the classification problem, converges to 1. See, for example, section 6.3.3.

Remark 6.1. *Sharpening the term $O_p(p^\epsilon)$ in* (6.9). If, as in the problem treated by [8], the distribution functions Ψ_{Wj} are all identical and the variables $X_{i_1j_1}$ and $Y_{i_2j_2}$, for $1 \le i_1 \le n_X$, $1 \le i_2 \le n_Y$ and $1 \le j_1, j_2 \le p$, are completely independent, then a refinement of the argument leading to (6.9) shows that the $O_p(p^\epsilon)$ term there can be reduced to $O_p(\sqrt{\log p})$. Here it is not necessary to assume that the common distributions are normal. Indeed, in that context [8] noted that the $O_p(p^\epsilon)$ term in (6.9) can be replaced by $O_p(\sqrt{\log\log p})$.

Remark 6.2. *Relaxing the monotonicity condition* (B2)(*b*). Assumption (B2)(b) asks that $G_{Wj}(u,v) = P(|V_{Wj} + v| \le u)$ be nonincreasing in $|v|$ for each u. However, if the distributions of X_{ij} and Y_{ij} are identical for all but at most q values of j then it is sufficient to ask that, for these particular j, it be possible to write, for each $\epsilon > 0$,

$$G_{Wj}(u,v) = H_{Wj}(u,v) + o\{p^\epsilon\, q^{-1}\, \psi_W(u)^{1/2}\},$$

uniformly in $1 \le j \le p$, $u \ge t$ and $W = X$ and Y, where each H_{Wj} has the monotonicity property asked of G_{Wj} in (B2)(b).

Remark 6.3. *Mixing at polynomial rate.* The exponential-like mixing rate implied by (B1) is a consequence of the fact that (a) and (b) in (B1) hold for each $C, \epsilon > 0$. If, instead, those properties apply only for a particular positive pair C, ϵ, then (6.9) continues to hold with p^ϵ there replaced by p^η, where $\eta > 0$ depends on C, ϵ from (B1), and decreases to zero as C increases and ϵ decreases.

6.3.3. *Delineating good performance*

Theorem 6.2 gives a simple representation of the higher-criticism statistic. It implies that, if Z is drawn from Π_Q where $(W,Q) = (X,Y)$ or (Y,X), and if $\overline{\mathrm{hc}}_{WZ}$ exceeds a constant multiple of p^ϵ for some $\epsilon > 0$, then the probability that we make the correct decision in either a hypothesis testing or classification problem, Z converges to 1 as $p \to \infty$. We shall use this result to determine a region where hypothesis testing, or classification, are possible. For simplicity, in this section we shall assume that each $\mu_{Xj} = 0$, and $\mu_{Yj} = 0$ for all but q values of j, for which $\mu_{Yj} = \nu > 0$ and $\nu = \nu(p)$ diverges with p and does not depend on j. The explicit form of $\overline{\mathrm{hc}}_{WZ}$, at (6.8), makes it possible to handle many other settings, but a zero-or-ν representation of each mean difference permits an insightful comparison with results discussed by [8].

In principle, two cases are of interest, where the tails of the distribution of V_{Wj} decrease polynomially or exponentially fast, respectively. However, in the polynomial case it can be proved using (6.8) that the hypothesis testing and classification problems are relatively simple. Therefore, we study only the exponential setting. In this context, and reflecting the discussion in section 6.2.2, we take the distribution of V_{Wj} to be that of the difference between two Studentised means, standardised by dividing by $\sqrt{2}$, and the distributions of X_{ij} and Y_{ij} to represent translations of that distribution. See (C1) and (C2) below. Alternatively we could work with the case where X_{ij} is a Studentised mean for a distribution with zero mean, and Y_{ij} is computed similarly but for the case where the expected value is shifted by $\pm\nu$. Theoretical arguments in the latter setting are almost identical to those given here, being based on results of [32].

Condition C

(C1) For each pair (W, j), where $W = X$ or Y and $1 \le j \le p$, let U_{Wjk}, for $1 \le k \le N_{Wj}$, denote random variables that are independent and identically distributed as U_{Wj}, where $E(U_{Wj}) = 0$, $E(U_{Wj}^4)$ is bounded uniformly in (W, j), $E(U_{Wj}^2)$ is bounded away from zero uniformly in (W, j), and $N_{Wj} \ge 2$. Let V_{Wj} have the distribution $2^{-1/2}$ times the difference between two independent copies of $N_{Wj}^{1/2}\, \bar{U}_{Wj}/S_{Wj}$, where $\bar{U}_{Wj}$ and S_{Wj} denote respectively the empirical mean and variance of the data $U_{Wj1}, \ldots, U_{WjN_{Wj}}$. Take $\mu_{Xj} = 0$ for each j, $\mu_{Yj} = 0$ for all but $q = q(p)$ values of j, say $j_1, \ldots, j_q$, and $|\mu_{Yj}| = \nu$ for these particular values of j.

(C2) The quantity ν in (C1) is given by $\nu = \sqrt{2w \log p}$, and the threshold, t, satisfies $B \le t \le \sqrt{2s \log p}$ for some $B, s > 0$, where $0 < w < 1$ and $0 < s < \min(4w, 1)$.

The setting described by (C1) is one where a working statistician would, in practice, generally take each distribution approximation $\widehat{\Psi}_{Wj}(u)$ to be simply $P(|\xi| \le u)$, where ξ has the standard normal distribution. The signal detection boundary in this setting is obtained using a polynomial model for the number of added shifts:

$$\text{for some} \quad \tfrac{1}{2} < \beta < 1, \quad q \sim \text{const.}\, p^{1-\beta} \tag{6.10}$$

(see [8]). The boundary is then determined by:

$$w \ge \begin{cases} \beta - \frac{1}{2} & \text{if } \frac{1}{2} < \beta \le \frac{3}{4} \\ (1 - \sqrt{1-\beta})^2 & \text{if } \frac{3}{4} < \beta < 1\,. \end{cases} \tag{6.11}$$

The inequality (6.11) is also sufficient in hypothesis testing and classification problems where the data are exactly normally distributed. Likewise it is valid if we use a normal approximation and if that approximation is good enough. The question we shall address is, "how good is good enough?" The following theorem answers this question in cases where N_{Wj} diverges at at least a logarithmic rate, as a function of p. The proof is given in Appendix A.2.

Theorem 6.3. *Assume* (C1), (C2), (6.10), *that* w *satisfies* (6.11), *and that, for* $W = X$ *or* Y *and* $1 \le j \le p$, N_{Wj}, *given in* (C1), *satisfies*

$$N_{Wj}^{-1} (\log p)^4 \to 0 . \tag{6.12}$$

Suppose too that Z *is from* Π_Q, *where* $(W, Q) = (X, Y)$ *or* (Y, X). *Then, for constants* $B, \eta > 0$,

$$\overline{\mathrm{hc}}_{WZ} \ge B\, p^{\eta} . \tag{6.13}$$

Condition (6.12) confirms that the samples on which the coordinate data are based need be only logarithmically large, as a function of p, in order for the higher-criticism classifier to be able to detect the difference between the W and Q populations.

6.4. Further Results

6.4.1. *Alternative constructions of* hc_W *and* $\widehat{\mathrm{hc}}_W$

There are several other ways of constructing higher-criticism statistics when the distribution functions Ψ_{Wj} depend on j and have to be estimated. For example, omitting for simplicity the threshold t, we could re-define $\widehat{\mathrm{hc}}_W$ as:

$$\widehat{\mathrm{hc}}_W = p^{1/2} \inf_{u\,:\,pu(1-u)\ge C} \{u\,(1-u)\}^{-1/2} \sum_{j=1}^{p} \big[I\{|Z_j - \bar{W}_{\cdot j}| \le \widehat{\Psi}_{Wj}^{-1}(u)\} - u \big] . \tag{6.14}$$

If Z were drawn from Π_W then the random variable $K = \sum_j I\{|Z_j - \bar{W}_{\cdot j}| \le \Psi_{Wj}^{-1}(u)\}$ would have exactly a binomial $\mathrm{Bi}(p, u)$ distribution. The normalisation in formula (6.14) for $\widehat{\mathrm{hc}}_W$ reflects this property. However, replacing K by $\widehat{K} = \sum_j I\{|Z_j - \bar{W}_{\cdot j}| \le \widehat{\Psi}_{Wj}^{-1}(u)\}$, as in (6.14), destroys the independence of the summands, and makes normalisation problematic. This is particularly true when, as would commonly be the case in practice, the estimators $\widehat{\Psi}_{Wj}$ are computed from data W_{ij_1} for values of j_1 that are local to j. In such cases the estimators $\widehat{\Psi}_{Wj}$ would not be root-p consistent for the respective distributions Ψ_{Wj}.

If the distribution of $|Z_j - \bar{W}_{.j}|$ were known up to its standard deviation, σ_{Wj}, and if we had an estimator, $\hat{\sigma}_{Wj}$, of σ_{Wj} for each W and j, then we could construct a third version of $\widehat{\text{hc}}_W$:

$$\widehat{\text{hc}}_W = \inf_{u\,:\,\phi_W(u)\geq C} \phi_W(u)^{-1/2} \sum_{j=1}^{p} \big\{I\big(|Z_j - \bar{W}_{.j}|/\hat{\sigma}_{Wj} \leq u\big) - \Phi_{Wj}(u)\big\},$$

where Φ_{Wj} denotes the distribution function of $|Z_j - \bar{W}_{.j}|/\sigma_{Wj}$ under the assumption that Z is drawn from Π_W, and $\phi_W(u) = \sum_j \Phi_{Wj}\,(1 - \Phi_{Wj})$. Again, however, the correlation induced through estimation, this time the estimation of σ_{Wj}, makes the normalisation difficult to justify.

In some problems there is good reason to believe that if the marginal means of the populations Π_X and Π_Y differ, then the differences are of a particular sign. For example, it might be known that $\mu_{Xj} \geq \mu_{Yj}$ for all j. In this case we would alter the construction of the higher-criticism statistics hc_W and $\widehat{\text{hc}}_W$, at (6.2) and (6.4), to:

$$\text{hc}_W^{\text{os}} = \inf_{u\in\mathcal{U}_W^{\text{os}}} \psi_W^{\text{os}}(u)^{-1/2} \sum_{j=1}^{p} \big\{I(Z_j - \bar{W}_{.j} \leq u) - \Psi_{Wj}^{\text{os}}(u)\big\}, \quad (6.15)$$

$$\widehat{\text{hc}}_W^{\text{os}} = \inf_{u\in\widehat{\mathcal{U}}_W^{\text{os}}} \hat{\psi}_W^{\text{os}}(u)^{-1/2} \sum_{j=1}^{p} \big\{I(Z_j - \bar{W}_{.j} \leq u) - \widehat{\Psi}_{Wj}^{\text{os}}(u)\big\}, \quad (6.16)$$

respectively, where

$$\psi_W^{\text{os}}(u) = \sum_{j=1}^{p} \Psi_{Wj}^{\text{os}}(u)\,\{1-\Psi_{Wj}^{\text{os}}(u)\}\,, \quad \hat{\psi}_W^{\text{os}}(u) = \sum_{j=1}^{p} \widehat{\Psi}_{Wj}^{\text{os}}(u)\,\{1-\widehat{\Psi}_{Wj}^{\text{os}}(u)\}\,,$$

$\widehat{\Psi}_{Wj}^{\text{os}}(u)$ is an empirical approximation to the probability $\Psi_{Wj}^{\text{os}}(u) = P(Z_j - \bar{W}_{.j} \leq u)$ when Z is drawn from Π_W, $\mathcal{U}_W^{\text{os}} = \mathcal{U}_W^{\text{os}}(C,t)$ is the set of u for which $u \geq t$, $\psi_W^{\text{os}}(u) \geq C$, $\widehat{\mathcal{U}}_W^{\text{os}}$ is defined analogously, and the superscript "os" denotes "one-sided." When using $\widehat{\text{hc}}_W^{\text{os}}$ we would classify Z as coming from Π_X if $\widehat{\text{hc}}_X^{\text{os}} \geq \widehat{\text{hc}}_Y^{\text{os}}$, and as coming from Π_Y otherwise.

Remark 6.4. *Adapting Theorems 6.1 and 6.2.* Theorems 6.1 and 6.2 have direct analogues, formulated in the obvious manner, for the one-sided classifiers hc_W^{os} and hc_W^{os} introduced above. In particular, the one-sided version of $\overline{\text{hc}}_{WZ}$, at (6.8), is obtained by removing the absolute value signs there. The regularity conditions too differ in only minor respects. For example, when formulating the appropriate version of (A1) we replace $\widehat{\Psi}_{Wj}$, Ψ_{Wj}, $\widehat{\psi}_W$ and ψ_W by $\widehat{\Psi}^{\rm os}_{Wj}$, $\Psi^{\rm os}_{Wj}$, $\widehat{\psi}^{\rm os}_W$ and $\psi^{\rm os}_W$, respectively. Part (b) of (B2) can be dropped on this occasion, since its analogue in the one-sided case follows directly from the monotonicity of a distribution function.

6.4.2. *Advantages of incorporating the threshold*

By taking the threshold, t, large we can construct the higher-criticism statistics hc_W and $\widehat{\mathrm{hc}}_W$, at (6.2) and (6.4), so that they emphasise relatively large values of $|Z_j - \bar{W}_{\cdot j}|$. This is potentially advantageous, especially when working with $\widehat{\mathrm{hc}}_W$, since we expect the value of u at which the infimum at (6.4) is achieved also to be large.

The most important reasons for thresholding are more subtle than this argument would suggest, however. They are founded on properties of relative errors in distribution approximations, and on the fact that the divisor in (6.2) is $\psi_W^{1/2}$, not simply ψ_W. To see why this is significant, consider the case where the distribution functions Ψ_{Wj} are all identical, to Ψ say. Then $\psi_W = p\,\Psi\,(1-\Psi)$, which we estimate by $\widehat{\psi}_W = p\,\widehat{\Psi}\,(1-\widehat{\Psi})$, say. In order for the effect of replacing each $\Psi_{Wj}(u)$ (appearing in (6.2)) by $\widehat{\Psi}_{Wj}(u)$ (in (6.4)) to be asymptotically negligible, we require the quantity

$$\psi_W(u)^{-1/2} \sum_{j=1}^{p} \big|\widehat{\Psi}_{Wj}(u) - \Psi_{Wj}(u)\big| = \frac{p^{1/2}|\widehat{\Psi}(u) - \Psi(u)|}{\Psi(u)^{1/2}\{1-\Psi(u)\}^{1/2}}$$

to be small. Equivalently, if u is in the upper tail of the distribution Ψ, we need the ratio

$$\frac{p^{1/2}\,|\widehat{\Psi}(u) - \Psi(u)|}{\{1-\Psi(u)\}^{1/2}} \tag{6.17}$$

to be small.

If the approximation of Ψ by $\widehat{\Psi}$ (or more particularly, of $1-\Psi$ by $1-\widehat{\Psi}$) is accurate in a relative sense, as it is (for example) if Ψ is the distribution of a Studentised mean, then, for large u,

$$\rho(u) \equiv \frac{|\widehat{\Psi}(u) - \Psi(u)|}{1-\Psi(u)} \tag{6.18}$$

is small for u in the upper tail as well as for u in the middle of the distribution. When u is in the upper tail, so that $1-\Psi(u)$ is small, then, comparing (6.17) and (6.18), we see that we do not require $\rho(u)$ to be as small as it would have to be in the middle of the distribution. By insisting that $u \geq t$, where the threshold t is relatively large, we force u to be in the upper tail, thus obtaining the advantage mentioned in the previous sentence.

Below, we show in more detail why, if thresholding is not undertaken, that is, if we do not choose t large when applying the higher-criticism classifier, substantial errors can occur when using the classifier. They arise through an accumulation of errors in the approximation $\widehat{\Psi}_{Wj} \approx \Psi_{Wj}$.

Commonly, the approximation of Ψ_{Wj} by $\widehat{\Psi}_{Wj}$ can be expressed as

$$\widehat{\Psi}_{Wj}(u) = \Psi_{Wj}(u) + \delta_p\, \alpha_{Wj}(u) + o(\delta_p)\,, \tag{6.19}$$

where δ_p decreases to zero as p increases and represents the accuracy of the approximation; α_{Wj} is a function, which may not depend on j; and the remainder, $o(\delta_p)$, denotes higher-order terms. Even if α_{Wj} depends on j, its contribution cannot be expected to "average out" of $\widehat{\mathrm{hc}}_W$, by some sort of law-of-large-numbers effect, as we sum over j.

In some problems the size of δ_p is determined by the number of data used to construct $\widehat{\Psi}_{Wj}$. For example, in the analysis of gene-expression data, $\widehat{\Psi}_{Wj}$ might be calculated by borrowing information from neighbouring values of j. In order for this method to be adaptive, only a small proportion of genes would be defined as neighbours for any particular j, and so a theoretical description of δ_p would take that quantity to be no smaller than $p^{-\eta}$, for a small constant $\eta > 0$. In particular, assuming that $\widehat{\Psi}_{Wj}$ was root-p consistent for Ψ_{Wj}, i.e. taking η as large as $\frac{1}{2}$, would be out of the question.

In other problems the coordinate data X_{ij} and Y_{ij} can plausibly be taken as approximately normally distributed, since they are based on Student's t statistics. See sections 6.2.2 and 6.3.3 for discussion. In such cases the size of δ_p is determined by the number of data in samples from which the t statistic is computed. This would also be much less than p, and so again a mathematical account of the size of δ_p would have it no smaller than $p^{-\eta}$, for $\eta > 0$ much less than $\frac{1}{2}$.

Against this background; and taking, for simplicity, $\Psi = \Psi_{Wj}$, $\alpha = \alpha_W$ and $\delta_p = p^{-\eta}$; we find that $\widehat{\psi}_W \sim \psi_W = p\,\Psi\,(1-\Psi)$ and $\sum_j (\widehat{\Psi}_{Wj} - \Psi_{Wj}) = p^{1-\eta}\,\alpha + o(p^{1-\eta})$. These results, and (6.19), lead to the conclusion that, for fixed u, the argument of the infimum in the definition of $\widehat{\mathrm{hc}}_W$, at (6.4), is given by

$$A(u) \equiv \widehat{\psi}_W(u)^{-1/2} \sum_{j=1}^{p} \{I(|Z_j - \bar{W}_{.j}| \le u) - \widehat{\Psi}_{Wj}(u)\}$$

$$= \{1 + o(1)\} \psi_W(u)^{-1/2} \sum_{j=1}^{p} \{I(|Z_j - \bar{W}_{.j}| \le u) - \Psi_{Wj}(u)\}$$

$$- p^{(1/2)-\eta} \gamma(u) + o\big(p^{(1/2)-\eta}\big), \tag{6.20}$$

where $\gamma = \alpha \{\Psi (1 - \Psi)\}^{-1/2}$.

Assume, again for simplicity, that Z is drawn from Π_W. Then, for fixed u, the series on the right-hand side of (6.20) has zero mean, and equals $O_p(p^{1/2})$. In consequence,

$$A(u) = O_p(1) - p^{(1/2)-\eta} \gamma(u) + o_p\big(p^{(1/2)-\eta}\big). \tag{6.21}$$

Referring to the definition of $\widehat{\text{hc}}_W$ at (6.4), we conclude from (6.21) that for fixed u,

$$\widehat{\text{hc}}_W \le O_p(1) - p^{(1/2)-\eta} \gamma(u) + o_p\big(p^{(1/2)-\eta}\big). \tag{6.22}$$

If u is chosen so that $\gamma(u) > 0$ then, since $\eta < \frac{1}{2}$, the subtracted term on the right-hand side of (6.22) diverges to $-\infty$ at a polynomial rate, and this behaviour is readily mistaken for detection of a value of Z that does not come from Π_W. (There, the rate of divergence to zero can be particularly small; see section 6.3.3 and [8].) This difficulty has arisen through the accumulation of errors in the distribution approximation.

6.5. Numerical Properties in the Case of Classification

We applied the higher-criticism classifier to simulated data. In each case, we generated $n_W = 10$ vectors of dimension $p = 10^6$, from $\Pi_W = \Pi_X$ or Π_Y; and one observation Z from Π_Y. We generated the data such that, for $i = 1, \ldots, n_W$ and $j = 1, \ldots, p$; and with W denoting X or Y; $W_{i,j} = (\bar{U}^W_{ij,1} - \bar{U}^W_{ij,2})/\sqrt{(S^2_{U,1} + S^2_{U,2})/N_U} + \mu_W$ where, for $k = 1$ and 2, $\bar{U}^W_{ij,k}$ was the empirical mean and $S^2_{U,k}$ was the empirical variance of: (1) in the case of independence, $N_U = 20$ independent and identically distributed random variables, having the distribution function of a N(0, 1) variable, a student T_{10} or a χ^2_6 random variable; and (2) in the case of dependence, $N_U = 20$ random variables of the type $V^W_{i,j,k}$, where, for $i = 1, \ldots, n_W$ and $j = 1, \ldots, p$, $V^W_{i,j,k} = \sum_{\ell=0}^{L} \theta^\ell \, \varepsilon^W_{i,j-\ell,k}$, with $\theta = 0.8$ and $\varepsilon^W_{i,j,k} \sim \text{N}(0, (1 + \theta^2)^{-1})$ denoting independent variables.

We set $\mu_{X,j} = 5(j-1)/(p-1)$ and, in compliance with (C2), (6.10), (6.11), took $\mu_X = \mu_Y$ for all but $q = \langle p^{1-\beta} \rangle$ randomly selected components, for which $\mu_{Y_j} = \mu_{X,j} + \sqrt{2w \log p}$, where $\langle \cdot \rangle$ denotes the integer-part function; and we considered different values of $\beta \in (\frac{1}{2}, 1)$ and $w \in (0, \frac{1}{2})$. Reflecting the results in sections 6.3.1 and 6.3.3, we estimated the unknown distribution function of the observed data as the standard normal distribution function. In all cases considered, we generated 500 samples in the manner described above, and we repeated the classification procedure 500 times. Below we discuss the percentages of those samples which led to correct classification.

Application of the method necessitated selection of the two parameters t and C defining $\mathcal{U}_W$. In view of condition (A2), we reformulated $\mathcal{U}_W$ as $\mathcal{U}_W = [t_1, t_2]$, and we replaced choice of t and C by choice of t_1 and t_2. If we have sufficient experience with the distributions of the data, t_1 and t_2 can be selected 'theoretically' to maximise the percentage of correct classifications.

In the tables below we compare the results obtained using three methods: the higher-criticism procedure for the optimal choice of (t_1, t_2), referring to it as HC_T; higher criticism without thresholding, i.e. for $(t_1, t_2) = (-\infty, \infty)$, to which we refer as simply HC; and the thresholded nearest-neighbour method, NN_T (see e.g. [15]),) i.e. the nearest-neighbour method applied to thresholded data $W_j \, I\{W_j > t\}$, where W denotes X, Y or Z and the threshold, t, is selected in a theoretically optimal way using the approach described above for choosing (t_1, t_2).

It is known ([15]) that, for normal variables, when the distribution of the observations is known, classification using HC_T is possible if w and β are above the boundary determined by (6.11), but classification using NN_T is possible only above the more restricted boundary determined by $w = 2\beta - 1$. Below, we show that these results hold in our context too, where the distribution of the data is known only approximately (more precisely, estimated by the distribution of a standard normal variable). We shall consider values of (β, w) that lie above, on or below the boundary $w = 2\beta - 1$, including values which lie between this boundary and that for higher criticism. Tables 1 and 2 summarise results for the independent case (1), when the observations were averages of, respectively, Student T_{10} variables and χ^2_6 variables. In all cases, including those where classification was possible for both methods, we see that the thresholded higher-criticism method performs significantly better than the thresholded nearest-neighbour approach. The results also show very clearly the improvement obtainable using the thresholded version of higher criticism.

Table 1: Percentage of correct classifications if case (1) with T_{10} variables, using the optimal values of t, t_1 and t_2.

	$w = 0.2$			$w = 0.3$			$w = 0.4$			$w = 0.5$		
β	NN_T	HC_T	HC	NN_T	HC_T	HC	NN_T	HC_T	HC	NN_T	HC_T	HC
0.5	99.8	100	95.8									
0.6	77.4	86.4	83.0	86.2	97.4	89.6	94.0	100	94.0			
0.65	63.8	72.8	70.6	74.4	85.4	73.4	77.2	97.8	82.6			
0.7				63.2	70.6	62.2	67.6	86.0	64.6	69.8	96.8	75.8
0.75							59.0	72.8	58.2	65.2	86.2	66.6

Table 2: Percentage of correct classifications if case (1) with χ^2_6 variables, using the optimal values of t, t_1 and t_2.

	$w = 0.2$			$w = 0.3$			$w = 0.4$			$w = 0.5$		
β	NN_T	HC_T	HC	NN_T	HC_T	HC	NN_T	HC_T	HC	NN_T	HC_T	HC
0.5	99.8	100	97.6									
0.6	75.8	85.4	78.6	86.2	98.4	89.0	93.6	100	91.8			
0.65	66.2	70.2	68.0	69.8	86.2	72.4	80.6	98.2	79.2			
0.7				64.4	74.8	64.2	68.4	88.4	64.4	75.0	97.0	69.8
0.75							60.0	73.8	56.0	64.2	85.4	59.4

Table 3: Percentage of correct classifications if case (1) with normal variables (line 1), case (2) with $L = 1$ (line 2) or $L = 3$ (line 3), using the optimal values of t, t_1 and t_2.

	$w = 0.2$			$w = 0.3$			$w = 0.4$			$w = 0.5$		
β	NN_T	HC_T	HC	NN_T	HC_T	HC	NN_T	HC_T	HC	NN_T	HC_T	HC
.5	99.8	100	94.0									
	98.0	100	93.8									
	95.6	100	93.6									
.6	77.2	83.4	79.2	85.2	98.0	88.6	93.0	100	95.2			
	73.0	83.2	80.0	81.8	97.2	85.6	90.0	100	92.4			
	67.4	82.0	77.2	75.6	97.6	84.0	85.6	100	92.0			
.65	63.6	70.2	65.6	71.0	83.8	74.2	79.4	97.3	82.4			
	61.8	72.0	68.4	68.2	84.0	72.4	76.0	97.8	82.8			
	58.2	67.8	64.4	62.6	83.0	72.0	73.2	96.4	79.2			
.7				64.2	70.2	65.0	69.0	83.0	66.4	72.8	95.2	75.6
				59.6	69.6	57.4	63.6	84.2	67.2	71.2	95.0	76.6
				59.4	70.8	63.6	63.0	82.8	63.8	66.4	94.4	70.2
.75							59.4	69.2	60.8	62.0	85.4	63.2
							59.0	71.4	59.2	59.8	85.0	62.6
							54.0	71.4	57.4	60.4	80.8	62.6

In Table 3 we compare the results of the independent case (1), where the data were Studentised means of independent $N(0,1)$ variables and so had Student's t distribution; and the dependent case (2), where the observations were Studentised means of correlated normal variables with either $L = 1$ or $L = 3$. Here we see that as the strength of correlation increases, the nearest-neighbour method seems to deteriorate more rapidly than higher criticism, which, as indicated in section 6.3.2, remains relatively unaffected by lack of independence.

If previous experience with data of the type being analysed is not sufficient to permit effective choice of threshold using that background, then a data-driven selection needs to be developed. We implemented a cross-validation procedure, described in Appendix A.1.

6.6. Technical Arguments

6.6.1. *Proof of Theorem 6.1*

Since $\psi_W(u) \geq C$ for each $u \in \mathcal{U}_W(C,t)$ then (A1) implies that $\psi_W(u)^{-1} \sum_j |\widehat{\Psi}_{Wj}(u) - \Psi_{Wj}(u)| = o_p(1)$ uniformly in $u \in \mathcal{U}_W(C,t)$, and hence that $\widehat{\psi}_W(u)/\psi_W(u) = 1 + o_p(1)$, uniformly in $u \in \mathcal{U}_W(C,t)$. Call this result R_1. That property and (A2) imply that with probability converging to 1 as $p \to \infty$, $\mathcal{U}_W(C_3,t) \subseteq \widehat{\mathcal{U}}_W(C_2,t) \subseteq \mathcal{U}_W(C_1,t)$; call this result R_2. (Since $\psi_W(t) \geq C_3$ then $t \in \mathcal{U}_W(C_3,t)$, and so the latter set is nonempty.) Results R_1, R_2 and (A1) together give (6.6). Property R_2 and (6.6) imply (6.7).

6.6.2. *Proof of Theorem 6.2*

Let V'_{Wj} be as in (B1). Since, in the case where Z is drawn from Π_W, V_{Wj}, for $1 \leq j \leq p$, have the joint distribution of $Z_j - \bar{W}_{\cdot j}$, for $1 \leq j \leq p$, then for Z from either Π_X or Π_Y we may write $Z_j - \bar{W}_{\cdot j} = V_{Wj} + \nu_{WZj}$, where $\nu_{WZj} = \mu_{Zj} - \mu_{Wj}$. Substituting this representation for $Z_j - \bar{W}_{\cdot j}$ into the definition of hc_W at (6.2), and defining $\Delta_{Wj} = V_{Wj} - V'_{Wj}$, we see that

$$\mathrm{hc}_W = \inf_{u \in \mathcal{U}_W} \psi_W(u)^{-1/2} \sum_{j=1}^{p} \{I(|V'_{Wj} + \Delta_{Wj} + \nu_{WZj}| \leq u) - \Psi_{Wj}(u)\}. \tag{6.23}$$

Given $D > 0$ and $v = 0$ or ± 1, define

$$\mathrm{hc}'_{WZ}(v) = \inf_{u \in \mathcal{U}_W} \psi_W(u)^{-1/2} \sum_{j=1}^{p} \{I(|V'_{Wj} + \nu_{WZj}| \leq u + vp^{-D}) - \Psi_{Wj}(u)\},$$

$$\mathrm{hc}''_{WZ} = \inf_{u \in \mathcal{U}_W} \psi_W(u)^{-1/2} \sum_{j=1}^{p} \left\{ I(|V'_{Wj} + \nu_{WZj}| \le u) - P(|V'_{Wj}| \le u) \right\}.$$

Let $\mathcal{E}_W$ denote the event that $|\Delta_{Wj}| \le p^{-D}$ for each $1 \le j \le p$. In view of (B1)(a),

$$\text{for all} \quad C_3 > 0\,, \quad P(\mathcal{E}_W) = 1 - O(p^{-C_3})\,. \tag{6.24}$$

Now, with probability 1,

$$\begin{aligned} &\mathrm{hc}'_{WZ}(-1) \le \mathrm{hc}'_{WZ}(0) \le \mathrm{hc}'_{WZ}(1) \qquad \text{and} \\ &\mathrm{hc}'_{WZ}(-1) \le \mathrm{hc}_W \le \mathrm{hc}'_{WZ}(1) \text{ if } \mathcal{E}_W \text{ holds}\,, \end{aligned} \tag{6.25}$$

where we used (6.23) to obtain the second set of inequalities. Furthermore,

$$\begin{aligned} 0 &\le \mathrm{hc}'_{WZ}(1) - \mathrm{hc}'_{WQ}(-1) \\ &\le \sup_{u \in \mathcal{U}_W} \sqrt{\psi_W(u)} \sum_{j=1}^{p} I\big(u - p^{-D} < |V'_{Wj} + \nu_{WZj}| \le u + p^{-D}\big) \\ &\le \sup_{u \in \mathcal{U}_W} \sqrt{\psi_W(u)} \sum_{j=1}^{p} I\big(u - 2p^{-D} < |V_{Wj} + \nu_{WZj}| \le u + 2p^{-D}\big) \end{aligned} \tag{6.26}$$

where the first inequality holds with probability 1 and the second holds almost surely on $\mathcal{E}_W$.

Let $1 \le j_1 \le p$, and take $C_4 > 0$. Using (B1) and (B2) it can be shown that the probability that there are no integers $j_2 \ne j_1$ with $1 \le j_2 \le p$ and

$$\Big| |V_{Wj_1} + \nu_{Zj_1}| - |V_{Wj_2} + \nu_{Zj_2}| \Big| \le C_4\, p^{-D}\,, \tag{6.27}$$

is bounded below by $1 - C_5\, p^{1-DC_1}$ uniformly in j_1, where $C_5 > 0$ and C_1 is the constant in (B2)(a). Adding over $1 \le j_1 \le p$, and choosing $D > 2C_1^{-1}$, we deduce that:

The probability that there is no pair (j_1, j_2) of distinct indices such that $|V_{Wj_1} + \nu_{Qj_1}|$ and $|V_{Wj_2} + \nu_{Qj_2}|$ are closer than $C_4\, p^{-D}$, converges to zero as $p \to \infty$. (6.28)

If, in the case $C_4 = 4$, the inequality (6.27) fails for all distinct integer pairs (j_1, j_2) with $1 \le j_1, j_2 \le p$, then the series on the far right-hand side of (6.26) can have no more than one nonzero term. That term, if it exists, must equal 1. In this case the far right-hand side of (6.26) cannot exceed $\sup_{u\in\mathcal{U}_W} \psi_W(u)^{-1/2}$, which in turn is bounded above by a constant, $C_6 = C^{-1/2}$. Hence, (6.26) and (6.28) imply that

$$P\{0 \le \mathrm{hc}'_{WZ}(1) - \mathrm{hc}'_{WZ}(-1) \le C_6\} \to 1 . \tag{6.29}$$

Combining (6.24), (6.25) and (6.29) we deduce that

$$P\{|\mathrm{hc}_W - \mathrm{hc}'_{WZ}(0)| \le C_6\} \to 1 . \tag{6.30}$$

Observe too that, uniformly in u,

$$\begin{aligned}\big|P(|V'_{Wj}| \le u) - \Psi_{Wj}(u)\big| &\le \big|\Psi_{Wj}(u+p^{-D}) - \Psi_{Wj}(u-p^{-D})\big| + P(\mathcal{E}_W)\\ &\le C_7\,(4p^{-D})^{C_1} + P(\mathcal{E}_W) = O\big(p^{-DC_1}\big),\end{aligned}$$

where we have used (B2) and (6.24). Hence,

$$\begin{aligned}|\mathrm{hc}''_{WZ} - \mathrm{hc}'_{WZ}(0)| &\le \sup_{u\in\mathcal{U}_W} \psi_W(u)^{-1/2} \sum_{j=1}^{p} \big|P(|V'_{Wj}| \le u) - \Psi_{Wj}(u)\big|\\ &= O\big(p^{-DC_1}\big).\end{aligned}$$

Combining this result and (6.30) we deduce that if $C_8 > 0$ is chosen sufficiently large,

$$P\big(|\mathrm{hc}_W - \mathrm{hc}''_{WZ}| \le C_8\big) \to 1. \tag{6.31}$$

Next we introduce further notation, defining $\Psi_{WZj}(u) = P(|V_{Wj} + \nu_{WZj}| \le u)$,

$$\Psi^{\text{dash}}_{Wj}(u) = P(|V'_{Wj}| \le u)\,, \qquad \Psi^{\text{dash}}_{WZj}(u) = P(|V'_{Wj} + \nu_{WZj}| \le u)\,,$$

$$\psi_{WZ} = \sum_{j=1}^{p} \Psi_{WZj}\,(1 - \Psi_{WZj})\,, \quad \psi^{\text{dash}}_{WZ} = \sum_{j=1}^{p} \Psi^{\text{dash}}_{WZj}\,(1 - \Psi^{\text{dash}}_{WZj})\,,$$

$$\phi_{WZ} = \sum_{j=1}^{p} (\Psi_{Wj} - \Psi_{WZj})\,, \qquad \omega_{WZ} = \psi_W + \phi_{WZ}\,,$$

$$\mathrm{hc}^{(3)}_{WZ} = \sup_{u\in\mathcal{U}_W} \omega_{WZ}(u)^{-1/2} \left| \sum_{j=1}^{p} \{I(|V'_{Wj} + \nu_{WZj}| \le u) - \Psi^{\text{dash}}_{WZj}(u)\} \right|,$$

$$\begin{aligned}\mathrm{hc}^{(4)}_{WZ} &= \sup_{u\in\mathcal{U}_W} \psi_W(u)^{-1/2} \sum_{j=1}^{p} \{P(|V'_{Wj}| \le u) - P(|V'_{Wj} + \nu_{WZj}| \le u)\}\\ &= \sup_{u\in\mathcal{U}_W} \psi_W(u)^{-1/2} \sum_{j=1}^{p} \{\Psi^{\text{dash}}_{Wj}(u) - \Psi^{\text{dash}}_{WZj}(u)\}\,.\end{aligned}$$

The remainder of the proof develops approximations to $\mathrm{hc}_{WZ}^{(3)}$ and $\mathrm{hc}_{WZ}^{(4)}$.

Using (B1)(a) and (B2)(a) it can be shown that, uniformly in u,

$$\left|\psi_{WZ} - \psi_{WZ}^{\mathrm{dash}}\right| = O(p^{1-DC_1}) \to 0\,, \tag{6.32}$$

provided $D > C_1^{-1}$. Also, if $D > C_1^{-1}$ then a similar argument can be used to show that, with $\overline{\mathrm{hc}}_{WZ}$ defined as at (6.8),

$$\left|\mathrm{hc}_{WZ}^{(4)} - \overline{\mathrm{hc}}_{WZ}\right| \to 0\,. \tag{6.33}$$

By (B2)(b), $0 \le \Psi_{WZj} \le \Psi_{Wj} \le 1$, from which it follows that

$$\begin{aligned}
&\Psi_{Wj}\,(1-\Psi_{Wj}) + \Psi_{Wj} - \Psi_{WZj} \\
&= \Psi_{WZj}\,(1-\Psi_{WZj}) + (\Psi_{Wj} - \Psi_{WZj})\,(2 - \Psi_{Wj} - \Psi_{WZj}) \\
&\ge \Psi_{WZj}\,(1-\Psi_{WZj})
\end{aligned}$$

for each j. Adding over j we deduce that $\omega_{WZ} \ge \psi_{WZ}$. Combining this result with (6.32), and noting that $\omega_{WZ}(u) \ge \psi_W(u) > C$ for $u \in \mathcal{U}_W = \mathcal{U}_W(C,t)$, we deduce that, for a constant $C_9 > 0$,

$$\text{for all } u \in \mathcal{U}_W\,, \quad \psi_{WZ}^{\mathrm{dash}}(u) \le C_9\,\omega_{WZ}(u)\,. \tag{6.34}$$

Write $\langle\cdot\rangle$ for the integer-part function. Given $\epsilon \in (0,1)$, use (B1)(b) to break the sum inside the absolute value in the definition of $\mathrm{hc}_{WZ}^{(3)}$, taken over $1 \le j \le p$, into $\langle p^\epsilon\rangle$ series, each consisting only of independent terms. Let $S_{WZk}(u)$, for $1 \le k \le \langle p^\epsilon\rangle$, denote the kth of these series. Now, $E\{S_{WZk}(u)\} = 0$ and, for $u \in \mathcal{U}_W$,

$$\mathrm{var}\{S_{WZk}(u)\} \le \psi_{WZ}^{\mathrm{dash}}(u) \le C_9\,\omega_{WZ}(u)\,, \tag{6.35}$$

where the variance is computed using the expression for $S_{WZk}(u)$ as a sum of independent random variables, and the second inequality comes from (6.34).

Employing (6.35), and noting again the independence property, standard arguments can be used to show that for each choice of $C_{10}, C_{11} > 0$,

$$\max_{1\le k\le\langle p^\epsilon\rangle} P\left\{\sup_{u\in\mathcal{U}_W} \omega_{WZ}(u)^{-1/2}\,|S_{WZk}(u)| > p^{C_{10}}\right\} = O\big(p^{-C_{11}}\big)\,. \tag{6.36}$$

In particular, using Rosenthal's inequality, Markov's inequality and the fact that $\omega_{WZ}(u) \ge \psi_W(u) \ge C$ for $u \in \mathcal{U}_W$, we may show that for all $B_1, B_2 > 0$,

$$\max_{1\le k\le\langle p^\epsilon\rangle}\ \sup_{u\in\mathcal{U}_W} P\big\{|S_{WZk}(u)| > \omega_{WZ}(u)^{1/2}\,p^{B_1}\big\} = O\big(p^{-B_2}\big)\,.$$

Therefore, if $\mathcal{V}_W = \mathcal{V}_W(p)$ denotes any subset of $\mathcal{U}_W$ that contains only $O(p^{B_3})$ elements, for some $B_3 > 0$, then for all $B_1, B_2 > 0$,

$$\max_{1\le k\le \langle p^\epsilon\rangle} P\Big\{ \max_{u\in\mathcal{V}_W} \omega_{WZ}(u)^{-1/2}\,|S_{WZk}(u)| > p^{B_1}\Big\} = O(p^{B_3-B_2}) = O(p^{-B_4}), \tag{6.37}$$

where $B_4 = B_2 - B_3$. Since B_3 and B_4 both can be taken arbitrarily large, then, using the monotonicity of the function $g(u) = I(v \le u)$, and also properties (B1) and (B2), it can be seen that $\max_{u\in\mathcal{V}_W}$ in (6.37) can be replaced by $\sup_{u\in\mathcal{U}_W}$, giving (6.36). In this context, condition (B2)(c) ensures that, with an error that is less than p^{-B_5}, for any given $B_5 > 0$, the distribution Ψ_{Wj} can be truncated at a point p^{B_6}, for sufficiently large B_6; and, within the interval $[0, p^{B_6}]$, the points in $\mathcal{V}_W$ can be chosen less than p^{-B_7} apart, where $B_7 > 0$ is arbitrarily large.

Result (6.36) implies that

$$P\bigg\{ \max_{1\le k\le\langle p^\epsilon\rangle} \sup_{u\in\mathcal{U}_W} \omega_{WZ}(u)^{-1/2}\,|S_{WZk}(u)| > p^{C_{10}}\bigg\} = O(p^{\epsilon-C_{11}}),$$

from which it follows that $P(\mathrm{hc}^{(3)}_{WZ} > p^{\epsilon+C_{10}}) = O(p^{\epsilon-C_{11}})$. Since ϵ, C_{10} and C_{11} are arbitrary positive numbers then we may replace ϵ here by zero, obtaining: for each $C_{10}, C_{11} > 0$,

$$P\big(\mathrm{hc}^{(3)}_{WZ} > p^{C_{10}}\big) = O\big(p^{-C_{11}}\big). \tag{6.38}$$

It can be deduced directly from the definitions of hc''_{WZ}, $\mathrm{hc}^{(3)}_{WZ}$ and $\mathrm{hc}^{(4)}_{WZ}$ that:

$$|\mathrm{hc}''_{WZ} + \mathrm{hc}^{(4)}_{WZ}| \le \mathrm{hc}^{(3)}_{WZ} \sup_{u\in\mathcal{U}_W} \sqrt{\frac{\omega_{WZ}(u)}{\psi_W(u)}} = \mathrm{hc}^{(3)}_{WZ} \sup_{u\in\mathcal{U}_W} \sqrt{1 + \frac{\phi_{WZ}(u)}{\psi_W(u)}}.$$

Combining this result with (6.31), (6.33) and (6.38); and noting that

$$\overline{\mathrm{hc}}_{WZ} = \sup_{u\in\mathcal{U}_W} \frac{\phi_{WZ}(u)}{\psi_W(u)^{1/2}},$$

and, since $\psi_W(u) \ge C$ for $u \in \mathcal{U}_W$,

$$\sup_{u\in\mathcal{U}_W} \bigg\{1 + \frac{\phi_{WZ}(u)}{\psi_W(u)}\bigg\}^{1/2} \le \big(1 + C^{-1/4}\big) \sup_{u\in\mathcal{U}_W} \bigg\{1 + \frac{\phi_{WZ}(u)}{\psi_W(u)^{1/2}}\bigg\}^{1/2};$$

we deduce that for each $\epsilon > 0$,

$$\mathrm{hc}_W + \overline{\mathrm{hc}}_{WZ} = O_p\Big\{p^\epsilon\,\big(1 + \overline{\mathrm{hc}}_{WZ}\big)^{1/2}\Big\}. \tag{6.39}$$

Theorem 6.2 follows directly from (6.39).

References

[1] Abramovich, F., Benjamini, Y., Donoho, D.L. and Johnstone, I.M. (2006). Adapting to unknown sparsity by controlling the false discovery rate. *Ann. Statist.* **34** 584–653.

[2] Benjamini, Y. and Hochberg, Y. (1995). Controlling the false discovery rate: a practical and powerful approach to multiple testing. *J. Roy. Statist. Soc.* Ser. B. **57** 289–300.

[3] Broberg, P. (2003). Statistical methods for ranking differentially expressed genes. *Genome Biol.* **4** R41 (electronic).

[4] Cai, T., Jin, J. and Low, M. (2007). Estimation and confidence sets for sparse normal mixtures. *Ann. Statist.*, to appear.

[5] Cayón, L., Banday, A. J., Jaffe, T., Eriksen, H. K. K., Hansen, F. K., Gorski, K. M. and Jin, J. (2006). No higher criticism of the Bianchi corrected WMAP data. *Mon. Not. Roy. Astron. Soc.* **369** 598–602.

[6] Cayón, L., Jin, J. and Treaster, A. (2005). Higher criticism statistic: Detecting and identifying non-Gaussianity in the WMAP first year data. *Mon. Not. Roy. Astron. Soc.* **362** 826–832.

[7] Cruz, M., Cayón, L., Martínez-González, E., Vielva, P. and Jin, J. (2007). The non-Gaussian cold spot in the 3-year WMAP data. *Astrophys. J.* **655** 11–20.

[8] Donoho, D. L. and Jin, J. (2004). Higher criticism for detecting sparse heterogeneous mixtures. *Ann. Statist.* **32** 962–994.

[9] Efron, B. (2004). Large-scale simultaneous hypothesis testing: the choice of a null hypothesis. *J. Amer. Statist. Assoc.* **99** 96–104.

[10] Fan, J., Chen, Y., Chan, H. M., Tam, P. and Ren, Y. (2005). Removing intensity effects and identifying significant genes for Affymetrix arrays in MIF-suppressed neuroblastoma cells. *Proc. Nat. Acad. Sci. USA* **102** 17751–17756.

[11] Fan, J. and Fan, Y. (2007). High dimensional classification using features annealed independence rules. *Ann. Statist.*, to appear.

[12] Fan, J., Peng, H. and Huang, T. (2005). Semilinear high-dimensional model for normalization of microarray data: a theoretical analysis and partial consistency. (With discussion.) *J. Amer. Statist. Assoc.* **100** 781–813.

[13] Fan, J., Tam, P., Vande Woude, G. and Ren, Y. (2004). Normalization and analysis of cDNA micro-arrays using within-array replications applied to neuroblastoma cell response to a cytokine. *Proc. Nat. Acad. Sci. USA* **101** 1135–1140.

[14] Hall, P. and Jin, J. (2006). Properties of higher criticism under long-range dependence. *Ann. Statist.*, to appear.

[15] Hall, P., Pittelkow, Y. and Ghosh, M. (2008). On relative theoretical performance of classifiers suitable for high-dimensional data and small sample sizes. *J. Roy. Statist. Soc.* Ser. B, to appear.

[16] Huang, J., Wang, D. and Zhang, C. (2003). A two-way semi-linear model for normalization and significant analysis of cDNA microarray data. Manuscript.

[17] Ingster, Yu. I. (1999). Minimax detection of a signal for l^n-balls. *Math.*

Methods Statist. **7** 401–428.
[18] Ingster, Yu. I. (2001). Adaptive detection of a signal of growing dimension. I. Meeting on Mathematical Statistics. *Math. Methods Statist. 10* 395–421.
[19] Ingster, Yu. I. (2002). Adaptive detection of a signal of growing dimension. II. *Math. Methods Statist.* **11** 37–68.
[20] Ingster, Yu. I. and Suslina, I. A. (2003). *Nonparametric Goodness-of-Fit Testing Under Gaussian Models.* Springer, New York.
[21] Jin, J. (2006). Higher criticism statistic: Theory and applications in non-Gaussian detection. In *Statistical Problems in Particle Physics, Astrophysics And Cosmology.* (Eds. L. Lyons and M.K. Ünel.) Imperial College Press, London.
[22] Jin, J., Starck, J.-L., Donoho, D. L., Aghanim, N. and Forni, O. (2004). Cosmological non-Gaussian signature detection: Comparing performance of different statistical tests. *Eurasip J. Appl. Signal Processing* **15** 2470–2485.
[23] Lehmann, E. L., Romano, J. P. and Shaffer, J. P. (2005). On optimality of stepdown and stepup multiple test procedures. *Ann. Statist.* **33** 1084–1108.
[24] Lönnstedt, I. and Speed, T. (2002). Replicated microarray data. *Statist. Sinica* **12** 31–46.
[25] Meinshausen, M. and Rice, J. (2006). Estimating the proportion of false null hypotheses among a large number of independent tested hypotheses. *Ann. Statist.* **34** 373–393.
[26] Scheffé, H. (1959). *The Analysis of Variance.* Wiley, New York.
[27] Storey, J. D. and Tibshirani, R. (2003). Statistical significance for genome-wide experiments. *Proc. Nat. Acad. Sci. USA* **100** 9440–9445.
[28] Storey, J. D., Taylor, J. E. and Siegmund, D. (2004). Strong control, conservative point estimation, and simultaneous conservative consistency of false discovery rates: A unified approach. *J. Roy. Stat. Soc.* Ser. B **66** 187–205.
[29] Tseng, G. C., Oh, M. K., Rohlin, L., Liao, J. C. and Wong, W. H. (2001). Issues in cDNA microarray analysis: Quality filtering, channel normalization, models of variations and assessment of gene effects. *Nucleic Acids Res.* **29**, 2549–2557.
[30] Tukey, J.W. (1953). The problem of multiple comparisons. Manuscript. Department of Statistics, Princeton University.
[31] Wang, Q. (2005). Limit theorems for self-nomalized large deviation. *Electronic J. Probab.* **38** 1260–1285.
[32] Wang, Q. and Hall, P. (2007). Relative errors in central limit theorem for Student's t statistic, with applications. Manuscript.

Appendix

A.1. Description of the Cross-Validation Procedure

If previous experience with data of the type being analysed is not sufficient to permit effective choice of threshold using that background, then a data-driven selection needs to be developed. This, however, is a challenging task, as the sample sizes are typically very small. As a first practical method, we implemented a cross-validation (CV) procedure where the basic idea was as follows. Create $n_X + n_Y$ cross-validation samples $(\mathcal{X}_{CV,k}, \mathcal{Y}_{CV,k}, Z_{CV,k}) = (\mathcal{W}^{(-j)}, \mathcal{T}, W_j)$, $k = j + n_X I(W = Y)$, $j = 1, \ldots, n_W$, $(W,T) = (X,Y)$ or (Y,X), where $\mathcal{W}^{(-j)}$ denotes the sample $\mathcal{W}$ with the jth observation W_j left out; apply the classification procedure to each CV sample, and then choose (t_1, t_2) to give a large number of correct classifications, but not too large so as to avoid 'overfitting' the data. We experimented with different ways of avoiding the overfitting problem, and found that the following gave quite good results.

(a) Here we describe how to choose the grid on which we search for (t_1, t_2). One of the problems in our context is that p is so large that removing one of the data values, as is done in cross-validation, has substantial impact on the range of the observed data. Therefore, and since we expect t_2 to be related to the extreme observed values, it would not be appropriate to choose a grid for (t_1, t_2) and keep it fixed over each iteration of the algorithm. Instead, at each step k, where $k = 1, \ldots, n_X + n_Y$, of the algorithm we define the grid in terms of a set of $K \in [2, 2p - 1]$ order statistics $U_{(i_1)} < U_{(i_2)} < \ldots < U_{(i_K)}$ of the vector $U = (|Z_{CV,k} - \bar{X}_{CV,k}|, |Z_{CV,k} - \bar{Y}_{CV,k}|)$. (To make notations less heavy, we omit the index k from U.) We keep fixed the vector $I = (i_1, \ldots, i_K)$ of K indices. At each step we define our grid for (t_1, t_2) as $U_{(I)} \times U_{(I)}$, where $U_{(I)}$ denotes $(U_{(i_1)}, \ldots, U_{(i_K)})$. The indices $1 \le i_1 < i_2 < \ldots < i_K \le 2p$ are chosen such that the last, say, $K - S$ order statistics $V_{(i_{S+1})} < V_{(i_{S+2})} < \ldots < V_{(i_K)}$ of the vector $V = (|Z_k - \bar{X}_k|, |Z_k - \bar{Y}_k|)$ consist of the extreme values of V, and the first S order statistics $V_{(i_1)} < V_{(i_2)} < \ldots < V_{(i_S)}$ are uniformly distributed over the interval $[V_{(1)}, V_{(i_{S+1}-1)}]$.

(b) For $k = 1, \ldots, n_X + n_Y$, apply the HC procedure to the kth cross-validation sample, for each (t_1, t_2) in the grid $U_{(I)} \times U_{(I)}$.

(c) For each $1 \le j, k \le K$, let $C_{j,k}$ denote the number of correct classifications out of the $n_X + n_Y$ cross-validation trials at (b), obtained by taking $(t_1, t_2) = (U_{(i_j)}, U_{(i_k)})$. Of course, since t_1 must be less than t_2, we set $C_{j,k} = 0$ for all $j > k$.

(d) Taking V as in (a), construct the vector t_2^* of all values $V_{(i_k)}$ for which $\sup_j C_{j,k} \ge M' = \sup_{j,k} C_{j,k} - (n_X + n_Y)/10$. The factor $(n_X + n_Y)/10$ was chosen heuristically and it is introduced to avoid overfitting the data. Take t_2 as the component of t_2^*, say $V_{(i_\ell)}$, for which $\#\{j \text{ s.t. } C_{j,\ell} \geq M'\}$ is the largest — in case of non uniqueness, take $V_{(i_\ell)}$ as the largest such component. Then take t_1 as the average of all $V_{(i_j)}$'s such that $C_{j,\ell} \geq M'$.

In most cases this method gave good results, with performance lying approximately midway between that using the theoretically optimal (t_1, t_2) or no thresholding, i.e. $(t_1, t_2) = (-\infty, \infty)$, respectively.

A.2. Proof of Theorem 6.3

For simplicity, we denote N_{Wj} by N. Recall that χ_{Wj} denotes the distribution of V_{Wj}, i.e. the distribution of $Z_j - \bar{W}_{.j}$ when Z is drawn from Π_W. It can be proved from results of [31] that, under (C1), uniformly in values of $u > 0$ that satisfy $u = o(N^{-1/6})$, and uniformly in W and in $1 \leq j \leq p$,

$$\chi_{Wj}(u) = \Phi(u) + O\big[N^{-1/2} |u|^3 \{1 - \Phi(u)\}\big],$$

where Φ denotes the standard normal distribution function. An analogous result for $u \leq 0$ also holds. Hence for $u > 0$ satisfying $u = o(N^{-1/6})$, we have uniformly in W and in $1 \leq j \leq p$,

$$P(|V_{Wj}| \leq u) = 2\,\Phi(u) - 1 + O\big[N^{-1/2}\, u^3 \{1 - \Phi(u)\}\big]. \tag{6.40}$$

Similarly it can be shown that, uniformly in $j = j_1, \ldots, j_q$, the latter as in (C1),

$$\begin{aligned} P(|V_{Wj} \pm \nu| \leq u) = \Phi(u + \nu) + \Phi(u - \nu) - 1 \\ + O\big[N^{-1/2}\, (u + \nu)^3 \{1 - \Phi(|u - \nu|)\}\big]. \end{aligned} \tag{6.41}$$

Let $a_{WQ}(u)$ denote the series in the definition of $\overline{\text{hc}}_{WQ}$, at (6.8). Combining (6.40) and (6.41) we deduce that, if $Q \neq W$,

$$\begin{aligned} a_{WQ}(u) &= q\,\{2\,\Phi(u) - \Phi(u + \nu) - \Phi(u - \nu)\} \\ &\quad + O\big[N^{-1/2}\, q\,(u + \nu)^3 \{1 - \Phi(|u - \nu|)\}\big], \qquad (6.42) \\ \psi_W(u) &= 2p\,\{2\Phi(u) - 1\}\,\{1 - \Phi(u)\} + O\big[N^{-1/2}\, p\, u^3 \{1 - \Phi(u)\}\big] \\ &= \{1 + o(1)\}\, 2p\,\{2\Phi(u) - 1\}\,\{1 - \Phi(u)\}, \qquad (6.43) \end{aligned}$$

uniformly in $u \in \mathcal{U}_W(C, t)$. To obtain the second identity in (6.43) we used the properties $t \geq B > 0$ and $N^{-1/2}\, (\log p)^{3/2} \to 0$, from (C2) and (6.12) respectively.

Take $u = \sqrt{2v \log p}$ where $0 < v = v(p) \le 1$, and recall that $\nu = \sqrt{2w \log p}$, where w and s are as in (C2). It can be shown, borrowing ideas from [8], that

$$2\,\Phi(u) - \Phi(u+\nu) - \Phi(u-\nu) = g_1(p)\, p^{-(\sqrt{v}-\sqrt{w})^2}, \tag{6.44}$$

$$\{2\Phi(u)-1\}\,\{1-\Phi(u)\} \sim C_1\,(\log p)^{-1/2}\, p^{-v}, \tag{6.45}$$

where, here and below, g_j denotes a function that is bounded above by C_2 and below by $C_3\,(\log p)^{-1/2}$, and C_1, C_2, C_3 denote positive constants. To derive (6.44), write $2\,\Phi(u) - \Phi(u+\nu) - \Phi(u-\nu)$ as

$$\{1-\Phi(u+\nu)\} + \{1-\Phi(u-\nu)\} - 2\,\{1-\Phi(u)\},$$

and use conventional approximations to $1 - \Phi(z)$, for moderate to large positive z, and, when $u - \nu < 0$, to $\Phi(-z)$, for z in the same range.

In view of (6.44) and (6.45),

$$\frac{2\,\Phi(u) - \Phi(u+\nu) - \Phi(u-\nu)}{[2p\,\{2\Phi(u)-1\}\,\{1-\Phi(u)\}]^{1/2}} = g_2(p)\,(\log p)^{1/4}\, p^{b_1}, \tag{6.46}$$

where $b_1 = \frac{1}{2}\,(v-1) - (\sqrt{v}-\sqrt{w})^2$. Similarly,

$$\frac{N^{-1/2}\,(u+\nu)^3\,\{1-\Phi(|u-\nu|)\}}{[p\,\{1-\Phi(u)\}]^{1/2}} = O\{N^{-1/2}\,(\log p)^{7/4}\, p^{b_1}\}. \tag{6.47}$$

Using (6.42), (6.43), (6.46) and (6.47) we deduce that, provided $N^{-1}\,(\log p)^4 \to 0$,

$$\frac{a_{WQ}(u)}{\psi_W(u)^{1/2}} \sim q\, g_2(p)\,(\log p)^{1/4}\, p^{b_1} = g_3(p)\,(\log p)^{1/4}\, p^{b_2}, \tag{6.48}$$

where $b_2 = \frac{1}{2}\,(v+1) - \beta - (\sqrt{v}-\sqrt{w})^2$.

Since s, in the definition of $t = \sqrt{2s \log p}$, satisfies $0 < s < \min(4w, 1)$, we can take

$$v = \begin{cases} 4w & \text{if } 0 < w < \frac{1}{4} \\ 1 - c(\log p)^{-1} \log\log p & \text{if } \frac{1}{4} \le w < 1, \end{cases}$$

where $c > \frac{1}{2}$, and have

$$u = \sqrt{2v \log p} = \min\left(2\nu, \sqrt{2\log p - 2c\log\log p}\right) \in \mathcal{U}(C,t).$$

For this choice of v, $b_2 = 2\eta$ where $\eta > 0$, and it follows from (6.48) that

$$\overline{\text{hc}}_{WQ} \ge \frac{a_{WQ}(u)}{\psi_W(u)^{1/2}} \ge C_4\, p^{\eta}.$$

Result (6.13) follows.

Chapter 7

Bayesian Nonparametric Approach to Multiple Testing

Subhashis Ghosal[1] and Anindya Roy[2]

[1] *Department of Statistics,*
North Carolina State University,
2501 Founders Drive,
Raleigh NC 27695-8203, USA
sghosal@stat.ncsu.edu

[2] *Department of Mathematics & Statistics,*
University of Maryland Baltimore County,
1000 Hilltop Circle,
Baltimore, MD 21250, USA
anindya@math.umbc.edu

Motivated by the problems in genomics, astronomy and some other emerging fields, multiple hypothesis testing has come to the forefront of statistical research in the recent years. In the context of multiple testing, new error measures such as the false discovery rate (FDR) occupy important roles comparable to the role of type I error in classical hypothesis testing. Assuming that a random mechanism decides the truth of a hypothesis, substantial gain in power is possible by estimating error measures from the data. Nonparametric Bayesian approaches are proven to be particularly suitable for estimation of error measure in multiple testing situation. A Bayesian approach based on a nonparametric mixture model for p-values can utilize special features of the distribution of p-values that significantly improves the quality of estimation. In this paper we describe the nonparametric Bayesian modeling exercise of the distribution of the p-values. We begin with a brief review of Bayesian nonparametric concepts of Dirichlet process and Dirichlet mixtures and classical multiple hypothesis testing. We then review recently proposed nonparametric Bayesian methods for estimating errors based on a Dirichlet mixture of prior for the p-value density. When the test statistics are independent, a mixture of beta kernels can adequately model the p-value density, whereas in the dependent case one can consider a Dirichlet mixture of multivariate skew-normal kernel prior for probit transforms of the p-values. We conclude the paper by illustrating the scope of these methods in some real-life applications.

7.1. Bayesian Nonparametric Inference

To make inference given an observed set of data, one needs to model how the data are generated. The limited knowledge about the mechanism often does not permit explicit description of the distribution given by a relatively few parameters. Instead, only very general assumptions leaving a large portion of the mechanism unspecified can be reasonably made. This nonparametric approach thus avoids possible gross misspecification of the model, and understandably is becoming the preferred approach to inference, especially when many samples can be observed. Nonparametric models are actually not parameter free, but they contain infinite dimensional parameters, which can be best interpreted as functions. In common applications, the cumulative distribution function (c.d.f.), density function, nonparametric regression function, spectral density of a time series, unknown link function in a generalized linear model, transition density of a Markov chain and so on can be the unknown function of interest. Classical approach to nonparametric inference has flourished throughout the last century. Estimation of c.d.f. is commonly done by the empirical c.d.f., which has attractive asymptotic properties. Estimation of density, regression function and similar objects in general needs smoothing through the use of a kernel or through a basis expansion. Testing problems are generally approached through ranks, which typically form the maximal invariant class under the action of increasing transformations.

Bayesian approach to inference offers a conceptually straightforward and operationally convenient method, since one needs only to compute the posterior distribution given the observations, on which the inference is based. In particular, standard errors and confidence sets are automatically obtained along with a point estimate. In addition, the Bayesian approach enjoys philosophical justification and often Bayesian estimation methods have attractive frequentist properties, especially in large samples. However, Bayesian approach to nonparametric inference is challenged by the issue of construction of prior distribution on function spaces. Philosophically, specifying a genuine prior distribution on an infinite dimensional space amounts to adding infinite amount of prior information about all fine details of the function of interest. This is somewhat contradictory to the motivation of nonparametric modeling where one likes to avoid specifying too much about the unknown functions. This issue can be resolved by considering the so called "automatic" or "default" prior distributions, where some tractable automatic mechanism constructs most part of the prior by spreading the mass all over the parameter space, while only a handful of key parameters may be chosen subjectively. Together with additional conditions, large support of the prior helps the posterior distribution concentrate around the true value of the unknown function of interest. This property, known as posterior consistency, validates a Bayesian procedure from the frequentist view, in that it ensures that, with sufficiently large amount of data, the truth can be discovered accurately and the data eventually overrides any prior information. Therefore, a frequentist will be more likely to agree to the inference based on a default nonparametric prior. Lack of consistency is thus clearly undesirable since this means that the posterior distribution is not directed toward the truth. For a consistent posterior, the speed of convergence to the true value, called the rate of convergence, gives a more refined picture of the accuracy of a Bayesian procedure in estimating the unknown function of interest.

For estimating an arbitrary probability measure (equivalently, a c.d.f.) on the real line, with independent and identically distributed (i.i.d.) observations from it, Ferguson ([19]) introduced the idea of a Dirichlet process — a random probability distribution P such that for any finite measurable partition $\{B_1, \dots, B_k\}$ of $\mathbb{R}$, the joint distribution of $(P(B_1), \dots, P(B_k))$ is a finite dimensional Dirichlet distribution with parameters $(\alpha(B_1), \dots, \alpha(B_k))$, where α is a finite measure called the base measure of the Dirichlet process $\mathcal{D}_\alpha$. Since clearly $P(A) \sim \text{Beta}(\alpha(A), \alpha(A^c))$, we have $\mathrm{E}(P(A)) = \alpha(A)/(\alpha(A) + \alpha(A^c)) = G(A)$, where $G(A) = \alpha(A)/M$, a probability measure called the center measure and $M = \alpha(\mathbb{R})$, called the precision parameter. This implies that if $X|P \sim P$ and $P \sim \mathcal{D}_\alpha$, then marginally $X \sim G$. Observe that $\text{var}(P(A)) = G(A)G(A^c)/(M+1)$, so that the prior is more tightly concentrated around its mean when M is larger. If P is given the measure $\mathcal{D}_\alpha$, we shall write $P \sim \text{DP}(M, G)$. The following give the summary of the most important facts about the Dirichlet process:

(i) If $\int |\psi| dG < \infty$, then $\mathrm{E}(\int \psi dP) = \int \psi dG$.
(ii) If $X_1, \dots, X_n | P \overset{iid}{\sim} P$ and $P \sim \mathcal{D}_\alpha$, then $P|X_1, \dots, X_n \sim \mathcal{D}_{\alpha + \sum_{i=1}^n \delta_{X_i}}$.
(iii) $\mathrm{E}(P|X_1, \dots, X_n) = \frac{M}{M+n} G + \frac{n}{M+n} \mathbb{P}_n$, a convex combination of the prior mean and the empirical distribution $\mathbb{P}_n$.
(iv) Dirichlet sample paths are a.s. discrete distributions.
(v) The topological support of $\mathcal{D}_\alpha$ is $\{P^* : \text{supp}(P^*) \subset \text{supp}(G)\}$.

(vi) The marginal joint distribution of $(X_1,\dots,X_n)$ from P, where $P \sim \mathcal{D}_\alpha$, can be described through the conditional laws

$$X_i|(X_l, l \neq i) \sim \begin{cases} \delta_{\phi_j}, & \text{with probability } \frac{n_j}{M+n-1},\ j=1,\dots,k_{-i}, \\ G, & \text{with probability } \frac{M}{M+n-1}, \end{cases}$$

where k_{-i} is the number of distinct observations in X_l, $l \neq i$ and $\phi_1,\dots,\phi_{k_{-i}}$ are those distinct values with multiplicities $n_1,\dots,n_{k_{-i}}$. Thus the number of distinct observations K_n in $X_1,\dots,X_n$, is generally much smaller than n with $\mathrm{E}(K_n) = M\sum_{i=1}^{n}(M+i-1)^{-1} \sim M\log(n/M)$, introducing sparsity.

(vii) Sethuraman's ([51]) stick-breaking representation: $P = \sum_{i=1}^{\infty} V_i\delta_{\theta_i}$, where $\theta_i \overset{iid}{\sim} G$, $V_i = [\prod_{j=1}^{i-1}(1-Y_j)]Y_i$, $Y_i \overset{iid}{\sim} \mathrm{Beta}(1,M)$. This allows us to approximately generate a Dirichlet process and is indispensable in various complicated applications involving the Dirichlet process, where posterior quantities can be simulated approximately with the help of a truncation and Markov chain Monte-Carlo (MCMC) techniques.

In view of (iii), clearly G should be elicited as the prior guess about P, while M should be regarded as the strength of this belief. Actual specification of these are quite difficult in practice, so we usually let G contain additional hyperparameters ξ, and some flat prior is put on ξ, leading to a mixture of Dirichlet process ([1]).

A widely different scenario occurs when one mixes parametric families nonparametrically. Assume that given a latent variable θ_i, the observations X_i follows a parametric density $\psi(\cdot;\theta_i)$, $i=1,\dots,n$, respectively, and the random effects $\theta_i \overset{iid}{\sim} P$, $P \sim \mathcal{D}_\alpha$ ([20], [33]). In this case, the density of the observation can be written as $f_P(x) = \int \psi(x;\theta)dP(\theta)$. The induced prior distribution on f_P through $P \sim \mathrm{DP}(M,G)$ is called a Dirichlet process mixture (DPM). Since $f_P(x)$ is a linear functional of P, the expressions of posterior mean and variance of the density $f_P(x)$ can be analytically expressed. However, these expressions contain enormously large number of terms. On the other hand, computable expressions can be obtained by MCMC methods by simulating the latent variables $(\theta_1,\dots,\theta_n)$ from their posterior distribution by a scheme very similar to (vi); see [18]. More precisely, given θ_j, $j \neq i$, only X_i affects the posterior distribution of θ_i. The observation X_i weighs the selection probability of an old θ_j by $\psi(X_i;\theta_j)$, and the fresh draw by $M\int\psi(X_i;\theta)dG(\theta)$, and a fresh draw, whenever obtained, is taken from the "baseline posterior" defined by

$dG_i(\theta) \propto \psi(X_i;\theta)dG(\theta)$. The procedure is known as the generalized Polya urn scheme.

The kernel used in forming DPM can be chosen in different ways depending on the sample space under consideration. A location-scale kernel is appropriate for densities on the line with unrestricted shape. In Section 7.3, we shall use a special type of beta kernels for decreasing densities on the unit interval modeling the density of p-values in multiple hypothesis testing problem.

To address the issue of consistency, let Π be a prior on the densities and let f_0 stand for the true density. Then the posterior probability of a set B of densities given observations $X_1, \ldots, X_n$ can be expressed as

$$\Pi(f \in B | X_1, \ldots, X_n) = \frac{\int_B \prod_{i=1}^n \frac{f(X_i)}{f_0(X_i)} d\Pi(f)}{\int \prod_{i=1}^n \frac{f(X_i)}{f_0(X_i)} d\Pi(f)}. \tag{7.1}$$

When B is the complement of a neighborhood U of f_0, consistency requires showing that the expression above goes to 0 as $n \to \infty$ a.s. $[P_{f_0}]$. This will be addressed by showing that the numerator in (7.1) converges to zero exponentially fast, while the denominator multiplied by $e^{\beta n}$ goes to infinity for all $\beta > 0$. The latter happens if $\Pi(f : \int f_0 \log(f_0/f) < \epsilon) > 0$ for all $\epsilon > 0$. The assertion about the numerator in (7.1) holds if a uniformly exponentially consistent test exists for testing the null hypothesis $f = f_0$ against the alternative $f \in U^c$. In particular, the condition holds automatically if U is a weak neighborhood, which is the only neighborhood we need to consider in our applications to multiple testing.

7.2. Multiple Hypothesis Testing

Multiple testing procedures are primarily concerned with controlling the number of incorrect significant results obtained while simultaneously testing a large number of hypothesis. In order to control such errors an appropriate error rate must be defined. Traditionally, the family-wise error rate (FWER) has been the error rate of choice until recently when the need was felt to define error rates that more accurately reflect the scientific goals of modern statistical applications in genomics, proteomics, functional magnetic resonance imaging (fMRI) and other biomedical problems. In order to define the FWER and other error rates we must first describe the different components of a typical multiple testing problem. Suppose $H_{10}, \ldots, H_{m0}$ are m null hypotheses whose validity is being tested simultaneously. Suppose m_0 of those hypotheses are true and after making

Table 7.1. Number of hypotheses accepted and rejected and their true status.

	Decision		
Hypothesis	Accept	Reject	Total
True	U	V	m_0
False	T	S	$m - m_0$
Total	Q	R	m

decisions on each hypothesis, R of the m hypotheses are rejected. Also, denote the m ordered p-values obtained from testing the m hypotheses as $X_{(1)} < X_{(2)} < \cdots < X_{(m)}$. Table 7.1 describes the components associated with this scenario.

The FWER is defined as the probability of making at least one false discovery, i.e. FWER $= P(V \geq 1)$. The most common FWER controlling procedure is the Bonferroni procedure where each hypotheses is tested at level α/m to meet an overall error rate of α; see [35]. When m is large, this measure is very conservative and may not yield any "statistical discovery", a term coined by [54] to describe a rejected hypothesis. Subsequently, several generalization of the Bonferroni procedure were suggested where the procedures depend on individual p-values, such as [52], [30], [31], [29] and [42]. In the context of global testing where one is interested in the significance of a set of hypotheses as a whole, [52] introduced a particular sequence of critical values, $\alpha_i = i\alpha/n$, to compare with each p-value. More recently, researchers proposed generalization of the FWER (such as the k-FWER) that is more suitable for modern applications; see [32].

While the FWER gives a very conservative error rate, at the other extreme of the spectrum of error rates is the per comparison error rate (PCER) where significance of any hypothesis is decided without any regard to the significance of the rest of the hypothesis. This is equivalent to testing each hypothesis at a fixed level α and looking at the average error over the m tests conducted, i.e. PCER $= \mathrm{E}(V/m)$. While the PCER is advocated by some ([53]) it is too liberal and may result in several false discoveries. A compromise was proposed by [7] where they described a sequential procedure to control the false discovery rate (FDR), defined as FDR $= \mathrm{E}(V/R)$. The ratio V/R is defined to be zero if there are no rejections. The FDR as an error rate has many desirable properties. First of all, as described in [7] and by many others, one can devise algorithms to control FDR in multiple testing situation under fairly general joint behavior of the test statistics for the hypotheses. Secondly, if all hypotheses are true, controlling FDR is equivalent to controlling the FWER. In general, FDR falls between the other two error rates, the FWER and PCER (cf. [24]).

The Benjamini-Hochberg (B-H) FDR control procedure is a sequential step-up procedure where the p-values (starting with the largest p-value) are sequentially compared with a sequence of critical values to find a critical p-value such that all hypotheses with p-values smaller than the critical value are rejected. Suppose $\hat{k} = \max\{i : X_{(i)} \leq \alpha_i\}$ where $\alpha_i = i\alpha/m$. The the B-H procedure rejects all hypotheses with p-values less than or equal to $X_{(\hat{k})}$. If no such $\hat{k}$ exists, then none of the hypotheses is rejected. Even though the algorithm sequentially steps down through the sequence of p-values, it is called a step-up procedure because this is equivalent to stepping up with respect to the associated sequence of test statistics to find a minimal significant test value. The procedure is also called a linear step-up procedure due to the linearity of the critical function α_i with respect to i. [9], [46], [59] among others have shown the FDR associated with this particular step-up procedure is exactly equal to $m_0\alpha/m$ in the case when the test statistics are independent and is less than $m_0\alpha/m$ if the test statistics have positive dependence: for every test function ϕ, the conditional expectation $\mathrm{E}[\phi(X_1, \ldots, X_m)|X_i]$ is increasing with X_i for each i. [46] has suggested an analogous step-down procedure where one fails to reject all hypotheses with p-values above a critical value α_i, that is, if $\hat{l} = \min\{i : X_{(i)} > \alpha_i\}$, none of the hypotheses associated with p-value $X_{(\hat{l})}$ and above is rejected. [46] used the same set of critical values $\alpha_i = i\alpha/m$ as in [7] which also controls the FDR at the desired level (see [47]). However, for the step-down procedure even in the independent case the actual FDR may be less than $m_0\alpha/m$.

Since in the independent case the FDR of the linear step-up procedure is exactly equal to $m_0\alpha/m$, if the proportion of true null hypotheses, $\pi = m_0/m$, is known then α can be adjusted to get FDR equal to any target level. Specifically, if $\alpha_i = i\alpha/(m\pi)$ then the FDR of the linear step-up procedure is exactly equal to α in the independent case. Unfortunately, in any realistic situation m_0 is not known. Thus, in situations where π is not very close to one, FDR can be significantly smaller than the desired level, and the procedure may be very conservative with poor power properties.

Another set of sequential FDR controlling procedures were introduced more recently, where π is adaptively estimated from the data and the critical values are modified as $\alpha_i = i\alpha/(m\hat{\pi})$. Heuristically, this procedure would yield an FDR close to $\pi\alpha\mathrm{E}(\hat{\pi}^{-1})$, and if $\hat{\pi}$ is an efficient estimator of π then the FDR for the adaptive procedure will be close to the target level

α. However, merely plugging-in an estimator of π in the expression for α_i may yield poor results due to the variability of the estimator of π^{-1}. [57] suggested using $\hat{\pi} = [m - R(\lambda) + 1]/[m(1-\lambda)]$, where $R(\lambda) = \sum \mathbb{1}\{X_i \le \lambda\}$ is the number of p-values smaller than λ and $0 < \lambda < 1$ is a constant; here and below $\mathbb{1}$ will stand for the indicator function. Similar estimators had been originally suggested by [56]. Then for any λ, choose the sequence of critical points as

$$\alpha_i = \min\left\{\lambda, \frac{i\alpha(1-\lambda)}{m - R(\lambda) + 1}\right\}.$$

The adaptive procedure generally yields tighter FDR control and hence can enhance the power properties of the procedures significantly ([8], [12]). Of course, the performance of the procedure will be a function of the choice of λ. [58] suggested various procedures for choosing λ. [11] suggested choosing $\lambda = \alpha/(1+\alpha)$ and they looked at the power properties of the adaptive procedure. [50] investigated theoretical properties of these two stage procedures and [22] suggested analogous adaptive step-down procedures.

The procedures described above for controlling FDR can be thought of as fixed-error rate approach where the individual hypotheses are tested at different significance level to maintain a constant overall error rate. [57, 58] introduced the fixed-rejection-region approach where $\alpha_i = \alpha$ for all i (i.e. the rejection region is fixed). The FDR given the rejection region is estimated from the data and then α is chosen to set the estimated FDR at a predetermined level. [57] also argued that since one becomes concerned about false discoveries only in the situation where there are some discoveries, one should look at the expected proportion of false discoveries conditional on the fact that there has been some discoveries. Thus the positive false discovery rate (pFDR) is defined as pFDR $= \mathrm{E}(V/R|R > 0)$. [57] showed that if we assume a mixture model for the hypotheses, i.e., if we can assume that the true null hypothesis are arising as a Bernoulli sequence with probability π, then the expression for pFDR reduces to

$$\mathrm{pFDR}(\alpha) = \frac{\pi\alpha}{F(\alpha)}, \tag{7.2}$$

where $F(\cdot)$ is the marginal c.d.f. of the p-values. Although it cannot be controlled in the situation when there are no discoveries, given its simple expression, pFDR is ideally suited for the estimation approach. Once an estimator for pFDR has been obtained, the error control procedure reduces to rejecting all p-values less than or equal to $\hat{\gamma}$ where

$$\hat{\gamma} = \max\{\gamma : \widehat{\mathrm{pFDR}}(\gamma) \le \alpha\}. \tag{7.3}$$

Storey (cf. [57]) showed that the B-H linear step-up procedure can be viewed as Storey's procedure where π is estimated by 1. Therefore, it is clear that using the procedure (7.3) will improve the power substantially unless π is actually very close to 1.

Storey (cf. [58]) also showed that the pFDR can be given a Bayesian interpretation as the posterior probability of a null hypothesis being true given that it has been rejected. This interpretation connects the frequentist and the Bayesian paradigms in the multiple testing situation. Given that p-values are fundamental quantities that can be interpreted in both paradigms, this connection in the context of a procedure based on p-values is illuminating. Several multiple testing procedures have resulted by substituting different estimators of pFDR in (7.3). Most of these procedures rely on the expression (7.2) and substitute the empirical c.d.f. for $F(\alpha)$ in the denominator. These procedures mainly differ in the way they estimate π. However, since $\pi\alpha$ is less than or equal to $F(\alpha)$, there is always a risk of violating the inequality if one estimates $F(\alpha)$ and π independently. [60] suggested a nonparametric Bayesian approach that simultaneously estimates π and $F(\alpha)$ within a mixture model framework that naturally constrain the estimators to maintain the relationship. This results in a more efficient estimator of pFDR.

The case when the test statistics (equivalently, p-values) are dependent is of course of great practical interest. A procedure that controls the FDR under positive regression dependence was suggested in [9] where the B-H critical values are replaced by $\alpha_i = \frac{i\alpha}{m\sum_{j=1}^{i} j^{-1}}$. The procedure is very conservative because the critical values are significantly smaller than the B-H critical values. [50] suggested an alternative set of critical values and investigated the performance under some special dependence structures. [21] and [17] suggested modeling the probit transform of the p-values as joint normal distribution to capture dependence among the p-values. A similar procedure to model the joint behavior of the p-values was suggested by [44] who used a mixture of skew-normal densities to incorporate dependence among the p-values. This mixing distribution is then estimated using nonparametric Bayesian techniques described in Section 7.1.

Other error measure such as the local FDR ([17]) were introduced to suit modern large dimensional datasets. While the FDR depends on the tail probability of the marginal p-value distribution, $F(\alpha)$, the local FDR depends on the marginal p-value density. Other forms of generalization can be found in ([48], [49]) and the references therein. Almost all error measures are functionals of the marginal p-value distribution, while few have been analyzed under the possibility of dependence among the p-values. A model based approach that estimates the components of the marginal distribution of the p-values has the advantage that once accurate estimates of the components of the marginal distribution are obtained, then it is possible to estimate several of these error measures and make a comparative study. Bayesian methodologies in multiple testing were discussed in [13], [60], [27] and [44]. [26] used a weighted p-value scheme that incorporates prior information about the hypothesis in the FDR controlling procedure. Empirical Bayes estimation of FDR was discussed in [15].

A particularly attractive feature of the Bayesian approach in the multiple testing situation is its ability to attach a posterior probability to an individual null hypothesis being actually false. In particular, it is easy to predict the false discovery proportion (FDP), V/R. Let $I_i(\alpha) = 1\!\!1\{X_i < \alpha\}$ denote that the ith hypothesis is rejected at a threshold level α and let H_i be the indicator that the ith alternative hypothesis is true. The FDP process evaluated at a threshold α (cf. [25]) is defined by

$$\mathrm{FDP}(\alpha) = \frac{\sum_{i=1}^{m} I_i(\alpha)(1 - H_i)}{\sum_{i=1}^{m} I_i(\alpha) + \prod_{i=1}^{m}(1 - I_i(\alpha))}.$$

Assuming that $(H_i, I_i(\alpha))$, $i = 1, \ldots, m$, are exchangeable, [44] showed that $\mathrm{FDR}(\alpha) = \pi b(\alpha)\mathrm{P}(\text{at least one rejection})$, where $b(\alpha)$ is the expected value of a function of the indicator functions. This implies that $\mathrm{pFDR}(\alpha) = \pi b(\alpha)$, which reduces to the old expression under independence. A similar expression was derived in [9] and also in [47]. In particular, [47] showed that the quantity $b(\alpha)/\alpha$ is the expectation of a jackknife estimator of $\mathrm{E}[(1 + R)^{-1}]$.

Thus the simple formula for pFDR as $\pi\alpha/F(\alpha)$ does not hold if the p-values are dependent, but the FDP with better conditional properties, seems to be more relevant to a Bayesian. Estimating the pFDR will generally involve computing high dimensional integrals, and hence will be difficult to obtain in reasonable time, but predicting the FDP is considerably simpler. Since the Bayesian methods are able to generate from the joint conditional distribution of $(H_1, \ldots, H_m)$ given data, we can predict the FDP by calculating its conditional expectation given data.

The theoretical model for the null distribution of the p-values is U[0, 1]. The theoretical null model may not be appropriate for the observed p-values in many real-life applications due to composite null hypothesis, complicated test statistic or dependence among the datasets used to test the multiple hypothesis. For a single hypothesis, the uniform null model may be approximately valid even for very complex hypothesis testing situations with composite null and complicated test statistics; see [4]. However, as argued by [16] and [6], if the multiple hypotheses tests are dependent then the m_0 null p-values collectively can behave very differently from a collection of independent uniform random variables. For example, the histogram of the probit transformed null p-values may be significantly skinnier than the standard normal, the theoretical null distribution of the probit p-values. [16] showed that a small difference between the theoretical null and an empirical null can have a significant impact on the conclusions of an error control procedure. Fortunately, large scale multiple testing situations provide one with the opportunity to empirically estimate the null distribution using a mixture model framework. Thus, validity of the theoretical null assumption can be tested from the data and if the observed values show significant departure from the assumed model, then the error control procedure may be built based on the empirical null distribution.

7.3. Bayesian Mixture Models for p-Values

As discussed in the previous section, p-values play an extremely important role in controlling the error in a multiple hypothesis testing problem. Therefore, it is a prudent strategy to base our Bayesian approach considering p-values as fundamental objects rather than as a product of some classical testing procedure. Consider the estimation approach of Storey ([57, 58]) discussed in the previous section. Here the false indicator H_i of the ith null hypothesis, is assumed to arise through a random mechanism, being distributed as independent Bernoulli variables with success probability $1-\pi$. Under this scenario, even though the original problem of multiple testing belongs to the frequentist paradigm, the probabilities that one would like to estimate are naturally interpretable in a Bayesian framework. In particular, the pFDR function can be written in the form of a posterior probability. There are other advantages of the Bayesian approach too. Storey's estimation method of π is based on the implicit assumption that the the density of p-values h under the alternative is concentrated near zero, and hence almost every p-value over the chosen threshold λ must arise from null hypotheses. Strictly speaking, this is incorrect because p-values bigger than λ can occur under alternatives as well. This bias can be addressed through elaborate modeling of the p-value density. Further, it is unnatural to assume that the value of the alternative distribution remains fixed when the hypotheses themselves are appearing randomly. It is more natural to assume that, given that the alternative is true, the value of the parameter under study is chosen randomly according to some distribution. This additional level of hierarchy is easily absorbed in the mixture model for the density of p-values proposed below.

7.3.1. *Independent case: Beta mixture model for p-values*

In this subsection, we assume that the test statistics, and hence the p-values, arising from different hypotheses are independent. Then the p-values $X_1, \dots, X_m$ may be viewed as i.i.d. samples from the two component mixture model: $f(x) = \pi g(x) + (1-\pi)h(x)$, where g stands for the density of p-values under the null hypothesis and h that under the alternative. The distribution of X_i under the corresponding null hypothesis $H_{0,i}$ may be assumed to be uniformly distributed on $[0,1]$, at least approximately. This happens under a number of scenarios:

(i) the test statistic is a continuous random variable and the null hypothesis is simple;
(ii) in situations like t-test or F-test, where the null hypothesis has been reduced to a simple one by considerations of similarity or invariance;
(iii) if a conditional predictive p-value or a partial predictive p-value ([4], [41]) is used.

Thus, unless explicitly stated, hereafter we assume that g is the uniform density. It is possible that this assumption fails to hold, which will be evident from the departure of the empirical null distribution from the theoretical null. However, even when this assumption fails to hold, generally the actual g is stochastically larger than the uniform. Therefore it can be argued that the error control procedures that assume the uniform density remain valid in the conservative sense. Alternatively, this difference can be incorporated in the mixture model by allowing the components of the mixture distribution that are stochastically larger than the uniform distribution to constitute the actual null distribution.

The density of p-values under alternatives is not only concentrated near zero, but usually has more features. In most multiple testing problems, individual tests are usually simple one-sided or two-sided z-test, χ^2-test, or more generally, tests for parameters in a monotone likelihood ratio (MLR) family. When the test is one-sided and the test statistic has the MLR property, it is easy to see that the density of p-values is decreasing (Proposition 1 of [27]). For two-sided alternatives, the null distribution of the test statistic is often symmetric, and in that case, a two-sided analog of the MLR property implies that the p-value density is decreasing (Proposition 2 of [27]). The p-value density for a one-sided hypothesis generally decays to zero as x tends to 1. For a two-sided hypothesis, the minimum value of the p-value density will be a (small) positive number. For instance, for the two-sided normal location model, the minimum value is $e^{-n\theta^2/2}$, where n is the sample size on which the test is based on. In either case, the p-value density looks like a reflected "J", a shape exhibited by a beta density with parameters $a < 1$ and $b \geq 1$. In fact, if we are testing for the scale parameter of the exponential distribution, it is easy to see that the p-value density is exactly beta with $a < 1$ and $b = 1$. In general, several distributions on $[0, 1]$ can be well approximated by mixtures of beta distributions (see [14], [40]). Thus it is reasonable to approximate the p-value density under the alternative by an arbitrary mixture of beta densities with parameters $a < 1$ and $b \geq 1$, that is, $h(x) = \int \mathrm{be}(x|a, b)dG(a, b)$, where $\mathrm{be}(x; a, b) = x^{a-1}(1-x)^{b-1}/B(a, b)$ is the beta density with parameters a and b, and $B(a, b) = \Gamma(a)\Gamma(b)/\Gamma(a+b)$ is the beta function. The mixing distribution can be regarded as a completely arbitrary distribution subject to the only restriction that G is concentrated in $(0, 1) \times [1, \infty)$. [60] took this approach and considered a Dirichlet process prior on the mixing distribution G. Note that, if the alternative values arise randomly from a population distribution and individual p-value densities conditional on the alternative are well approximated by mixtures of beta densities, then the beta mixture model continues to approximate the overall p-value density. Thus, the mixture model approach covers much wider models and has a distinct advantage over other methods proposed in the literature. The resulting posterior can be computed by an appropriate MCMC method, as described below. The resulting Bayesian estimator, because of shrinkage properties, offers a reduction in the mean squared error and is generally more stable than its empirical counterpart considered by Storey ([57, 58]). [60] ran extensive simulation to demonstrate the advantages of the Bayesian estimator.

The DPM model is equivalent to the following hierarchical model, where associated with each X_i there is a latent variable $\theta_i = (a_i, b_i)$,

$$X_i|\theta_i \sim \pi + (1 - \pi)\,\mathrm{be}(x_i|\theta_i), \quad \theta_1, \ldots, \theta_m|G \overset{iid}{\sim} G \quad \text{and} \quad G \sim \mathrm{DP}(M, G_0).$$

The random measure G can be integrated out from the prior distribution to work with only finitely many latent variables $\theta_1, \ldots, \theta_m$.

In application to beta mixtures, it is not possible to choose G_0 to be conjugate with the beta likelihood. Therefore it is not possible to obtain closed-form expressions for the weights and the baseline posterior distribution in the generalized Polya urn scheme for sampling from the posterior distribution of $(\theta_1, \ldots, \theta_n)$. To overcome this difficulty, the no-gaps algorithm ([34]) may be used, which can bypass the problems of evaluating the weights and sampling from the baseline posterior. For other alternative MCMC schemes, consult [36].

[60] gave detailed description of how the no-gaps algorithm can be implemented to generate samples from the posterior of $(\theta_1, \ldots, \theta_m, \pi)$. Once MCMC sample values of $(\theta_1, \ldots, \theta_m, \pi)$ are obtained, the posterior mean is approximately given by the mean of the sample π-values. Since the pFDR functional is not linear in (G, π), evaluation of the posterior mean of pFDR(α) requires generating posterior samples of the infinite dimensional parameter h using Sethuraman's representation of G. This is not only cumbersome, but also requires truncating the infinite series to finitely many terms and controlling the error resulting from the truncation. We avoid this path by observing that, when m is large (which is typical in multiple testing applications), the "posterior distribution" of G given $\theta_1, \ldots, \theta_m$ is essentially concentrated at the "posterior mean" of G given $\theta_1, \ldots, \theta_m$, which is given by $\mathrm{E}(G|\theta_1, \ldots, \theta_m) = (M+m)^{-1}MG_0 + (M+m)^{-1}\sum_{i=1}^m \delta_{\theta_i}$, where $\delta_\theta(x) = \mathbb{1}\{\theta \leq x\}$ now stands for the c.d.f. of the distribution degenerate at θ. Thus the approximate posterior mean of pFDR(α) can be obtained by the averaging the values of $\pi\alpha/[(M+m)^{-1}MG_0(\alpha) + (M+m)^{-1}\sum_{i=1}^m \delta_{\theta_i}(\alpha)]$ realized in the MCMC samples. In the simulations of [60], it turned out that the sensitivity of the posterior to prior parameters is minimal.

In spite of the success of the no gaps algorithm in computing the Bayes estimators of π and pFDR(α), the computing time is exorbitantly high in large scale applications. In many applications, real-time computing giving instantaneous results is essential. Newton's algorithm ([38], [39], [37]) is a computationally fast way of solving general deconvolution problems in mixture models, but it can also be used to compute density estimates.

For a general kernel mixture, Newton's algorithm may be described as follows: Assume that $Y_1, \ldots, Y_m \overset{iid}{\sim} h(y) = \int k(y;\theta)\psi(\theta)d\nu(\theta)$, where the mixture density $\psi(\theta)$ with respect to the dominating measure $\nu(\theta)$ is to be estimated. Start with an initial estimate $\psi_0(\theta)$, such as the prior mean, of $\psi(\theta)$. Fix weights $1 \geq w_1 \geq w_2 \geq \cdots w_m > 0$ such as $w_i = i^{-1}$. Recursively compute

$$\psi_i(\theta) = (1 - w_i)\psi_{i-1}(\theta) + w_i \frac{k(Y_i;\theta)\psi_{i-1}(\theta)}{\int k(Y_i;t)\psi_{i-1}(t)d\nu(t)}, \quad i = 2, \dots, m,$$

and declare $\psi_m(\theta)$ as the final estimate $\hat{\psi}(\theta)$. The estimate is not a Bayes estimate (it depends on the ordering of the observations), but it closely mimics the Bayes estimate with respect to a DPM prior with kernel $k(x;\theta)$ and center measure with density $\psi_0(\theta)$. If $\sum_{i=1}^{\infty} w_i = \infty$ and $\sum_{i=1}^{\infty} w_i^2 < \infty$ then the mixing density is consistently estimated ([37], [28]).

In the multiple testing context, ν is the sum of point mass at 0 of size 1 and the Lebesgue measure on $(0,1)$. Then π is identified as $\psi(0)$ and $F(\alpha)$ as $\psi(0) + \int_{(0,\alpha]} \psi(\theta)d\theta$. Then a reasonable estimate is obtained by $\hat{\psi}(0)\alpha/[\hat{\psi}(0)\alpha + \int_{(0,\alpha]} \hat{\psi}(\theta)d\theta]$. The computation is extremely fast and the performance of the estimator is often comparable to that of the Bayes estimator.

Since π takes the most important role in the expression for the pFDR function, it is important to estimate π consistently. However, a conceptual problem arises because π is not uniquely identifiable from the mixture representation $F(x) = \pi x + (1-\pi)H(x)$, where $H(\cdot)$ is another c.d.f. on [0,1]. Note that the class of such distributions is weakly closed. The components π and H can be identified by imposing the additional condition that H cannot be represented as a mixture with another uniform component, which, for the case when H has a continuous density h, translates into $h(1) = 0$. Letting $\pi(F)$ be the largest possible value of π in the representation, it follows that $\pi(F)$ upper bounds the actual proportion of null hypothesis and hence the actual pFDR is bounded by $\overline{\text{pFDR}}(F;\alpha) := \pi(F)\alpha/F(\alpha)$. This serves the purpose from a conservative point of view. The functional $\pi(F)$ and the $\overline{\text{pFDR}}$ are upper semicontinuous with respect to the weak topology in the sense that if $F_n \to_w F$, then $\limsup_{n\to\infty} \pi(F_n) \leq \pi(F)$ and $\limsup_{n\to\infty} \overline{\text{pFDR}}(F_n;\alpha) \leq \overline{\text{pFDR}}(F;\alpha)$.

Full identifiability of the components π and H in the mixture representation is possible under further restriction on F if $H(x)$ has a continuous density h with $h(1) = 0$ or the tail of H at 1 is bounded by $C(1-x)^{1+\epsilon}$ for some $C, \epsilon > 0$. The second option is particularly attractive since it also yields continuity of the map taking F to π under the weak topology. Thus posterior consistency of estimating F under the weak topology in this case will imply consistency of estimating π and the pFDR function, uniformly on compact subsets of $(0,1]$. The class of distributions satisfying the latter condition will be called $\mathcal{B}$ and $\mathcal{D}$ will stand for the class of continuous decreasing densities on $(0, 1]$.

Consider a prior Π for H supported in $\mathcal{B} \cap \mathcal{D}$ and independently a prior μ for π with full support on $[0, 1]$. Let the true value of π and h be respectively π_0 and h_0 where $0 < \pi_0 < 1$ and $H_0 \in \mathcal{B} \cap \mathcal{D}$. In order to show posterior consistency under the weak topology, we apply Schwartz's result [55]. Clearly we need the true p-value density to be in the support of the beta mixture prior. A density h happens to be a pointwise mixture of be(a, b) with $a < 1$ and $b \geq 1$ if $H(e^{-y})$ or $1 - H(1 - e^{-y})$ is completely monotone, that is, has all derivatives which are negative for odd orders and positive for even orders. Since pointwise approximation is stronger than L_1-approximation by Scheffe's theorem, densities pointwise approximated by beta densities are in the L_1-support of the prior in the sense that $\Pi(h : \|h - h_0\|_1 < \epsilon) > 0$ for all $\epsilon > 0$. Because both the true and the random mixture densities contain a uniform component, both densities are bounded below. Then a relatively simple analysis shows that the Kullback–Leibler divergence is essentially bounded by the L_1-distance up to a logarithmic term, and hence $f_0 = \pi_0 + (1 - \pi_0)h_0$ is in the Kullback–Leibler support of the prior on $f = \pi + (1 - \pi)h$ induced by Π and μ. Thus by the consistency result discussed in Section 7.1 applies so that the posterior for F is consistent under the weak topology. Hence under the tail restriction on H described above, posterior consistency for π and pFDR follows. Even if the tail restriction does not hold, a one-sided form of consistency, which may be called "upper semi-consistency", holds: For any $\epsilon > 0$, $\Pr(\pi < \pi_0 + \epsilon | X_1, \ldots, X_m) \to 1$ a.s. and that the posterior mean $\hat{\pi}_m$ satisfies $\limsup_{m \to \infty} \hat{\pi}_m \leq \pi_0$ a.s.

Unfortunately, the latter has limited significance since typically one would not like to underestimate the true π_0 (and the pFDR) while overestimation is less serious. When the beta mixture prior is used on h with the center measure of the Dirichlet process G_0 supported in $(0, 1) \times (1 + \epsilon, \infty)$ and h_0 is in the L_1-support of the Dirichlet mixture prior, then full posterior consistency for estimating π and pFDR holds. Since the Kullback–Leibler property is preserved under mixtures by Fubini's theorem, the result continues to hold even if the precision parameter of the Dirichlet process is obtained from a prior and the center measure G_0 contains hyperparameters.

7.3.2. *Dependent case: Skew-normal mixture model for probit p-values*

Due to the lack of a suitable multivariate model for the joint distribution of the p-values, most applications assume that the data associated with the family of tests are independent. However, empirical evidence obtained in many important applications such as fMRI, proteomics (two-dimensional gel electrophoresis, mass-spectroscopy) and microarray analysis, shows that the data associated with the different tests for multiple hypotheses are more likely to be dependent. In an fMRI example, tests regarding the activation of different voxels are spatially correlated. In diffusion tensor imaging problems, the diffusion directions are correlated and generate dependent observations over a spatial grid. Hence, a grid-by-grid comparison of such images across patient groups will generate several p-values that are highly dependent.

The p-values, X_i, take values in the unit interval on which it is hard to formulate a flexible multivariate model. It is advantageous to transform X_i to a real-valued random variable Y_i, through a strictly increasing smooth mapping $\Psi : [0,1] \to \mathbb{R}$. A natural choice for Ψ is the probit link function, Φ^{-1}, the quantile function of the standard normal distribution. Let $Y_i = \Phi^{-1}(X_i)$ be referred to as the *probit p-values.* We shall build flexible nonparametric mixture models for the joint density of $(Y_1, \dots, Y_m)$.

The most obvious choice of a kernel is an m-variate normal density. Efron (cf. [17]) advocated in favor of this kernel. This can automatically include the null component, which is the standard normal density after the probit transformation of the uniform. However, the normal mixture has a shortcoming. As in the previous subsection, marginal density of a p-value is often decreasing. Thus the model on the probit p-values should conform to this restriction whenever it is desired so. The transformed version of a normal mixture is not decreasing for any choice of the mixing distribution unless all components have variance exactly equal to one. This prompts for a generalization of the normal kernel which still includes the standard normal as a special case but can reproduce the decreasing shape of the p-value density by choosing the mixing distribution appropriately. [44] suggested using the multivariate skew-normal kernel as a generalization of the normal kernel. The mixture of skew-normal distribution does provide decreasing p-value densities for a large subset of parameter configurations.

To understand the point, it is useful to look at the unidimensional case. Let

$$q(y; \mu, \omega, \lambda) = 2\phi(y; \mu, \omega^2)\Phi(\lambda\omega^{-1}y)$$

denote the skew-normal density (cf. [2]) with location parameter μ, scale parameter ω and shape parameter λ, where $\phi(y; \mu, \omega^2)$ denotes the $N(\mu, \omega^2)$ density and $\Phi(\cdot)$ denotes the standard normal c.d.f. The skew-normal family has got a lot of recent attention due to its ability to naturally generalize the normal family to incorporate skewness and form a much more flexible class. The skewness of the distribution is controlled by λ and when $\lambda = 0$, it reduces to the normal distribution. If Y has density $q(y; \mu, \omega, \lambda)$, then [44] showed that the density of $X = \Phi(Y)$ is decreasing in $0 < x < 1$ if and only if

$$\omega^2 \geq 1,\ \lambda > \sqrt{(\omega^2 - 1)/\omega^2} \text{ and } \mu < \lambda H^*(\beta_1(\omega^2, \lambda)),$$

where $\beta_1(\omega^2, \lambda) = (\omega^2 - 1)/(\lambda^2 \omega^2)$, $0 \leq \beta_1 \leq 1$, $H^*(\beta_1) = \inf_x [H(x) - \beta_1 x]$ and $H(\cdot)$ is the hazard function of the standard normal distribution. Now, since the class of decreasing densities forms a convex set, it follows that the decreasing nature of the density of the original p-value X will be preserved even when a mixture of skew-normal density $q(y; \mu, \omega, \lambda)$ is considered, provided that the mixing measure K is supported on

$$\{(\mu, \omega, \lambda) : \mu \leq m(\beta_1(\omega, \lambda)), \quad \omega \geq 1, \lambda \geq \sqrt{(\omega^2 - 1)/\omega^2}\}.$$

Location-shape mixtures of skew-normal family holding the scale fixed at $\omega = 1$ can be restricted to produce decreasing p-value densities if the location parameter is negative and shape parameter is positive. For scale-shape mixtures with the location parameter set to zero, the induced p-value densities are decreasing if the mixing measure has support on $\{(\omega, \lambda) : \omega \geq 1, \lambda \geq \sqrt{1 - \omega^{-2}}\}$. Location-scale mixtures with the shape parameter set to zero is the same as location-scale mixtures of normal family. It is clear from the characterization that the normal density is unable to keep the shape restriction. This is the primary reason why we do not work with normal mixtures.

By varying the location parameter μ and the scale parameter ω in the mixture, we can generate all possible densities. The skew-normal kernel automatically incorporates skewness even before taking mixtures, and hence it is expected to lead to a parsimonious mixture representation in presence of skewness, commonly found in the target density. Therefore we can treat the mixing measure K to be a distribution on μ and ω only and treat λ as a hyperparameter. The nonparametric nature of K can be maintained by putting a prior with large weak support, such as the Dirichlet process. A recent result of [61] shows that nonparametric Bayesian density estimation based on a skew-normal kernel is consistent under the weak topology, adding a strong justification for the use of this kernel. Interestingly, if the theoretical standard normal null distribution is way off from the empirical one, then one can incorporate this feature in the model by allowing K to assign weights to skew-normal components stochastically larger than the standard normal.

In the multidimensional case, [44] suggested replacing the univariate skew-normal kernel by a multivariate analog. [3] introduced the multivariate skew-normal density

$$SN_m(\boldsymbol{y}; \boldsymbol{\mu}, \boldsymbol{\Omega}, \boldsymbol{\alpha}) \equiv 2\phi_m(\boldsymbol{y}; \boldsymbol{\mu}, \boldsymbol{\Omega})\Phi(\boldsymbol{\alpha}^T \boldsymbol{\Omega}^{-1}(\boldsymbol{y} - \boldsymbol{\mu})),$$

where ϕ_m is the m-variate normal density. Somewhat more flexibility in separating skewness and correlation is possible with the version of [45].

[44] considered a scale-shape mixture under restriction to illustrate the capability of the skew-normal mixture model. Most commonly arising probit p-value densities can be well approximated by such mixtures. Analogous analysis is possible with mixtures of location, scale and shape. Consider an $m \times m$ correlation matrix $\boldsymbol{R}$ with possibly a very sparse structure. Let $\boldsymbol{\omega} = (\omega_1, \ldots, \omega_m)^T$, $\boldsymbol{\alpha} = (\alpha_1, \ldots, \alpha_m)^T$ and $\boldsymbol{\lambda} = (\lambda_1, \ldots, \lambda_m)^T$. Let H_i denote the indicators that the ith null hypothesis H_{i0} is false and let $\boldsymbol{H} = (H_1, \ldots, H_m)^T$. Then a multivariate mixture model for $\boldsymbol{Y} = (Y_1, \ldots, Y_m)^T$ is $(Y|\boldsymbol{\omega}, \boldsymbol{\lambda}, \boldsymbol{H}, \boldsymbol{R}) \sim SN_m(0; \boldsymbol{\Omega}, \boldsymbol{\alpha})$ where $\boldsymbol{\Omega} = \boldsymbol{\Delta}_\omega \boldsymbol{R} \boldsymbol{\Delta}_\omega$, $\boldsymbol{\Delta}_\omega = \mathrm{diag}(\boldsymbol{\omega})$ is the diagonal matrix of scale parameters and $\boldsymbol{\alpha} = \boldsymbol{R}^{-1}\boldsymbol{\lambda}$ is the vector of shape parameters. Let H_i be i.i.d. Bernoulli$(1 - \pi)$, and independently

$$(\omega_i, \lambda_i)|\boldsymbol{H} \sim \begin{cases} \delta_{1,0}, & \text{if } H_i = 0, \\ K_0, & \text{if } H_i = 1. \end{cases}$$

The skew-mixture model is particularly suitable for Bayesian estimation. [44] described an algorithm for obtaining posterior samples. Using a result from [3], one can represent $Y_i = \omega_i \delta_i |U| + \omega_i (1 - \delta_i^2) V_i$, where $\delta_i = \lambda_i/(1 + \lambda_i^2)$, U is standard normal and $\boldsymbol{V} = (V_1, \ldots, V_n)^T$ is distributed as n-variate normal with zero mean and dispersion matrix $\boldsymbol{R}$ independently of U. This representation naturally lends itself to an iterative MCMC scheme. The posterior sample for the parameters in $\boldsymbol{R}$ can be used to validate the assumption of independence. Also, using the posterior samples it is possible to predict the FDP.

It is not obvious how to formulate an analog of Newton's estimate for dependent observations, but we outline the sketch of a strategy below. If the joint density under the model can be factorized as

$Y_1|\theta \sim k_1(Y_1;\theta)$, $Y_2|(Y_1,\theta) \sim k_2(Y_2;Y_1,\theta)$, ..., $Y_m|(X_1,\ldots,Y_{m-1},\theta) \sim k_m(Y_m;Y_1,\ldots,Y_{m-1},\theta)$, then the most natural extension would be to use

$$\psi_i(\theta) = (1-w_i)\psi_{i-1}(\theta) + w_i \frac{k_i(Y_i;Y_1,\ldots,Y_{i-1}\theta)\psi_{i-1}(\theta)}{\int k_i(Y_i;Y_1,\ldots,Y_{i-1},t)\psi_{i-1}(t)d\nu(t)}. \quad (7.4)$$

Such factorizations are often available if the observations arise sequentially. On the other hand, if m is small and $(Y_i|Y_j, j \neq i)$ are simple, we may use the kernel $k_i(y_i|\theta, y_j, j \neq i)$. More generally, if the observations can be associated with a decomposable graphical model, we can proceed by fixing a perfect order of cliques and then reducing to the above two special cases through the decomposition.

7.4. Areas of Application

Multiple testing procedures have gained increasing popularity in statistical research in view of their wide applicability in biomedical applications. Microarray experiments epitomize the applicability of multiple testing procedures because in microarray we are faced with a severe multiplicity problem where the error rapidly accumulates as one tests for significance over thousands of gene locations. We illustrate this point using a dataset obtained from the National Center for Biotechnology Information (NCBI) database. The data comes from an analysis of isografted kidneys from brain dead donors. Brain death in donors triggers inflammatory events in recipients after kidney transplantation. Inbred male Lewis rats were used in the experiment as both donors and recipients, with the experimental group receiving kidneys from brain dead donors and the control group receiving kidneys from living donors. Gene expression profiles of isografts from brain dead donors and grafts from living donors were compared using a high-density oligonucleotide microarray that contained approximately 25,000 genes. [6] analyzed this dataset using a finite skew-mixture model where the mixing measure is supported on only a finite set of parameter values. Due to the high multiplicity of the experiment, even for a single step procedure with a very small α, the FDR can be quite large. [6] estimated that the pFDR for testing for the difference between brain dead donors and living donors at each of the 25,000 gene locations at a fixed level $\alpha = 0.0075$ is about 0.2. The mixture model framework also naturally provides estimates of effect size among the false null. While [6] looked at one sided t-test at each location to generate the p-values, they constructed the histogram of the 25,000 p-values generated from two-sided tests. The left panel of Figure 7.1 gives a default MATLAB kernel-smoother estimate of the observed p-value histogram. The density shows a general decreasing shape except for local variation and at the edges. The spikes at the edges are artifacts of the smoothing mechanism. The bumpy nature of the smoothed histogram motivates a mixture approach to modeling. The histogram in the probit scale is shown as the jagged line in the right panel in Figure 7.1. The smoothed curve is an estimate of the probit p-value density based on a skew-normal mixture model. Empirical investigation reveals the possibility of correlation among the gene locations. Thus, the multivariate skew-normal mixture would yield more realistic results by incorporating flexible dependence structure.

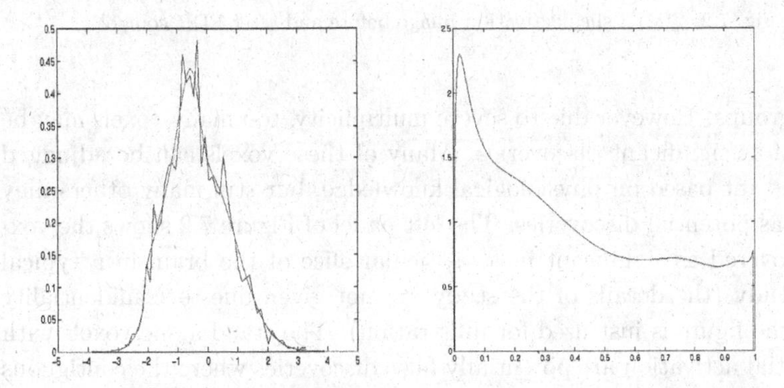

Fig. 7.1. Density of p-values obtained from the ratdata: original scale (left) and probit scale (right).

Another important application area of the FDR control procedure is fMRI. In fMRI data, one is interested in testing for brain activation in thousands of brain voxels simultaneously. In a typical experiment designed to determine the effect of covariate (say a drug or a disease status) on brain activation during a specific task (say eye movement), the available subjects will be divided into the treatment group (individual taking the drug or having a particular disease) and the control group (individuals taking a placebo or not having a disease) and their brain activation (blood oxygen level dependent signal) will be recorded at each voxel in a three dimensional grid in the brain. Then for each of the thousands of voxels, the responses for the individuals in both groups are recorded and then two sample tests are carried out voxel-by-voxel to determine the voxels with significant signal difference

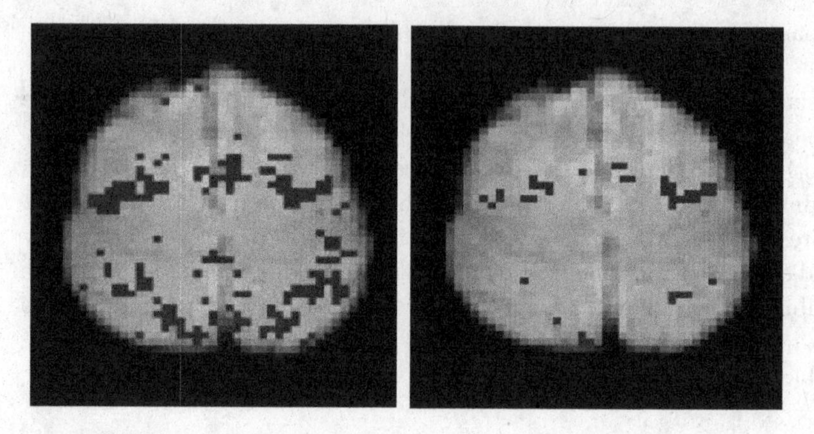

Fig. 7.2. fMRI slice activation image before and after FDR control.

across groups. However due to severe multiplicity, too many voxels may be declared as significant discoveries. Many of these voxels can be adjudged unimportant based on physiological knowledge, but still many others may remain as potential discoveries. The left panel of Figure 7.2 shows the voxels discovered as significant in a particular slice of the brain in a typical fMRI study (the details of the study are not given due to confidentiality issues, the figure is just used for illustration). The stand alone voxels with differential activation are potentially false discoveries where the contiguous clusters of voxels with significant activation pattern are potentially more meaningful findings. However, one needs to use statistical procedures to determine this as there will be tens of thousands of voxels and determining the validity of the findings manually is an infeasible task and a source of potential subjective bias. FDR control has been advocated by [23] to control for false discoveries in fMRI experiments. An application of the B-H procedure removes most of the voxels as false discoveries while keeping only a few with strong signal difference among the two groups. Thus the B-H procedure for this application turns out to be very conservative, and conflicts with scientific goal of finding anatomically rich activation patterns. An FDR control procedure that takes the dependence among voxels into account will be be more appropriate for this application. Work is underway to evaluate the merits of the dependent skew-mixture procedure in a typical fMRI dataset.

[6] also gave an illustration of the pitfalls of constraining the p-value model to have a theoretical null component. In their example, the null com-

ponents were made up of two components, one which is slightly stochastically smaller than the theoretical null and the other which is slightly bigger. With a single theoretical null distribution fitted to the data, both components were poorly estimated while the unconstrained fit with no prespecified theoretical null distribution gave an adequate approximation of both components.

Of course the applicability of multiple testing procedures is not restricted to biomedical problems. While the biomedical problems have been the primary motivation for developing false discovery control procedures, FDR control procedures are equally important in other fields, such as astronomy, where one may be interested in testing significance of findings of several celestial bodies simultaneously. There are important applications in reliability, meteorology and other disciplines as well.

References

[1] Antoniak, C. (1974). Mixtures of Dirichlet processes with application to Bayesian non-parametric problems. *Ann. Statist.* **2** 1152–1174.

[2] Azzalini, A. (1985). A class of distributions which includes the normal ones. *Scand. J. Statist.* **12** 171–178.

[3] Azzalini, A. and Dalla Valle, A. (1996). The multivariate skew-normal distribution. *Biometrika* **83** 715–726.

[4] Bayarri, M. J. and Berger, J. O. (2000). *p*-values for composite null models. *J. Amer. Statist. Asooc.* **95** 1127–1142.

[5] Bazan, J. L., Branco, M. D. and Bolfarine, H. (2006). A skew item response model. *Bayesian Analysis* **1** 861–892.

[6] Bean, G. J., DeRose, E. A., Mercer, L. D., Thayer, L. K. and Roy, A. (2008). Finite skew-mixture models for estimation of false discovery rates. Preprint.

[7] Benjamini, Y. and Hochberg, Y. (1995). Controlling the false discovery rate: A practical and powerful approach to multiple testing. *J. Roy. Statist. Soc. Ser. B* **57** 289–300.

[8] Benjamini, Y. and Hochberg, Y. (2000). On the adaptive control of the false discovery rate in multiple testing with independent statistics. *J. Behav. Educ. Statist.* **25** 60–83.

[9] Benjamini, Y. and Yekutieli, D. (2001). The control of the false discovery rate in multiple testing under dependency. *Ann. Statist.* **29** 1165–1188.

[10] Benjamini, Y. and Yekutieli, D. (2005). False discovery rate-adjusted multiple confidence intervals for selected parameters. *J. Amer. Statist. Assoc.* **100** 71–93.

[11] Benjamini, Y., Krieger, A. M. and Yekutieli, D. (2006). Adaptive linear step-up false discovery rate controlling procedures. *Biometrika* **93** 491–507.

[12] Black, M. A. (2004). A note on adaptive control of false discovery rates. *J. R. Statist. Soc. Ser. B* **66** 297–304.

[13] Chen, J. and Sarkar, S. K. (2004). Multiple testing of response rates with a control: A Bayesian stepwise approach. *J. Statist. Plann. Inf.* **125** 3–16.

[14] Diaconis, P. and Ylvisaker, D. (1985). Quantifying prior opinion. In *Bayesian Statistics* **2** (J. M. Bernardo, et al., eds.) North-Holland, Amsterdam, 133–156.

[15] Efron, B., Tibshirani, R., Storey, J. D. and Tusher, V. (2001). Empirical Bayes analysis of a microarray experiment. *J. Amer. Statist. Assoc.* **96** 1151–1160.

[16] Efron, B. (2004). Large-scale simultaneous hypothesis testing: the choice of a null hypothesis. *J. Amer. Statist. Assoc.* **99** 96–104.

[17] Efron, B. (2005). Local false discovery rates. Available at http://www-stat.stanford.edu/ brad/papers/False.pdf.

[18] Escobar, M. and West, M. (1995). Bayesian density estimation and inference using mixtures. *J. Amer. Statist. Assoc.* **90** 577–588.

[19] Ferguson, T. S. (1973). A Bayesian analysis of some nonparametric problems. *Ann. Statist.* **1** 209–230.

[20] Ferguson, T. S. (1983). Bayesian density estimation by mixtures of Normal distributions. In *Recent Advances in Statistics* (Rizvi M., Rustagi, J. and Siegmund, D., Eds.) 287–302.

[21] Finner, H. and Roters, M. (2002). Multiple hypothesis testing and expected number of type I errors. *Ann. Statist.* **30** 220–238.

[22] Gavrilov, Y., Benjamini, Y. and Sarkar, S. K. (2008). An adaptive step-down procedure with proven FDR control. *Ann. Statist.* (in press).

[23] Genovese, C. R., Lazar, N. A. and Nichols, T. E. (2002). Thresholding of statistical maps in functional neuroimaging using the false discovery rate. *NeuroImage* **15** 870–878.

[24] Genovese, C. and Wasserman, L. (2002). Operating characteristics and extensions of the false discovery rate procedure. *J. R. Stat. Soc. Ser. B* **64** 499–517.

[25] Genovese, C. R. and Wasserman, L. (2004). A stochastic process approach to false discovery rate. *Ann. Statist.* **32** 1035–1063.

[26] Genovese, C. R., Roeder, K. and Wasserman, L. (2006). False discovery control with p-value weighting. *Biometrika* **93** 509–524.

[27] Ghosal, S., Roy, A. and Tang, Y. (2008). Posterior consistency of Dirichlet mixtures of beta densities in estimating positive false discovery rates. In *Beyond Parametrics in Interdisciplinary Research: Festschrift in Honor of Professor Pranab K. Sen* (E. Pena *et al.*, eds.), Institute of Mathematical Statistics Collection **1** 105–115.

[28] Ghosh, J. K. and Tokdar, S. T. (2006). Convergence and consistency of Newton's algorithm for Estimating Mixing Distribution. *The Frontiers in Statistics* (J. Fan and H. Koul, eds.), Imperial College Press.

[29] Hochberg, Y. (1988). A sharper Bonferroni procedure for multiple tests of significance. *Biometrika* **75** 800–802.

[30] Holm, S. (1979). A simple sequentially rejective multiple test procedure. *Scand. J. Statist.* **6** 65–70.

[31] Hommel, G. (1988). A stagewise rejective multiple test procedure based on a modfied Bonferroni test. *Biometrika* **75** 383–386.

[32] Lehmann, E. L. and Romano, J. (2005). Generalizations of the familywise error rate. *Ann. Statist.* **33** 1138–1154.

[33] Lo, A. Y. (1984). On a class of Bayesian nonparametric estimates I: Density estimates. *Ann. Statist.* **12** 351–357.

[34] MacEachern, S. N. and Müller, P. (1998). Estimating mixture of Dirichlet process models. *J. Comput. Graph. Statist.* **7** 223–228.

[35] Miller, R. G. Jr. (1966). *Simultaneous Statistical Inference.* McGraw Hill, New York.

[36] Neal, R. M. (2000). Markov chain sampling methods for Dirichlet process mixture models. *J. Comput. Graph. Statist.* **9** 249–265.

[37] Newton, M. A. (2002). On a nonparametric recursive estimator of the mixing distribution. *Sankhya Ser. A* **64** 1–17.

[38] Newton, M. A., Quintana, F. A. and Zhang, Y. (1998). Nonparametric Bayes methods using predictive updating. In *Practical Nonparametric and Semiparametric Bayesian Statistics* (D. Dey *et al.*, eds.) 45–61. Springer-Verlag, New York.

[39] Newton, M. A. and Zhang, Y. (1999). A recursive algorithm for nonparametric analysis with missing data. *Biometrika* **86** 15–26.

[40] Parker, R. A. and Rothenberg, R. B. (1988). Identifying important results from multiple statistical tests. *Statistics in Medicine* **7** 1031–1043.

[41] Robins, J. M., van der Vaart, A. W. and Ventura, V. (2000). Asymptotic distribution of p-values in composite null models. *J. Amer. Statist. Asooc.* **95** 1143–1167.

[42] Rom, D. M. (1990). A sequentially rejective test procedure based on a modified Bonferroni inequality. *Biometrika* **77** 663–665.

[43] Romano, J. P. and Shaikh, A. M. (2006). Step-up procedures for control of generalizations of the familywise error rate. *Ann. Statist.* **34** (to appear).

[44] Roy, A. and Ghosal, S. (2008). Estimating false discovery rate under dependence; a mixture model approch. Preprint.

[45] Sahu, S. K., Dey, D. K. and Branco, M. D. (2003). A new class of multivariate skew distributions with applications to Bayesian regression models. *Canad. J. Statist.* **31** 129–150.

[46] Sarkar, S. K. (2002). Some results on false discovery rate in stepwise multiple testing procedures. *Ann. Statist.* **30** 239–257

[47] Sarkar, S. K. (2006). False discovery and false non-discovery rates in single-step multiple testing procedures. *Ann. Statist*, **34** 394–415.

[48] Sarkar, S. K. (2007). Step-up procedures controlling generalized FWER and generalized FDR. *Ann. Statist.* **35** 2405–2420.

[49] Sarkar, S. K. (2008). Generalizing Simes' test and Hochberg's step-up procedure. *Ann. Statist.* **36** 337–363.

[50] Sarkar, S. K. (2008).Two-stage step-up procedures controlling FDR. *J. Statist. Plann. Inf.* **138** 1072–1084.

[51] Sethuraman, J. (1994). A constructive definition of Dirichlet priors. *Statistica Sinica* **4** 639–650.

[52] Simes, R. J. (1986). An improved Bonferroni procedure for multiple tests of significance. *Biometrika* **73** 751–754.

[53] Saville, D. J. (1990). Multiple comparison procedures: the practical solution. *American Statistician* **44** 174–180.

[54] Soric, B. (1989). Statistical "discoveries" and effect size estimation. *J. Amer. Statist. Assoc.* **84** 608–610.

[55] Schwartz, L. (1965). On Bayes Procedures. *Z. Wahrsch. Verw. Gebiete*, **4** 10–26.

[56] Schweder, T. and Spjøtvoll, E. (1982). Plots of p-values to evaluate many test simultaneously. *Biometrika* **69** 493–502.

[57] Storey, J. D. (2002). A direct approach to false discovery rates. *J. Roy. Statist. Soc., Ser. B* **64** 479–498.

[58] Storey, J. D. (2003). The positive false discovery rate: A Bayesian interpretation and the q-value. *Ann. Statist.* **31** 2013–2035.

[59] Storey, J. D., Taylor, J. E. and Siegmund, D. (2004). Strong control, conservative point estimation and simultaneous conservative consistency of false discovery rates: A unified approach. *J. Roy. Statist. Soc. Ser. B* **66** 187–205.

[60] Tang, Y., Ghosal, S. and Roy, A. (2007). Nonparametric Bayesian estimation of positive false discovery rates. *Biometrics* **63** 1126–1134.

[61] Wu, Y. and Ghosal, S. (2008). Kullback–Leibler property of general kernel mixtures in Bayesian density estimation. *Electronic J. Statist.* **2** 298–331.

Chapter 8

Bayesian Inference on Finite Mixtures of Distributions

Kate Lee[1,*], Jean-Michel Marin[2], Kerrie Mengersen[1] and Christian Robert[3]

[1] *School of Mathematical Sciences, Queensland University of Technology, GPO Box 2434, Brisbane, QLD 4001, Australia*

[2] *Institut de Mathématiques et Modélisation, Université Montpellier 2, Case Courrier 51, place Eugène Bataillon, 34095 Montpellier cedex 5, France*

[3] *CEREMADE, Université Paris Dauphine, 75775 Paris cedex 16, France*

This survey covers state-of-the-art Bayesian techniques for the estimation of mixtures. It complements the earlier work [31] by studying new types of distributions, the multinomial, latent class and t distributions. It also exhibits closed form solutions for Bayesian inference in some discrete setups. Lastly, it sheds a new light on the computation of Bayes factors via the approximation of [8].

8.1. Introduction

Mixture models are fascinating objects in that, while based on elementary distributions, they offer a much wider range of modeling possibilities than their components. They also face both highly complex computational challenges and delicate inferential derivations. Many statistical advances have stemmed from their study, the most spectacular example being the EM algorithm. In this short review, we choose to focus solely on the Bayesian approach to those models (cf. [42]). [20] provides a book-long and in-depth coverage of the Bayesian processing of mixtures, to which we refer the reader whose interest is woken by this short review, while [29] give a broader perspective.

*Kate Lee is a PhD candidate at the Queensland University of Technology, Jean-Michel Marin is professor in Université Montpellier 2, Kerrie Mengersen is professor at the Queensland University of Technology, and Christian P. Robert is professor in Université Paris Dauphine and head of the Statistics Laboratory of CREST.

Without opening a new debate about the relevance of the Bayesian approach in general, we note that the Bayesian paradigm (cf. [41]) allows for probability statements to be made directly about the unknown parameters of a mixture model, and for prior or expert opinion to be included in the analysis. In addition, the latent structure that facilitates the description of a mixture model can be naturally aggregated with the unknown parameters (even though latent variables are *not* parameters) and a global posterior distribution can be used to draw inference about both aspects at once.

This survey thus aims to introduce the reader to the construction, prior modelling, estimation and evaluation of mixture distributions within a Bayesian paradigm. Focus is on both Bayesian inference and computational techniques, with light shed on the implementation of the most common samplers. We also show that exact inference (with no Monte Carlo approximation) is achievable in some particular settings and this leads to an interesting benchmark for testing computational methods.

In Section 8.2, we introduce mixture models, including the missing data structure that originally appeared as an essential component of a Bayesian analysis, along with the precise derivation of the exact posterior distribution in the case of a mixture of Multinomial distributions. Section 8.3 points out the fundamental difficulty in conducting Bayesian inference with such objects, along with a discussion about prior modelling. Section 8.4 describes the appropriate MCMC algorithms that can be used for the approximation to the posterior distribution on mixture parameters, followed by an extension of this analysis in Section 8.5 to the case in which the number of components is unknown and may be derived from approximations to Bayes factors, including the technique of [8] and the robustification of [2].

8.2. Finite Mixtures

8.2.1. *Definition*

A mixture of distributions is defined as a convex combination

$$\sum_{j=1}^{J} p_j f_j(x), \quad \sum_{j=1}^{J} p_j = 1, \quad p_j > 0, \quad J > 1,$$

of standard distributions f_j. The p_j's are called *weights* and are most often unknown. In most cases, the interest is in having the f_j's parameterised, each with an unknown parameter θ_j, leading to the generic parametric mixture model

$$\sum_{j=1}^{J} p_j f(x|\theta_j)\,. \tag{8.1}$$

The dominating measure for (8.1) is arbitrary and therefore the nature of the mixture observations widely varies. For instance, if the dominating measure is the counting measure on the simplex of $\mathbb{R}^m$

$$\mathcal{S}_{m,\ell} = \left\{ (x_1, \ldots, x_m); \sum_{i=1}^{m} x_i = \ell \right\},$$

the f_j's may be the product of ℓ independent Multinomial distributions, denoted "$\mathcal{M}_m(\ell; q_{j1}, ..., q_{jm}) = \otimes_{i=1}^{\ell} \mathcal{M}_m(1; q_{j1}, ..., q_{jm})$", with m modalities, and the resulting mixture

$$\sum_{j=1}^{J} p_j \mathcal{M}_m(\ell; q_{j1}, \ldots, q_{jm}) \tag{8.2}$$

is then a possible model for repeated observations taking place in $\mathcal{S}_{m,\ell}$. Practical occurrences of such models are repeated observations of *contingency tables.* In situations when contingency tables tend to vary more than expected, a mixture of Multinomial distributions should be more appropriate than a single Multinomial distribution and it may also contribute to separation of the observed tables in homogeneous classes. In the following, we note $q_{j\cdot} = (q_{j1}, \ldots, q_{jm})$.

Example 8.1. *For* $J = 2$, $m = 4$, $p_1 = p_2 = .5$, $q_{1\cdot} = (.2, .5, .2, .1)$, $q_{2\cdot} = (.3, .3, .1, .3)$ *and* $\ell = 20$, *we simulate* $n = 50$ *independent realisations from model (8.2). That corresponds to simulating some* 2×2 *contingency tables whose total sum is equal to* 20. *Figure 8.1 gives the histograms for the four entries of the contingency tables.* ◀

Another case where mixtures of Multinomial distributions occur is the *latent class model* where d discrete variables are observed on each of n individuals ([30]). The observations $(1 \le i \le n)$ are $\boldsymbol{x}_i = (x_{i1}, \ldots, x_{id})$, with x_{iv} taking values within the m_v modalities of the v-th variable. The distribution of $\boldsymbol{x}_i$ is then

$$\sum_{j=1}^{J} p_j \prod_{i=1}^{d} \mathcal{M}_{m_i}\left(1; q_1^{ij}, \ldots, q_{m_i}^{ij}\right),$$

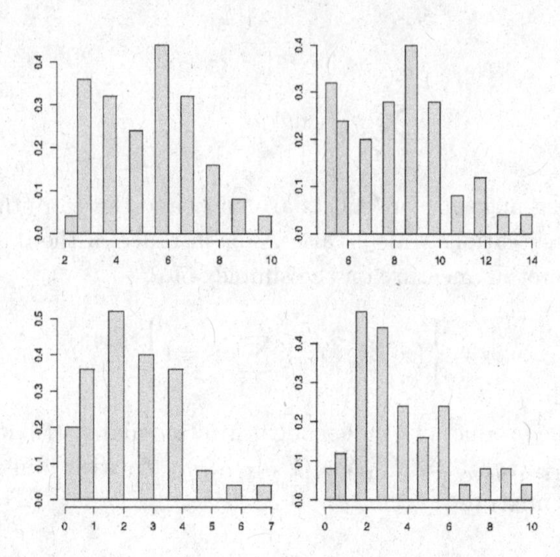

Fig. 8.1. For $J = 2$, $p_1 = p_2 = .5$, $q_{1\cdot} = (.2, .5, .2, .1)$, $q_{2\cdot} = (.3, .3, .1, .3)$, $\ell = 20$ and $n = 50$ independent simulations: histograms of the $m = 4$ entries.

so, strictly speaking, this is a mixture of *products* of Multinomials. The applications of this peculiar modelling are numerous: in medical studies, it can be used to associate several symptoms or pathologies; in genetics, it may indicate that the genes corresponding to the variables are not sufficient to explain the outcome under study and that an additional (unobserved) gene may be influential. Lastly, in marketing, variables may correspond to categories of products, modalities to brands, and components of the mixture to different consumer behaviours: identifying to which group a customer belongs may help in suggesting sales, as on Web-sale sites.

Similarly, if the dominating measure is the counting measure on the set of integers $\mathbb{N}$, the f_j's may be Poisson distributions $\mathcal{P}(\lambda_j)$ ($\lambda_j > 0$). We aim then to make inference about the parameters (p_j, λ_j) from a sequence $(x_i)_{i=1,\ldots,n}$ of integers.

The dominating measure may as well be the Lebesgue measure on $\mathbb{R}$, in which case the $f(x|\theta)$'s may all be normal distributions or Student's t distributions (or even a mix of both), with θ representing the unknown mean and variance, or the unknown mean and variance and degrees of freedom, respectively. Such a model is appropriate for datasets presenting multimodal or asymmetric features, like the aerosol dataset from [38] presented below.

Example 8.2. *The estimation of particle size distribution is important in understanding the aerosol dynamics that govern aerosol formation, which is of interest in environmental and health modelling. One of the most important physical properties of aerosol particles is their size; the concentration of aerosol particles in terms of their size is referred to as the* particle size distribution.

The data studied by [38] *and represented in Figure 8.2 is from Hyytiälä, a measurement station in Southern Finland. It corresponds to a full day of measurement, taken at ten minute intervals.* ◀

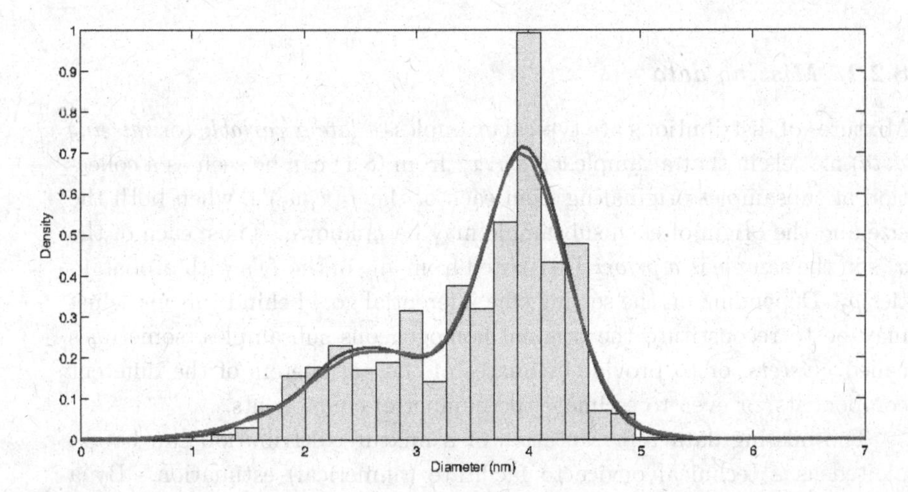

Fig. 8.2. Histogram of the aerosol diameter dataset, along with a normal (*red*) and a t (*blue*) modelling.

While the definition (8.1) of a mixture model is elementary, its simplicity does not extend to the derivation of either the maximum likelihood estimator (when it exists) or of Bayes estimators. In fact, if we take n iid observations $\mathbf{x} = (x_1, \ldots, x_n)$ from (8.1), with parameters

$$\mathbf{p} = (p_1 \ldots, p_J) \quad \text{and} \quad \boldsymbol{\theta} = (\theta_1, \ldots, \theta_J),$$

the full computation of the posterior distribution and in particular the explicit representation of the corresponding posterior expectation involves the expansion of the likelihood

$$\mathrm{L}(\boldsymbol{\theta}, \mathbf{p}|\mathbf{x}) = \prod_{i=1}^{n} \sum_{j=1}^{J} p_j f(x_i|\theta_j) \tag{8.3}$$

into a sum of J^n terms, with some exceptions (see, for example Section 8.3). This is thus computationally too expensive to be used for more than a few observations. This fundamental computational difficulty in dealing with the models (8.1) explains why those models have often been at the forefront for applying new technologies (such as MCMC algorithms, see Section 8.4).

8.2.2. *Missing data*

Mixtures of distributions are typical examples of *latent variable* (or *missing data*) models in that a sample $x_1, \ldots, x_n$ from (8.1) can be seen as a collection of subsamples originating from each of the $f(x_i|\theta_j)$'s, when both the size and the origin of each subsample may be unknown. Thus, each of the x_i's in the sample is *a priori* distributed from any of the f_j's with probabilities p_j. Depending on the setting, the inferential goal behind this modeling may be to reconstitute the original homogeneous subsamples, sometimes called *clusters*, or to provide estimates of the parameters of the different components, or even to estimate the number of components.

The missing data representation of a mixture distribution can be exploited as a technical device to facilitate (numerical) estimation. By a demarginalisation argument, it is always possible to associate to a random variable x_i from a mixture (8.1) a second (finite) random variable z_i such that

$$x_i|z_i = z \sim f(x|\theta_z), \quad \mathbb{P}(z_i = j) = p_j . \tag{8.4}$$

This auxiliary variable z_i identifies to which component the observation x_i belongs. Depending on the focus of inference, the z_i's may [or may not] be part of the quantities to be estimated. In any case, keeping in mind the availability of such variables helps into drawing inference about the "true" parameters. This is the technique behind the EM algorithm of [11] as well as the "data augmentation" algorithm of [51] that started MCMC algorithms.

8.2.3. *The necessary but costly expansion of the likelihood*

As noted above, the likelihood function (8.3) involves J^n terms when the n inner sums are expanded, that is, when all the possible values of the missing variables z_i are taken into account. While the likelihood at a given value $(\boldsymbol{\theta}, \mathbf{p})$ can be computed in $\mathrm{O}(nJ)$ operations, the computational difficulty in using the expanded version of (8.3) precludes analytic solutions via maximum likelihood or Bayesian inference. Considering n iid observations from model (8.1), if $\pi(\boldsymbol{\theta}, \mathbf{p})$ denotes the prior distribution on $(\boldsymbol{\theta}, \mathbf{p})$, the posterior distribution is naturally given by

$$\pi(\boldsymbol{\theta}, \mathbf{p}|\mathbf{x}) \propto \left(\prod_{i=1}^{n}\sum_{j=1}^{J} p_j f(x_i|\theta_j)\right) \pi(\boldsymbol{\theta}, \mathbf{p}).$$

It can therefore be computed in $\mathrm{O}(nJ)$ operations up to the normalising [marginal] constant, but, similar to the likelihood, it does not provide an intuitive distribution unless expanded.

Relying on the auxiliary variables $\mathbf{z} = (z_1, \ldots, z_n)$ defined in (8.4), we take $\mathcal{Z}$ to be the set of all J^n allocation vectors $\mathbf{z}$. For a given vector $(n_1, \ldots, n_J)$ of the simplex $\{n_1 + \ldots + n_J = n\}$, we define a subset of $\mathcal{Z}$,

$$\mathcal{Z}_j = \left\{\mathbf{z} : \sum_{i=1}^{n} \mathbb{I}_{z_i=1} = n_1, \ldots, \sum_{i=1}^{n} \mathbb{I}_{z_i=J} = n_J\right\};$$

that consists of all allocations $\mathbf{z}$ with the given allocation sizes $(n_1, \ldots, n_J)$, relabelled by $j \in \mathbb{N}$ when using for instance the lexicographical ordering on $(n_1, \ldots, n_J)$. The number of nonnegative integer solutions to the decomposition of n into J parts such that $n_1 + \ldots + n_J = n$ is equal to (see [17])

$$r = \binom{n+J-1}{n}.$$

Thus, we have the partition $\mathcal{Z} = \cup_{j=1}^{r} \mathcal{Z}_j$. Although the total number of elements of $\mathcal{Z}$ is the typically unmanageable J^n, the number of partition sets is much more manageable since it is of order $n^{J-1}/(J-1)!$. It is thus possible to envisage an exhaustive exploration of the $\mathcal{Z}_j$'s. ([6] did take advantage of this decomposition to propose a more efficient importance sampling approximation to the posterior distribution.)

The posterior distribution can then be written as

$$\pi(\boldsymbol{\theta}, \mathbf{p}|\mathbf{x}) = \sum_{i=1}^{r} \sum_{\mathbf{z} \in \mathcal{Z}_i} \omega(\mathbf{z})\, \pi(\boldsymbol{\theta}, \mathbf{p}|\mathbf{x}, \mathbf{z}), \tag{8.5}$$

where $\omega(\mathbf{z})$ represents the posterior probability of the given allocation $\mathbf{z}$. (See Section 8.2.4 for a derivation of $\omega(\mathbf{z})$.) Note that with this representation, a Bayes estimator of $(\boldsymbol{\theta}, \mathbf{p})$ can be written as

$$\sum_{i=1}^{r} \sum_{\mathbf{z} \in \mathcal{Z}_i} \omega(\mathbf{z}) \, \mathbb{E}^{\pi} \left[\boldsymbol{\theta}, \mathbf{p} | \mathbf{x}, \mathbf{z}\right] . \tag{8.6}$$

This decomposition makes a lot of sense from an inferential point of view: the Bayes posterior distribution simply considers each possible allocation $\mathbf{z}$ of the dataset, allocates a posterior probability $\omega(\mathbf{z})$ to this allocation, and then constructs a posterior distribution $\pi(\boldsymbol{\theta}, \mathbf{p}|\mathbf{x}, \mathbf{z})$ for the parameters conditional on this allocation. Unfortunately, the computational burden is of order $\mathrm{O}(J^n)$. This is even more frustrating when considering that the overwhelming majority of the posterior probabilities $\omega(\mathbf{z})$ will be close to zero for any sample.

8.2.4. *Exact posterior computation*

In a somewhat paradoxical twist, we now proceed to show that, in some very special cases, there exist exact derivations for the posterior distribution! This surprising phenomenon only takes place for discrete distributions under a particular choice of the component densities $f(x|\theta_i)$. In essence, the $f(x|\theta_i)$'s must belong to the natural exponential families, i.e.

$$f(x|\theta_i) = h(x) \exp\left\{\theta_i \cdot R(x) - \Psi(\theta_i)\right\} ,$$

to allow for sufficient statistics to be used. In this case, there exists a *conjugate prior* ([41]) associated with each θ in $f(x|\theta)$ as well as for the weights of the mixture. Let us consider the complete likelihood

$$\begin{aligned}
\mathrm{L}^c(\boldsymbol{\theta}, \mathbf{p}|\boldsymbol{x}, \boldsymbol{z}) &= \prod_{i=1}^{n} p_{z_i} \exp\left\{\theta_{z_i} \cdot R(x_i) - \Psi(\theta_{z_i})\right\} \\
&= \prod_{j=1}^{J} p_j^{n_j} \exp\left\{\theta_j \cdot \sum_{z_i=j} R(x_i) - n_j \Psi(\theta_j)\right\} \\
&= \prod_{j=1}^{J} p_j^{n_j} \exp\left\{\theta_j \cdot S_j - n_j \Psi(\theta_j)\right\} ,
\end{aligned}$$

where $S_j = \sum_{z_i=j} R(x_i)$. It is easily seen that we remain in an exponential family since there exist sufficient statistics with fixed dimension,

$(n_1, \ldots, n_J, S_1, \ldots, S_J)$. Using a Dirichlet prior

$$\pi(p_1, \ldots, p_J) = \frac{\Gamma(\alpha_1 + \ldots + \alpha_J)}{\Gamma(\alpha_1) \cdots \Gamma(\alpha_J)} p_1^{\alpha_1 - 1} \cdots p_J^{\alpha_J - 1}$$

on the vector of the weights $(p_1, \ldots, p_J)$ defined on the simplex of $\mathbb{R}^J$ and (independent) conjugate priors on the θ_j's,

$$\pi(\theta_j) \propto \exp\left\{\theta_j \cdot \tau_j - \delta_j \Psi(\theta_j)\right\},$$

the posterior associated with the complete likelihood $\mathrm{L}^c(\boldsymbol{\theta}, \mathbf{p}|\boldsymbol{x}, \boldsymbol{z})$ is then of the same family as the prior:

$$\begin{aligned}
\pi(\boldsymbol{\theta}, \mathbf{p}|\boldsymbol{x}, \boldsymbol{z}) &\propto \pi(\boldsymbol{\theta}, \mathbf{p}) \times \mathrm{L}^c(\boldsymbol{\theta}, \mathbf{p}|\boldsymbol{x}, \boldsymbol{z}) \\
&\propto \prod_{j=1}^{J} p_j^{\alpha_j - 1} \exp\left\{\theta_j \cdot \tau_j - \delta_j \Psi(\theta_j)\right\} \\
&\quad \times p_j^{n_j} \exp\left\{\theta_j \cdot S_j - n_j \Psi(\theta_j)\right\} \\
&= \prod_{j=1}^{J} p_j^{\alpha_j + n_j - 1} \exp\left\{\theta_j \cdot (\tau_j + S_j) - (\delta_j + n_j)\Psi(\theta_j)\right\};
\end{aligned}$$

the parameters of the prior get transformed from α_j to $\alpha_j + n_j$, from τ_j to $\tau_j + S_j$ and from δ_j to $\delta_j + n_j$.

If we now consider the observed likelihood (instead of the complete likelihood), it is the sum of the complete likelihoods over all possible configurations of the partition space of allocations, that is, a sum over J^n terms,

$$\sum_{\boldsymbol{z}} \prod_{j=1}^{J} p_j^{n_j} \exp\left\{\theta_j \cdot S_j - n_j \Psi(\theta_j)\right\}.$$

The associated posterior is then, up to a constant,

$$\begin{aligned}
&\sum_{\boldsymbol{z}} \prod_{j=1}^{J} p_j^{n_j + \alpha_j - 1} \exp\left\{\theta_j \cdot (\tau_j + S_j) - (n_j + \delta_j)\Psi(\theta_j)\right\} \\
&\quad = \sum_{\boldsymbol{z}} \omega(\boldsymbol{z})\, \pi(\boldsymbol{\theta}, \mathbf{p}|\boldsymbol{x}, \boldsymbol{z}),
\end{aligned}$$

where $\omega(\boldsymbol{z})$ is the normalising constant that is missing in

$$\prod_{j=1}^{J} p_j^{n_j + \alpha_j - 1} \exp\left\{\theta_j \cdot (\tau_j + S_j) - (n_j + \delta_j)\Psi(\theta_j)\right\}.$$

The weight $\omega(\mathbf{z})$ is therefore

$$\omega(z) \propto \frac{\prod_{j=1}^{J} \Gamma(n_j + \alpha_j)}{\Gamma(\sum_{j=1}^{J} \{n_j + \alpha_j\})} \times \prod_{j=1}^{J} K(\tau_j + S_j, n_j + \delta_j),$$

if $K(\tau, \delta)$ is the normalising constant of $\exp\{\theta_j \cdot \tau - \delta\Psi(\theta_j)\}$, i.e.

$$K(\tau, \delta) = \int \exp\{\theta_j \cdot \tau - \delta\Psi(\theta_j)\}\, \mathrm{d}\theta_j .$$

Unfortunately, except for very few cases, like Poisson and Multinomial mixtures, this sum does not simplify into a smaller number of terms because there exist no summary statistics. From a Bayesian point of view, the complexity of the model is therefore truly of magnitude $\mathrm{O}(J^n)$.

We process here the cases of both the Poisson and Multinomial mixtures, noting that the former case was previously exhibited by [16].

Example 8.3. *Consider the case of a two component Poisson mixture,*

$$x_1, \ldots, x_n \overset{iid}{\sim} p\,\mathcal{P}(\lambda_1) + (1-p)\,\mathcal{P}(\lambda_2),$$

with a uniform prior on p and exponential priors $\mathcal{E}xp(\tau_1)$ and $\mathcal{E}xp(\tau_2)$ on λ_1 and λ_2, respectively. For such a model, $S_j = \sum_{z_i=j} x_i$ and the normalising constant is then equal to

$$\begin{aligned} K(\tau, \delta) &= \int_{-\infty}^{\infty} \exp\{\lambda_j \tau - \delta \log(\lambda_j)\}\, d\lambda_j \\ &= \int_0^{\infty} \lambda_j^{\tau-1} \exp\{-\delta\lambda_j\}\, d\lambda_j = \delta^{-\tau}\,\Gamma(\tau). \end{aligned}$$

The corresponding posterior is (up to the overall normalisation of the weights)

$$\begin{aligned} &\sum_{z} \frac{\prod_{j=1}^{2} \Gamma(n_j+1)\Gamma(1+S_j)/(\tau_j+n_j)^{S_j+1}}{\Gamma(2+\sum_{j=1}^{2} n_j)}\, \pi(\boldsymbol{\theta}, \mathbf{p}|\boldsymbol{x}, \boldsymbol{z}) \\ &= \sum_{z} \frac{\prod_{j=1}^{2} n_j!\, S_j!/(\tau_j+n_j)^{S_j+1}}{(n+1)!}\, \pi(\boldsymbol{\theta}, \mathbf{p}|\boldsymbol{x}, \boldsymbol{z}) \\ &\propto \sum_{z} \prod_{j=1}^{2} \frac{n_j!\, S_j!}{(\tau_j+n_j)^{S_j+1}}\, \pi(\boldsymbol{\theta}, \mathbf{p}|\boldsymbol{x}, \boldsymbol{z}). \end{aligned}$$

$\pi(\boldsymbol{\lambda}, \mathbf{p}|\boldsymbol{x}, \boldsymbol{z})$ corresponds to a $\mathcal{B}(1+n_j, 1+n-n_j)$ (Beta distribution) on p_j and to a $\mathcal{G}(S_j+1, \tau_j+n_j)$ (Gamma distribution) on λ_j, $(j=1,2)$. An important feature of this example is that the above sum does not involve all of the 2^n terms, simply because the individual terms factorise in (n_1, n_2, S_1, S_2) that act like local sufficient statistics. Since $n_2 = n - n_1$ and $S_2 = \sum x_i - S_1$, the posterior only requires as many distinct terms as there are distinct values of the pair (n_1, S_1) in the completed sample. For instance, if the sample is $(0,0,0,1,2,2,4)$, the distinct values of the pair (n_1, S_1) are $(0,0), (1,0), (1,1), (1,2), (1,4), (2,0), (2,1), (2,2), (2,3), (2,4), (2,5), (2,6), \ldots, (6,5), (6,7), (6,8), (7,9)$. Hence there are 41 distinct terms in the posterior, rather than $2^8 = 256$. ◀

Let $\mathbf{n} = (n_1, \ldots, n_J)$ and $\mathbf{S} = (S_1, \ldots, S_J)$. The problem of computing the number (or cardinal) $\mu_n(\mathbf{n}, \mathbf{S})$ of terms in the sum with an identical statistic $(\mathbf{n}, \mathbf{S})$ has been tackled by [16], who proposes a recurrent formula to compute $\mu_n(\mathbf{n}, \mathbf{S})$ in an efficient book-keeping technique, as expressed below for a k component mixture:

If $\mathbf{e}_j$ denotes the vector of length J made of zeros everywhere except at component j where it is equal to one, if

$$\mathbf{n} = (n_1, \ldots, n_J), \quad \text{and} \quad \mathbf{n} - \mathbf{e}_j = (n_1, \ldots, n_j - 1, \ldots, n_J),$$

then

$$\mu_1(\mathbf{e}_j, R(x_1)\,\mathbf{e}_j) = 1, \quad \forall j \in \{1, \ldots, J\}, \quad \text{and}$$
$$\mu_n(\mathbf{n}, \mathbf{S}) = \sum_{j=1}^{J} \mu_{n-1}(\mathbf{n} - \mathbf{e}_j, \mathbf{S} - R(x_n)\,\mathbf{e}_j).$$

Example 8.4. *Once the $\mu_n(\mathbf{n}, \mathbf{S})$'s are all recursively computed, the posterior can be written as*

$$\sum_{(\mathbf{n},\mathbf{S})} \mu_n(\mathbf{n}, \mathbf{S}) \prod_{j=1}^{2} n_j!\, S_j! / (\tau_j + n_j)^{S_j+1}\, \pi(\boldsymbol{\theta}, \mathbf{p}|\boldsymbol{x}, \mathbf{n}, \mathbf{S}),$$

up to a constant, and the sum only depends on the possible values of the "sufficient" statistic $(\mathbf{n}, \mathbf{S})$.

This closed form expression allows for a straightforward representation of the marginals. For instance, up to a constant, the marginal in λ_1 is given *by*

$$\sum_{z} \prod_{j=1}^{2} n_j!\, S_j!/(\tau_j + n_j)^{S_j+1}\, (n_1+1)^{S_1+1} \lambda^{S_1} \exp\{-(n_1+1)\lambda_1\}/n_1!$$

$$= \sum_{(\mathbf{n},\mathbf{S})} \mu_n(\mathbf{n},\mathbf{S}) \prod_{j=1}^{2} n_j!\, S_j!/(\tau_j + n_j)^{S_j+1}$$

$$\times\, (n_1+\tau_1)^{S_1+1} \lambda_1^{S_1} \exp\{-(n_1+\tau_1)\lambda_1\}/n_1!\,.$$

The marginal in λ_2 *is*

$$\sum_{(\mathbf{n},\mathbf{S})} \mu_n(\mathbf{n},\mathbf{S}) \prod_{j=1}^{2} n_j!\, S_j!/(\tau_j + n_j)^{S_j+1}$$

$$(n_2+\tau_2)^{S_2+1} \lambda_2^{S_2} \exp\{-(n_2+\tau_2)\lambda_2\}/n_2!\,,$$

again up to a constant.

Another interesting outcome of this closed form representation is that marginal densities can also be computed in closed form. The marginal distribution of $\boldsymbol{x}$ *is directly related to the unnormalised weights in that*

$$m(\mathbf{x}) = \sum_{\mathbf{z}} \omega(\mathbf{z}) = \sum_{(\mathbf{n},\mathbf{S})} \mu_n(\mathbf{n},\mathbf{S}) \frac{\prod_{j=1}^{2} n_j!\, S_j!/(\tau_j+n_j)^{S_j+1}}{(n+1)!}$$

up to the product of factorials $1/x_1! \cdots x_n!$ *(but this product is irrelevant in the computation of the Bayes factor).* ◀

Now, even with this considerable reduction in the complexity of the posterior distribution, the number of terms in the posterior still explodes fast both with n and with the number of components J, as shown through a few simulated examples in Table 8.1. The computational pressure also increases with the range of the data, that is, for a given value of (J, n), the number of values of the sufficient statistics is much larger when the observations are larger, as shown for instance in the first three rows of Table 8.1: a simulated Poisson $\mathcal{P}(\lambda)$ sample of size 10 is mostly made of 0's when $\lambda = .1$ but mostly takes different values when $\lambda = 10$. The impact on the number of sufficient statistics can be easily assessed when $J = 4$. (Note that the simulated dataset corresponding to $(n, \lambda) = (10, .1)$ in Table 8.1 happens to correspond to a simulated sample made only of 0's, which explains the $n + 1 = 11$ values of the sufficient statistic $(n_1, S_1) = (n_1, 0)$ when $J = 2$.)

Table 8.1. Number of pairs $(\mathbf{n}, \mathbf{S})$ for simulated datasets from a Poisson $\mathcal{P}(\lambda)$ and different numbers of components. (*Missing terms are due to excessive computational or storage requirements.*)

(n, λ)	$J = 2$	$J = 3$	$J = 4$
(10, .1)	11	66	286
(10, 1)	52	885	8160
(10, 10)	166	7077	120,908
(20, .1)	57	231	1771
(20, 1)	260	20,607	566,512
(20, 10)	565	100,713	—
(30, .1)	87	4060	81,000
(30, 1)	520	82,758	—
(30, 10)	1413	637,020	—

Example 8.5. *If we have n observations $\boldsymbol{n}_i = (n_{i1}, \ldots, n_{im})$ from the Multinomial mixture*

$$\boldsymbol{n}_i \sim p\mathcal{M}_m(d_i; q_{11}, \ldots, q_{1m}) + (1-p)\mathcal{M}_m(d_i; q_{21}, \ldots, q_{2m})$$

where $n_{i1} + \cdots + n_{im} = d_i$ and $q_{11} + \cdots + q_{1m} = q_{21} + \cdots + q_{2m} = 1$, the conjugate priors on the q_{jv}'s are Dirichlet distributions, $(j = 1, 2)$

$$(q_{j1}, \ldots, q_{jm}) \sim \mathcal{D}(\alpha_{j1}, \ldots, \alpha_{jm}),$$

and we use once again the uniform prior on p. (A default choice for the α_{jv}'s is $\alpha_{jv} = 1/2$.) Note that the d_j's may differ from observation to observation, since they are irrelevant for the posterior distribution: given a partition $\boldsymbol{z}$ of the sample, the complete posterior is indeed

$$p^{n_1}(1-p)^{n_2} \prod_{j=1}^{2} \prod_{z_i=j} q_{j1}^{n_{i1}} \cdots q_{jm}^{n_{im}} \times \prod_{j=1}^{2} \prod_{v=1}^{m} q_{jv}^{-1/2},$$

(where n_j is the number of observations allocated to component j) up to a normalising constant that does not depend on $\boldsymbol{z}$. ◄

More generally, considering a Multinomial mixture with m components,

$$\boldsymbol{n}_i \sim \sum_{j=1}^{J} p_j \mathcal{M}_m(d_i; q_{j1}, \ldots, q_{jm}),$$

the complete posterior is also directly available, as

$$\prod_{j=1}^{J} p_j^{n_j} \times \prod_{j=1}^{J} \prod_{z_i=j} q_{j1}^{n_{i1}} \cdots q_{jm}^{n_{im}} \times \prod_{j=1}^{J} \prod_{v=1}^{m} q_{jv}^{-1/2},$$

once more up to a normalising constant.

Since the corresponding normalising constant of the Dirichlet distribution is

$$\frac{\prod_{v=1}^{m}\Gamma(\alpha_{jv})}{\Gamma(\alpha_{j1}+\cdots+\alpha_{jm})},$$

the overall weight of a given partition $\boldsymbol{z}$ is

$$n_1!n_2!\frac{\prod_{v=1}^{m}\Gamma(\alpha_{1v}+S_{1v})}{\Gamma(\alpha_{11}+\cdots+\alpha_{1m}+S_{1\cdot})}\times\frac{\prod_{v=1}^{m}\Gamma(\alpha_{2v}+S_{2v})}{\Gamma(\alpha_{21}+\cdots+\alpha_{2m}+S_{2\cdot})} \tag{8.7}$$

where S_{ji} is the sum of the n_{ji}'s for the observations i allocated to component j and

$$S_{ji}=\sum_{z_i=j}n_{ji} \quad\text{and}\quad S_{j\cdot}=\sum_i S_{ji}\,.$$

Given that the posterior distribution only depends on those "sufficient" statistics S_{ij} and n_i, the same factorisation as in the Poisson case applies, namely we simply need to count the number of occurrences of a particular local sufficient statistic $(n_1, S_{11}, \ldots, S_{Jm})$ and then sum over all values of this sufficient statistic. The book-keeping algorithm of [16] applies. Note however that the number of different terms in the closed form expression is growing extremely fast with the number of observations, with the number of components and with the number k of modalities.

Example 8.6. *In the case of the latent class model, consider the simplest case of two variables with two modalities each, so observations are products of Bernoulli's,*

$$\boldsymbol{x}\sim p\mathcal{B}(q_{11})\mathcal{B}(q_{12})+(1-p)\mathcal{B}(q_{21})\mathcal{B}(q_{22})\,.$$

We note that the corresponding statistical model is not identifiable beyond the usual label switching issue detailed in Section 8.3.1. Indeed, there are only two dichotomous variables, four possible realizations for the $\boldsymbol{x}$*'s, and five unknown parameters. We however take advantage of this artificial model to highlight the implementation of the above exact algorithm, which can then easily uncover the unidentifiability features of the posterior distribution.*

The complete posterior distribution is the sum over all partitions of the terms

$$p^{n_1}(1-p)^{n_2}\prod_{j=1}^{2}\prod_{v=1}^{2}q_{jv}^{s_{jv}}(1-q_{jv})^{n_j-s_{jv}}\times\prod_{j=1}^{2}\prod_{v=1}^{2}q_{jv}^{-1/2}$$

where $s_{jv} = \sum_{z_i=j} x_{iv}$, *the sufficient statistic is thus* $(n_1, s_{11}, s_{12}, s_{21}, s_{22})$, *of order* $O(n^5)$. *Using the benchmark data of* [50], *made of 216 sample points involving four binary variables related with a sociological questionnaire, we restricted ourselves to both first variables and 50 observations picked at random. A recursive algorithm that eliminated replicates gives the results that (a) there are* 5,928 *different values for the sufficient statistic and (b) the most common occurrence is the middle partition* (26, 6, 11, 5, 10), *with* 7.16×10^{12} *replicas (out of* 1.12×10^{15} *total partitions). The posterior weight of a given partition is*

$$\frac{\Gamma(n_1+1)\Gamma(n-n_1+1)}{\Gamma(n+2)} \prod_{j=1}^{2} \prod_{v=1}^{2} \frac{\Gamma(s_{jv}+1/2)\Gamma(n_j - s_{jv}+1/2)}{\Gamma(n_j+1)}$$

$$= \prod_{j=1}^{2} \prod_{v=1}^{2} \Gamma(s_{jv}+1/2)\Gamma(n_j - s_{jv}+1/2) \Big/ n_1!\,(n-n_1)!\,(n+1)!\,,$$

multiplied by the number of occurrences. In this case, it is therefore possible to find exactly the most likely partitions, namely the one with $n_1 = 11$ *and* $n_2 = 39$, $s_{11} = 11$, $s_{12} = 8$, $s_{21} = 0$, $s_{22} = 17$, *and the symmetric one, which both only occur once and which have a joint posterior probability of* 0.018. *It is also possible to eliminate all the partitions with very low probabilities in this example.* ◀

8.3. Mixture Inference

Once again, the apparent simplicity of the mixture density should not be taken at face value for inferential purposes; since, for a sample of arbitrary size n from a mixture distribution (8.1), there always is a non-zero probability $(1 - p_j)^n$ that the jth subsample is empty, the likelihood includes terms that do not bring any information about the parameters of the i-th component.

8.3.1. *Nonidentifiability, hence label switching*

A mixture model (8.1) is *senso stricto* never identifiable since it is invariant under permutations of the indices of the components. Indeed, unless we introduce some restriction on the range of the θ_i's, we cannot distinguish component number 1 (i.e., θ_1) from component number 2 (i.e., θ_2) in the likelihood, because they are exchangeable. This apparently benign feature has consequences on both Bayesian inference and computational implementation. First, exchangeability implies that in a J component mixture, the number of modes is of order $O(J!)$. The highly multimodal posterior surface is therefore difficult to explore via standard Markov chain Monte Carlo techniques. Second, if an exchangeable prior is used on $\boldsymbol{\theta} = (\theta_1, \ldots, \theta_J)$, all the marginals of the θ_j's are identical. Other and more severe sources of unidentifiability could occur as in Example 8.6.

Example 8.7. (Example 8.6 continued) *If we continue our assessment of the latent class model, with two variables with two modalities each, based on subset of data extracted from [50], under a Beta, $\mathcal{B}(a,b)$, prior distribution on p the posterior distribution is the weighted sum of Beta $\mathcal{B}(n_1+a, n-n_1+b)$ distributions, with weights*

$$\mu_n(\mathbf{n},\mathbf{s}) \prod_{j=1}^{2}\prod_{v=1}^{2} \Gamma(s_{jv}+1/2)\Gamma(n_j - s_{jv}+1/2) \Big/ n_1!\,(n-n_1)!\,(n+1)!\,,$$

where $\mu_n(\mathbf{n},\mathbf{s})$ denotes the number of occurrences of the sufficient statistic. Figure 8.3 provides the posterior distribution for a subsample of the dataset of [50] and $a=b=1$. Since p is not identifiable, the impact of the prior distribution is stronger than in an identifying setting: using a Beta $\mathcal{B}(a,b)$ prior on p thus produces a posterior [distribution] that reflects as much the influence of (a,b) as the information contained in the data. While a $\mathcal{B}(1,1)$ prior, as in Figure 8.3, leads to a perfectly symmetric posterior with three modes, using an asymmetric prior with $a \ll b$ strongly modifies the range of the posterior, as illustrated by Figure 8.4. ◀

Identifiability problems resulting from the exchangeability issue are called "label switching" in that the output of a properly converging MCMC algorithm should produce no information about the component labels (a feature which, incidentally, provides a fast assessment of the performance of MCMC solutions, as proposed in [7]). A naive answer to the problem proposed in the early literature is to impose an *identifiability constraint* on the parameters, for instance by ordering the means (or the variances or the weights) in a normal mixture. From a Bayesian point of view, this amounts to truncating the original prior distribution, going from $\pi(\boldsymbol{\theta}, \mathbf{p})$ to

$$\pi(\boldsymbol{\theta}, \mathbf{p})\, \mathbb{I}_{\mu_1 \le \ldots \le \mu_J}\,.$$

While this device may seem innocuous (because indeed the sampling distribution is the same with or without this constraint on the parameter

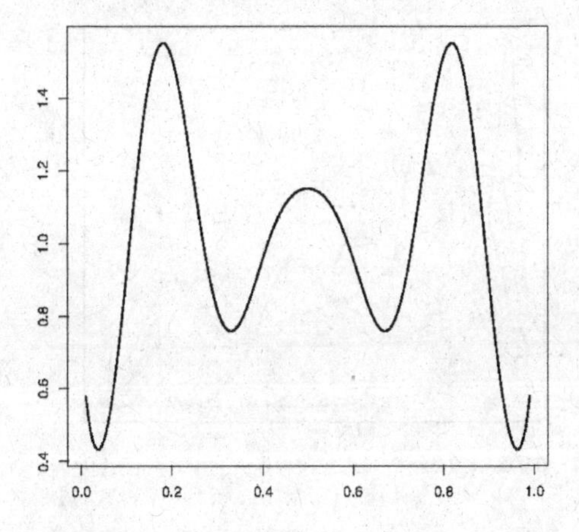

Fig. 8.3. Exact posterior distribution of p for a sample of 50 observations from the dataset of [50] and $a = b = 1$.

space), it is not without consequences on the resulting inference. This can be seen directly on the posterior surface: if the parameter space is reduced to its constrained part, there is no agreement between the above notation and the topology of this surface. Therefore, rather than selecting a single posterior mode and its neighbourhood, the constrained parameter space will most likely include parts of several modal regions. Thus, the resulting posterior mean may well end up in a very low probability region and be unrepresentative of the estimated distribution.

Note that, once an MCMC sample has been simulated from an unconstrained posterior distribution, any ordering constraint can be imposed on this sample, that is, after the simulations have been completed, for estimation purposes as stressed by [48]. Therefore, the simulation (if not the estimation) hindrance created by the constraint can be completely bypassed.

Once an MCMC sample has been simulated from an unconstrained posterior distribution, a natural solution is to identify one of the $J!$ modal regions of the posterior distribution and to operate the relabelling in terms of proximity to this region, as in [31]. Similar approaches based on clustering algorithms for the parameter sample are proposed in [48] and [7], and they achieve some measure of success on the examples for which they have been tested.

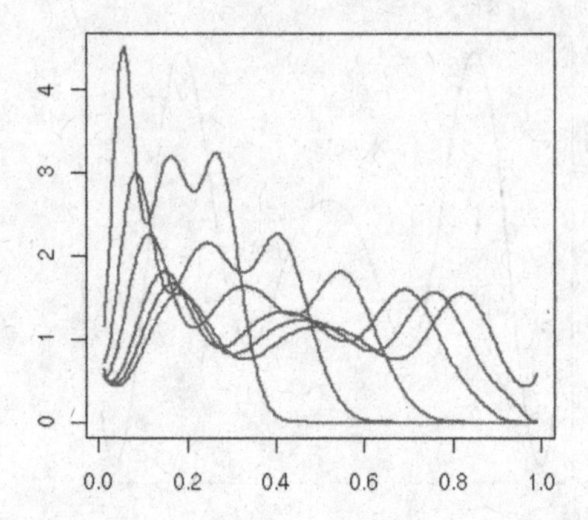

Fig. 8.4. Exact posterior distributions of p for a sample of 50 observations from the dataset of [50] under Beta $\mathcal{B}(a,b)$ priors when $a = .01, .05, .1, .05, 1$ and $b = 100, 50, 20, 10, 5, 1$.

An alternative approach is to eliminate the label switching problem by removing the labels altogether. This is done for instance in [7] by defining a loss function for the pairwise allocation of observations to clusters. From another perspective, [33] propose to work directly on the clusters associated with a mixture by defining the problem as an exchangeable process on the clusters: all that matters is then how data points are grouped together and this is indeed label-free.

8.3.2. *Restrictions on priors*

From a Bayesian point of view, the fact that few or no observation in the sample is (may be) generated from a given component has a direct and important drawback: this prohibits the use of independent improper priors,

$$\pi(\boldsymbol{\theta}) = \prod_{j=1}^{J} \pi(\theta_j),$$

since, if

$$\int \pi(\theta_j) \mathrm{d}\theta_j = \infty$$

then for any sample size n and any sample $\mathbf{x}$,

$$\int \pi(\boldsymbol{\theta}, \mathbf{p}|\mathbf{x}) \mathrm{d}\boldsymbol{\theta} \mathrm{d}\mathbf{p} = \infty.$$

The ban on using improper priors can be considered by some as being of little importance, since proper priors with large variances could be used instead. However, since mixtures are ill-posed problems, this difficulty with improper priors is more of an issue, given that the influence of a particular proper prior, no matter how large its variance, cannot be truly assessed.

There exists, nonetheless, a possibility of using improper priors in this setting, as demonstrated for instance by [34], by adding some degree of dependence between the component parameters. In fact, a Bayesian perspective makes it quite easy to argue *against* independence in mixture models, since the components are only properly defined in terms of one another. For the very reason that exchangeable priors lead to identical marginal posteriors on all components, the relevant priors must contain some degree of information that components are *different* and those priors must be explicit about this difference.

The proposal of [34], also described in [31], is to introduce first a common reference, namely a scale, location, or location-scale parameter (μ, τ), and then to define the original parameters in terms of *departure* from those references. Under some conditions on the reparameterisation, expressed in [43], this representation allows for the use of an improper prior on the reference parameter (μ, τ). See [35, 39, 53] for different approaches to the use of default or non-informative priors in the setting of mixtures.

8.4. Inference for Mixtures with a Known Number of Components

In this section, we describe different Monte Carlo algorithms that are customarily used for the approximation of posterior distributions in mixture settings when the number of components J is known. We start in Section 8.4.1 with a proposed solution to the label-switching problem and then discuss in the following sections Gibbs sampling and Metropolis-Hastings algorithms, acknowledging that a diversity of other algorithms exist (tempering, population Monte Carlo...), see [42].

8.4.1. *Reordering*

Section 8.3.1 discussed the drawbacks of imposing identifiability ordering constraints on the parameter space for estimation performances and there are similar drawbacks on the computational side, since those constraints decrease the explorative abilities of a sampler and, in the most extreme cases, may even prevent the sampler from converging ([7]). We thus consider samplers that evolve in an unconstrained parameter space, with the specific feature that the posterior surface has a number of modes that is a multiple of $J!$. Assuming that this surface is properly visited by the sampler (and this is not a trivial assumption), the derivation of point estimates of the parameters of (8.1) follows from an *ex-post* reordering proposed by [31] which we describe below.

Given a simulated sample of size M, a starting value for a point estimate is the naive approximation to the Maximum a Posteriori (MAP) estimator, that is the value in the sequence $(\boldsymbol{\theta}, \mathbf{p})^{(l)}$ that maximises the posterior,

$$l^* = \arg \max_{l=1,\ldots,M} \pi((\boldsymbol{\theta}, \mathbf{p})^{(l)}|\mathbf{x}).$$

Once an approximated MAP is computed, it is then possible to reorder all terms in the sequence $(\boldsymbol{\theta}, \mathbf{p})^{(l)}$ by selecting the reordering that is the closest to the approximate MAP estimator for a specific distance in the parameter space. This solution bypasses the identifiability problem without requiring a preliminary and most likely unnatural ordering with respect to one of the parameters (mean, weight, variance) of the model. Then, after the reordering step, an estimation of θ_j is given by

$$\sum_{l=1}^{M} (\theta_j)^{(l)}/M.$$

8.4.2. *Data augmentation and Gibbs sampling approximations*

The Gibbs sampler is the most commonly used approach in Bayesian mixture estimation ([12, 13, 15, 27, 52]) because it takes advantage of the missing data structure of the z_i's uncovered in Section 8.2.2.

The Gibbs sampler for mixture models (8.1) (cf. [13]) is based on the successive simulation of $\mathbf{z}$, $\mathbf{p}$ and $\boldsymbol{\theta}$ conditional on one another and on the data, using the full conditional distributions derived from the conjugate structure of the complete model. (Note that $\mathbf{p}$ only depends on the missing data $\boldsymbol{z}$.)

Gibbs sampling for mixture models

0. **Initialization:** choose $\mathbf{p}^{(0)}$ and $\boldsymbol{\theta}^{(0)}$ arbitrarily
1. **Step t.** For $t = 1, \ldots$
   1.1 Generate $z_i^{(t)}$ $(i = 1, \ldots, n)$ from $(j = 1, \ldots, J)$
   $$\mathbb{P}\left(z_i^{(t)} = j | p_j^{(t-1)}, \theta_j^{(t-1)}, x_i\right) \propto p_j^{(t-1)} f\left(x_i | \theta_j^{(t-1)}\right)$$
   1.2 Generate $\mathbf{p}^{(t)}$ from $\pi(\mathbf{p}|\mathbf{z}^{(t)})$
   1.3 Generate $\boldsymbol{\theta}^{(t)}$ from $\pi(\boldsymbol{\theta}|\mathbf{z}^{(t)}, \mathbf{x})$.

As always with mixtures, the convergence of this MCMC algorithm is not as easy to assess as it seems at first sight. In fact, while the chain is uniformly geometrically ergodic from a theoretical point of view, the severe augmentation in the dimension of the chain brought by the completion stage may induce strong convergence problems. The very nature of Gibbs sampling may lead to "trapping states", that is, concentrated local modes that require an enormous number of iterations to escape from. For example, components with a small number of allocated observations and very small variance become so tightly concentrated that there is very little probability of moving observations in or out of those components, as shown in [31]. As discussed in Section 8.2.3, [7] show that most MCMC samplers for mixtures, including the Gibbs sampler, fail to reproduce the permutation invariance of the posterior distribution, that is, that they do not visit the $J!$ replications of a given mode.

Example 8.8. *Consider a mixture of normal distributions with common variance σ^2 and unknown means and weights*

$$\sum_{j=1}^{J} p_j \mathcal{N}(\mu_j, \sigma^2).$$

This model is a particular case of model (8.1) and is not identifiable. Using conjugate exchangeable priors

$$\mathbf{p} \sim \mathcal{D}(1, \ldots, 1), \quad \mu_j \sim \mathcal{N}(0, 10\sigma^2), \quad \sigma^{-2} \sim \mathcal{E}xp(1/2),$$

it is straightforward to implement the above Gibbs sampler:

- *the weight vector $\mathbf{p}$ is simulated as the Dirichlet variable*
$$\mathcal{D}(1 + n_1, \ldots, 1 + n_J);$$

- *the inverse variance as the Gamma variable*

$$\mathcal{G}\left\{(n+2)/2, (1/2)\left[1+\sum_{j=1}^{J}\left(\frac{0.1 n_j \bar{x}_j^2}{n_j+0.1}+s_j^2\right)\right]\right\};$$

- *and, conditionally on σ, the means μ_j are simulated as the Gaussian variable*

$$\mathcal{N}(n_j \bar{x}_j/(n_j+0.1), \sigma^2/(n_j+0.1));$$

where $n_j = \sum_{z_i=j}$, $\bar{x}_j = \sum_{z_i=j} x_i$ *and* $s_j^2 = \sum_{z_i=j}(x_i - \bar{x}_j)^2/n_j$.
Note that this choice of implementation allows for the block simulation of the means-variance group, rather than the more standard simulation of the means conditional on the variance and of the variance conditional on the means ([13]). ◀

Consider the benchmark dataset of the galaxy radial speeds described for instance in [44]. The output of the Gibbs sampler is summarised on Figure 8.5 in the case of $J = 3$ components. As is obvious from the comparison of the three first histograms (and of the three following ones), label switching does not occur with this sampler: the three components remain isolated during the simulation process. ◀

Note that [22] (among others) dispute the relevance of asking for proper mixing over the $k!$ modes, arguing that on the contrary the fact that the Gibbs sampler sticks to a single mode allows for an easier inference. We obviously disagree with this perspective: first, from an algorithmic point of view, given the unconstrained posterior distribution as the target, a sampler that fails to explore all modes clearly fails to converge. Second, the idea that being restricted to a single mode provides a proper representation of the posterior is naively based on an intuition derived from mixtures with few components. As the number of components increases, modes on the posterior surface get inextricably mixed and a standard MCMC chain cannot be garanteed to remain within a single modal region. Furthermore, it is impossible to check in practice whether or not this is the case.

In his defence of "simple" MCMC strategies supplemented with post-processing steps, [22] states that

> *Celeux et al.'s ([7]) argument is persuasive only to the extent that there are mixing problems beyond those arising from permutation invariance of the posterior distribution. [7] does not make this argument,*

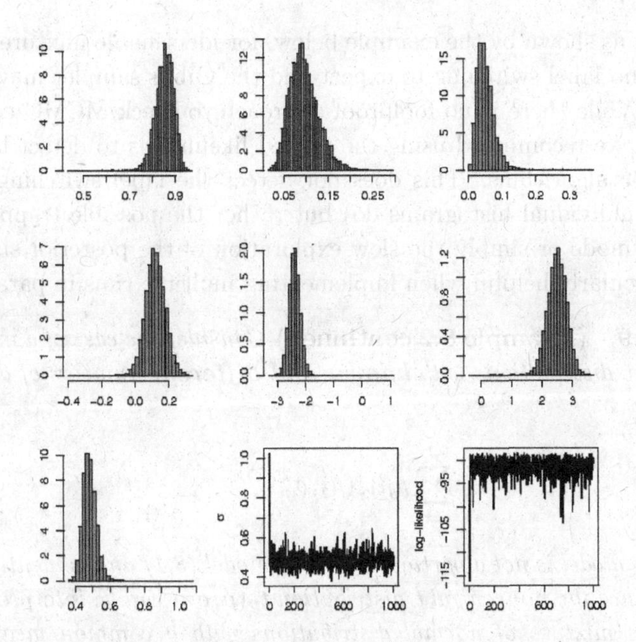

Fig. 8.5. From the left to the right, histograms of the parameters (p_1, p_2, p_3, μ_1, μ_2, μ_3, σ) of a normal mixture with $k = 3$ components based on 10^4 iterations of the Gibbs sampler and the galaxy dataset, evolution of the σ and of the log-likelihood.

> *indeed stating "The main defect of the Gibbs sampler from our perspective is the ultimate attraction of the local modes" (p. 959). That article produces no evidence of additional mixing problems in its examples, and we are not aware of such examples in the related literature. Indeed, the simplicity of the posterior distributions conditional on state assignments in most mixture models leads one to expect no irregularities of this kind.*

There are however clear irregularities in the convergence behaviour of Gibbs and Metropolis–Hastings algorithms as exhibited in [31] and [32] (Figure 6.4) for an identifiable two-component normal mixture with both means unknown. In examples such as those, there exist secondary modes that may have much lower posterior values than the modes of interest but that are nonetheless too attractive for the Gibbs sampler to visit other modes. In such cases, the posterior inference derived from the MCMC output is plainly incoherent. (See also [24] for another illustration of a multimodal posterior distribution in an identifiable mixture setting.)

However, as shown by the example below, for identifiable mixture models, there is no label switching to expect and the Gibbs sampler may work quite well. While there is no foolproof approach to check MCMC convergence ([42]), we recommend using the visited likelihoods to detect lack of mixing in the algorithms. This does not detect the label switching difficulties (but individual histograms do) but rather the possible trapping of a secondary mode or simply the slow exploration of the posterior surface. This is particularly helpful when implementing multiple runs in parallel.

Example 8.9. (Example 8.2 continued) *Consider the case of a mixture of Student's t distributions with* ***known and different*** *numbers of degrees of freedom*

$$\sum_{j=1}^{J} p_j t_{\nu_j}(\mu_j, \sigma_j^2).$$

This mixture model is not a particular case of model (8.1) and is identifiable. Moreover, since the noncentral t distribution $t_\nu(\mu, \sigma^2)$ can be interpreted as a continuous mixture of normal distributions with a common mean and with variances distributed as scaled inverse χ^2 random variable, a Gibbs sampler can be easily implemented in this setting by taking advantage of the corresponding latent variables: $x_i \sim t_\nu(\mu, \sigma^2)$ is the marginal of

$$x_i | V_i \sim \mathcal{N}\left(\mu, \frac{\sigma^2 \nu}{V_i}\right), \qquad V_i \sim \chi^2_\nu.$$

Once these latent variables are included in the simulation, the conditional posterior distributions of all parameters are available when using conjugate priors like

$$\mathbf{p} \sim \mathcal{D}(1, \ldots, 1), \quad \mu_j \sim \mathcal{N}(\mu_0, 2\sigma_0^2), \quad \sigma_j^2 \sim \mathcal{IG}(\alpha_\sigma, \beta_\sigma).$$

The full conditionals for the Gibbs sampler are a Dirichlet $\mathcal{D}(1+n_1, \ldots, 1+n_J)$ distribution on the weight vector, inverse Gamma

$$\mathcal{IG}\left\{\alpha_\sigma + \frac{n_j}{2}, \beta_\sigma + \sum_{z_i=j} \frac{(x_i - \mu_j)^2 V_i}{2\nu_j}\right\}$$

distribution on the variances σ_j^2, normal

$$\mathcal{N}\left(\frac{\mu_0 \sigma_j^2 + 2\sigma_0^2 \sum_{z_i=j} x_i V_i / \nu_j}{\sigma_j^2 + 2\sigma_0^2 \sum_{z_i=j} V_i/\nu_j}, \frac{2\sigma_0^2 \sigma_j^2}{\sigma_j^2 + 2\sigma_0^2 \sum_{z_i=j} V_i/\nu_j}\right)$$

distribution on the means μ_j, and Gamma

$$\mathcal{G}\left(\frac{1}{2}+\frac{\nu_j}{2}, \frac{1}{2}+\frac{(x_i-\mu_j)^2}{2\sigma_j^2\nu_j}\right)$$

distributions on the V_i.

In order to illustrate the performance of the algorithm, we simulated 2,000 *observations from the two-component t mixture with $\mu_1 = 0$, $\mu_2 = 5$, $\sigma_1^2 = \sigma_2^2 = 1$, $\nu_1 = 5$, $\nu_2 = 11$ and $p_1 = 0.3$. The output of the Gibbs sampler is summarized in Figure 8.6. The mixing behaviour of the Gibbs chains seems to be excellent, as they explore neighbourhoods of the true values.* ◀

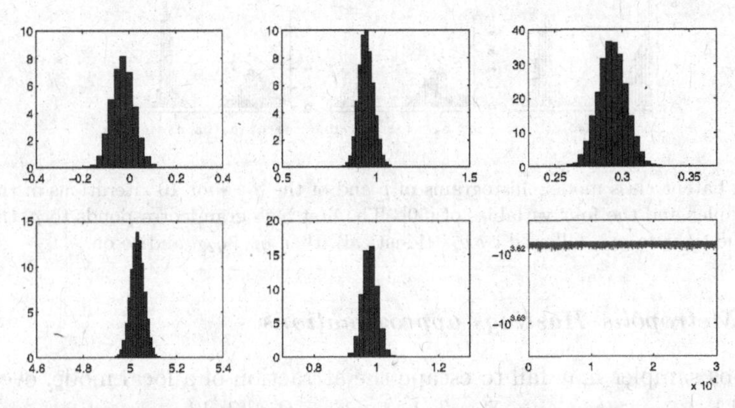

Fig. 8.6. Histograms of the parameters, $\mu_1, \sigma_1, p_1, \mu_2, \sigma_2$, and evolution of the (observed) log-likelihood along 30,000 iterations of the Gibbs sampler and a sample of 2,000 observations.

The example below shows that, for specific models and a small number of components, the Gibbs sampler may recover the symmetry of the target distribution.

Example 8.10. (Example 8.6 continued) *For the latent class model, if we use all four variables with two modalities each in* [50], *the Gibbs sampler involves two steps: the completion of the data with the component labels, and the simulation of the probabilities p and q_{tj} from Beta $\mathcal{B}(s_{tj} + .5, n_j - s_{tj} + .5)$ conditional distributions. For the* 216 *observations, the Gibbs sampler seems to converge satisfactorily since the output in Figure 8.7 exhibits the perfect symmetry predicted by the theory. We can note that, in this special case, the modes are well separated, and hence values can be crudely estimated for q_{1j} by a simple graphical identification of the modes.* ◀

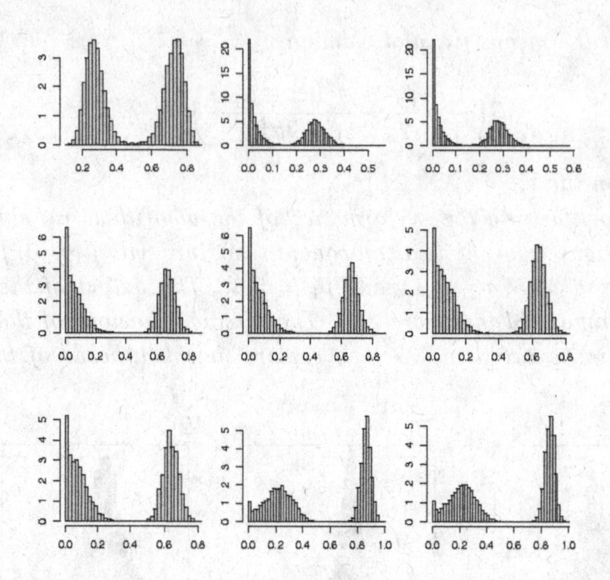

Fig. 8.7. Latent class model: histograms of p and of the q_{tj}'s for 10^4 iterations of the Gibbs sampler and the four variables of [50]. The first histogram corresponds to p, the next on the right to q_{11}, followed by q_{21} (identical), then q_{21}, q_{22}, and so on.

8.4.3. *Metropolis–Hastings approximations*

The Gibbs sampler may fail to escape the attraction of a local mode, even in a well-behaved case as in Example 1 where the likelihood and the posterior distributions are bounded and where the parameters are identifiable. Part of the difficulty is due to the completion scheme that increases the dimension of the simulation space and that reduces considerably the mobility of the parameter chain. A standard alternative that does not require completion and an increase in the dimension is the Metropolis–Hastings algorithm. In fact, the likelihood of mixture models is available in closed form, being computable in $\mathrm{O}(Jn)$ time, and the posterior distribution is thus available up to a multiplicative constant.

General Metropolis–Hastings algorithm for mixture models

0. **Initialization.** Choose $\mathbf{p}^{(0)}$ and $\boldsymbol{\theta}^{(0)}$
1. **Step t.** For $t = 1, \ldots$

   1.1 Generate $(\widetilde{\boldsymbol{\theta}}, \widetilde{\mathbf{p}})$ from $q\left(\boldsymbol{\theta}, \mathbf{p} | \boldsymbol{\theta}^{(t-1)}, \mathbf{p}^{(t-1)}\right)$,

1.2 Compute

$$r = \frac{f(\mathbf{x}|\widetilde{\boldsymbol{\theta}}, \widetilde{\mathbf{p}})\pi(\widetilde{\boldsymbol{\theta}}, \widetilde{\mathbf{p}})q(\boldsymbol{\theta}^{(t-1)}, \mathbf{p}^{(t-1)}|\widetilde{\boldsymbol{\theta}}, \widetilde{\mathbf{p}})}{f(\mathbf{x}|\boldsymbol{\theta}^{(t-1)}, \mathbf{p}^{(t-1)})\pi(\boldsymbol{\theta}^{(t-1)}, \mathbf{p}^{(t-1)})q(\widetilde{\boldsymbol{\theta}}, \widetilde{\mathbf{p}}|\boldsymbol{\theta}^{(t-1)}, \mathbf{p}^{(t-1)})},$$

1.3 Generate $u \sim \mathscr{U}_{[0,1]}$
If $r > u$ then $(\boldsymbol{\theta}^{(t)}, \mathbf{p}^{(t)}) = (\widetilde{\boldsymbol{\theta}}, \widetilde{\mathbf{p}})$
else $(\boldsymbol{\theta}^{(t)}, \mathbf{p}^{(t)}) = (\boldsymbol{\theta}^{(t-1)}, \mathbf{p}^{(t-1)})$.

The major difference with the Gibbs sampler is that we need to choose the proposal distribution q, which can be *a priori* anything, and this is a mixed blessing! The most generic proposal is the random walk Metropolis–Hastings algorithm where each unconstrained parameter is the mean of the proposal distribution for the new value, that is,

$$\widetilde{\theta}_j = \theta_j^{(t-1)} + u_j$$

where $u_j \sim \mathcal{N}(0, \zeta^2)$. However, for constrained parameters like the weights and the variances in a normal mixture model, this proposal is not efficient.

This is indeed the case for the parameter $\mathbf{p}$, due to the constraint that $\sum_{j=1}^{J} p_j = 1$. To solve this difficulty, [5] propose to overparameterise the model (8.1) as

$$p_j = w_j \Big/ \sum_{l=1}^{J} w_l \,, \quad w_j > 0 \,,$$

thus removing the simulation constraint on the p_j's. Obviously, the w_j's are not identifiable, but this is not a difficulty from a simulation point of view and the p_j's remain identifiable (up to a permutation of indices). Perhaps paradoxically, using overparameterised representations often helps with the mixing of the corresponding MCMC algorithms since they are less constrained by the dataset or the likelihood. The proposed move on the w_j's is $\log(\widetilde{w_j}) = \log(w_j^{(t-1)}) + u_j$ where $u_j \sim \mathcal{N}(0, \zeta^2)$.

Example 8.11. (Example 8.2 continued) *We now consider the more realistic case when the degrees of freedom of the t distributions are unknown. The Gibbs sampler cannot be implemented as such given that the distribution of the ν_j's is far from standard. A common alternative ([42]) is to introduce a Metropolis step within the Gibbs sampler to overcome this difficulty. If we use the same Gamma prior distribution with hyperparameters*

(α_ν, β_ν) for all the ν_js, the density of the full conditional distribution of ν_j is proportional to

$$\left(\frac{(1/2)^{\nu_j/2}}{\Gamma(\nu_j/2)}\nu_j^{-1/2}\right)^{n_j} \nu_j^{\alpha_\nu-1}\exp(-\beta_\nu\nu_j)\prod_{z_i=j} V_i^{\nu_j/2}\exp\left\{-\frac{(x_i-\mu_j)^2V_i}{2\sigma_j^2\nu_j^2}\right\}.$$

Therefore, we resort to a random walk proposal on the $\log(\nu_j)$*'s with scale $\varsigma = 5$. (The hyperparameters are $\alpha_\nu = 5$ and $\beta_\nu = 2$.)*

In order to illustrate the performances of the algorithm, two cases are considered: (i) all parameters except variances ($\sigma_1^2 = \sigma_2^2 = 1$) are unknown and (ii) all parameters are unknown. For a simulated dataset, the results are given on Figure 8.8 and Figure 8.9, respectively. In both cases, the posterior distributions of the ν_j's exhibit very large variances, which indicates that the data is very weakly informative about the degrees of freedom. The Gibbs sampler does not mix well-enough to recover the symmetry in the marginal approximations. The comparison between the estimated densities for both cases with the setting is given in Figure 8.10. The estimated mixture densities are indistinguishable and the fit to the simulated dataset is quite adequate. Clearly, the corresponding Gibbs samplers have recovered correctly one and only one of the 2 symmetric modes.

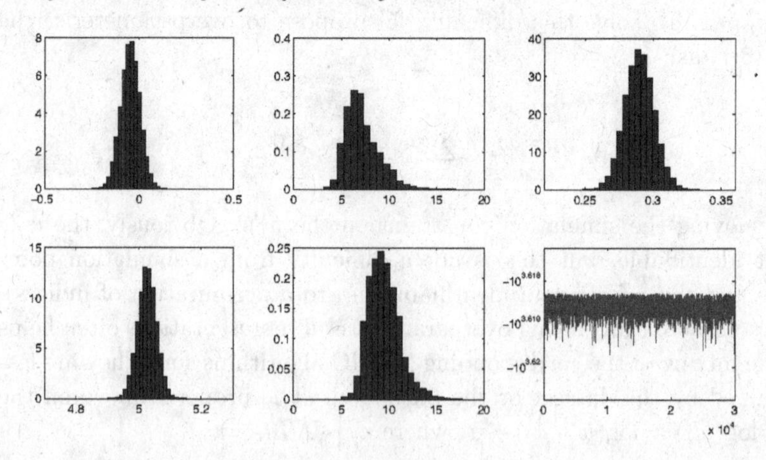

Fig. 8.8. Histograms of the parameters $\mu_1, \nu_1, p_1, \mu_2, \nu_2$ when the variance parameters are known, and evolution of the log-likelihood for a simulated t mixture with 2,000 points, based on 3×10^4 MCMC iterations.

We now consider the aerosol particle dataset described in Example 8.2. We use the same prior distributions on the ν_j's as before, that is $\mathcal{G}(5, 2)$. Figure 8.11 summarises the output of the MCMC algorithm. Since there

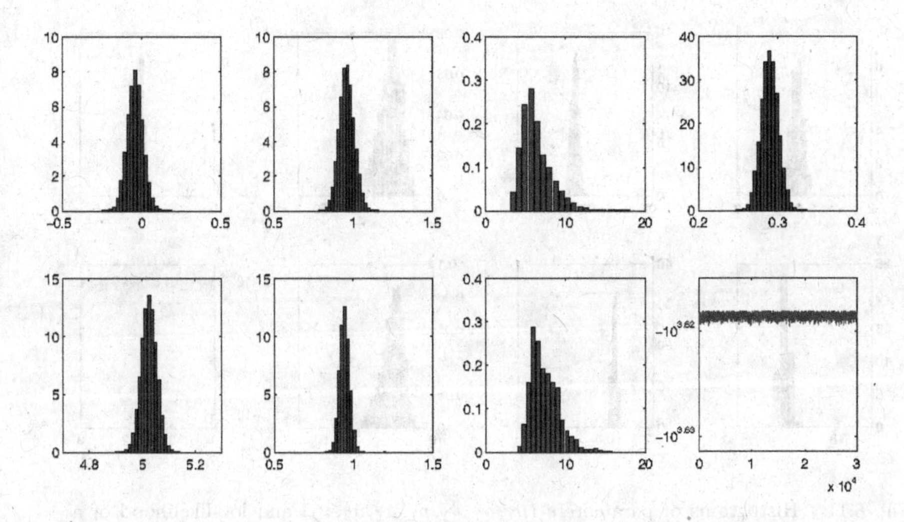

Fig. 8.9. Histograms of the parameters $\mu_1, \sigma_1, \nu_1, p_1, \mu_2, \sigma_2, \nu_2$, and evolution of the log-likelihood for a simulated t mixture with 2,000 points, based on 3×10^4 MCMC iterations.

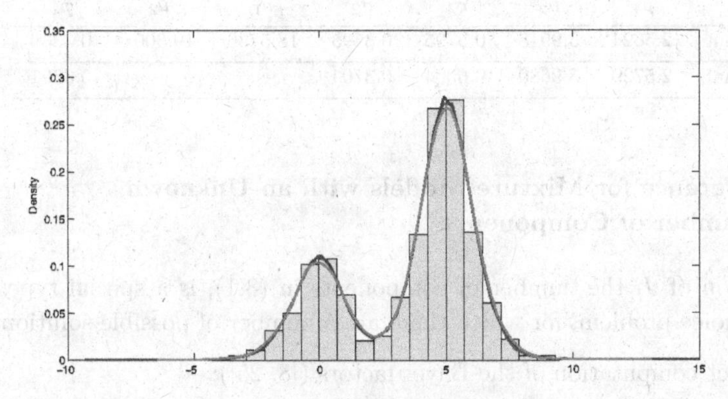

Fig. 8.10. Histogram of the simulated dataset, compared with estimated t mixtures with known σ^2 (*red*), known ν (*green*), and when all parameters are unknown (*blue*).

is no label switching and only two components, we choose to estimate the parameters by the empirical averages, as illustrated in Table 8.2. As shown by Figure 8.2, both t mixtures and normal mixtures fit the aerosol data reasonably well. ◀

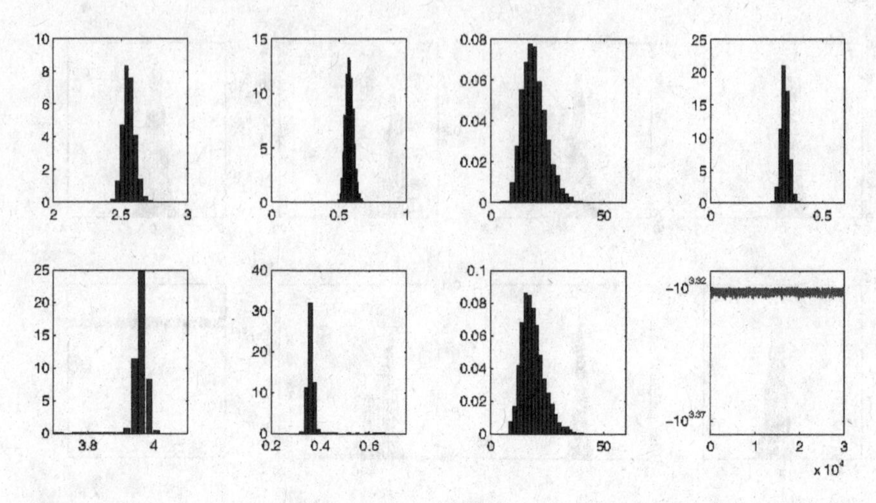

Fig. 8.11. Histograms of parameters $(\mu_1, \sigma_1, \nu_1, p_1, \mu_1, \sigma_2, \nu_2)$ and log-likelihood of a mixture of t distributions based on 30,000 iterations and the aerosol data.

Table 8.2. Estimates of the parameters for the aerosol dataset compared for t and normal mixtures.

	μ_1	μ_2	σ_1	σ_2	ν_1	ν_2	p_1
Student	2.5624	3.9918	0.5795	0.3595	18.5736	19.3001	0.3336
Normal	2.5729	3.9680	0.6004	0.3704	–	–	0.3391

8.5. Inference for Mixture Models with an Unknown Number of Components

Estimation of J, the number of components in (8.1), is a special type of model choice problem, for which there are a number of possible solutions:

(i) direct computation of the Bayes factors ([8, 25]);
(ii) evaluation of an entropy distance ([34, 45]);
(iii) generation from a joint distribution across models via reversible jump MCMC ([40]) or via birth-and-death processes ([49]);
(iv) indirect inference though Dirichlet process mixture models.

We refer to [31] for a short description of the reversible jump MCMC solution, a longer survey being available in [42] and a specific description for mixtures–including an R package—being provided in [32]. The alternative birth-and-death processes proposed in [49] has not generated as much follow-up, except for [5] who showed that the essential mechanism in this approach was the same as with reversible jump MCMC algorithms.

Dirichlet processes ([1, 18]) are often advanced as alternative to the estimation of the number of components for mixtures because they naturally embed a clustering mechanism. A Dirichlet process is a nonparametric object that formally involves a countably infinite number of components. Nonetheless, inference on Dirichlet processes for a finite sample size produces a random number of clusters. This can be used as an estimate of the number of components. From a computational point of view, see [37] for a MCMC solution and [3] for a variational alternative. We note that the proposal of [33] mentioned earlier also involves a Dirichlet cluster process modeling that leads to a posterior on the number of components.

We focus here on the first two approaches, because, first, the description of reversible jump MCMC algorithms require much care and therefore more space than we can allow to this paper and, second, this description exemplifies recent advances in the derivation of Bayes factors. These solutions pertain more strongly to the testing perspective, the entropy distance approach being based on the Kullback–Leibler divergence between a J component mixture and its projection on the set of $J-1$ mixtures, in the same spirit as in [14]. Given that the calibration of the Kullback divergence is open to various interpretations ([14, 23, 34]), we will only cover here some proposals regarding approximations of the Bayes factor oriented towards the direct exploitation of outputs from single model MCMC runs.

In fact, the major difference between approximations of Bayes factors based on those outputs and approximations based on the output from the reversible jump chains is that the latter requires a sufficiently efficient choice of proposals to move around models, which can be difficult despite significant recent advances ([4]). If we can instead concentrate the simulation effort on single models, the complexity of the algorithm decreases (a lot) and there exist ways to evaluate the performance of the corresponding MCMC samples. In addition, it is often the case that few models are in competition when estimating J and it is therefore possible to visit the whole range of potential models in an exhaustive manner.

We have

$$f_J(\boldsymbol{x}|\boldsymbol{\lambda}_J) = \prod_{i=1}^{n} \sum_{j=1}^{J} p_j f(x_i|\theta_j)$$

where $\boldsymbol{\lambda}_J = (\boldsymbol{\theta}, \boldsymbol{p}) = (\theta_1, \ldots, \theta_J, p_1, \ldots, p_J)$. Most solutions [20] revolve around an importance sampling approximation to the marginal likelihood integral

$$m_J(x) = \int f_J(\boldsymbol{x}|\boldsymbol{\lambda}_J)\,\pi_J(\boldsymbol{\lambda}_J)\,\mathrm{d}\boldsymbol{\lambda}_J$$

where J denotes the model index (that is the number of components in the present case). For instance, [28] use bridge sampling with simulated annealing scenarios to overcome the label switching problem. [47] rely on defensive sampling and the use of conjugate priors to reduce the integration to the space of latent variables (as in [6]) with an iterative construction of the importance function. [19] also centers her approximation of the marginal likelihood on a bridge sampling strategy, with particular attention paid to identifiability constraints. A different possibility is to use the representation in [21]: representation: starting from an arbitrary density g_J, the equality

$$\begin{aligned} 1 &= \int g_J(\boldsymbol{\lambda}_J)\,\mathrm{d}\boldsymbol{\lambda}_J = \int \frac{g_J(\boldsymbol{\lambda}_J)}{f_J(\boldsymbol{x}|\boldsymbol{\lambda}_J)\,\pi_J(\boldsymbol{\lambda}_J)} f_J(\boldsymbol{x}|\boldsymbol{\lambda}_J)\,\pi_J(\boldsymbol{\lambda}_J)\,\mathrm{d}\boldsymbol{\lambda}_J \\ &= m_J(\boldsymbol{x}) \int \frac{g_J(\boldsymbol{\lambda}_J)}{f_J(\boldsymbol{x}|\boldsymbol{\lambda}_J)\,\pi_J(\boldsymbol{\lambda}_J)} \pi_J(\boldsymbol{\lambda}_J|\boldsymbol{x})\,\mathrm{d}\boldsymbol{\lambda}_J \end{aligned}$$

implies that a potential estimate of $m_J(\boldsymbol{x})$ is

$$\hat{m}_J(\boldsymbol{x}) = 1 \Bigg/ \frac{1}{T} \sum_{t=1}^{T} \frac{g_J(\boldsymbol{\lambda}_J^{(t)})}{f_J(\boldsymbol{x}|\boldsymbol{\lambda}_J^{(t)})\,\pi_J(\boldsymbol{\lambda}_J^{(t)})}$$

when the $\boldsymbol{\lambda}_J^{(t)}$'s are produced by a Monte Carlo or an MCMC sampler targeted at $\pi_J(\boldsymbol{\lambda}_J|\boldsymbol{x})$. While this solution can be easily implemented in low dimensional settings ([10]), calibrating the auxiliary density g_k is always an issue. The auxiliary density could be selected as a non-parametric estimate of $\pi_k(\boldsymbol{\lambda}_J|x)$ based on the sample itself but this is very costly. Another difficulty is that the estimate may have an infinite variance and thus be too variable to be trustworthy, as experimented by [19].

Yet another approximation to the integral $m_J(\boldsymbol{x})$ is to consider it as the expectation of $f_J(\boldsymbol{x}|\boldsymbol{\lambda}_J)$, when $\boldsymbol{\lambda}_J$ is distributed from the prior. While a brute force approach simulating $\boldsymbol{\lambda}_J$ from the prior distribution is requiring a huge number of simulations ([36]), a Riemann based alternative is proposed by [46] under the denomination of *nested sampling*; however, [10] have shown in the case of mixtures that this technique could lead to uncertainties about the quality of the approximation.

We consider here a further solution, first proposed by [8], that is straightforward to implement in the setting of mixtures (see [9] for extensions).

Although it came under criticism by [36] (see also [19]), we show below how the drawback pointed by the latter can easily be removed. Chib's ([8]) method is directly based on the expression of the marginal distribution (loosely called *marginal likelihood* in this section) in Bayes' theorem:

$$m_J(\boldsymbol{x}) = \frac{f_J(\boldsymbol{x}|\boldsymbol{\lambda}_J)\,\pi_J(\boldsymbol{\lambda}_J)}{\pi_J(\boldsymbol{\lambda}_J|\boldsymbol{x})}$$

and on the property that the rhs of this equation is constant in $\boldsymbol{\lambda}_J$. Therefore, if an arbitrary value of $\boldsymbol{\lambda}_J$, $\boldsymbol{\lambda}_J^*$ say, is selected and if a good approximation to $\pi_J(\boldsymbol{\lambda}_J|\boldsymbol{x})$ can be constructed, $\hat{\pi}_J(\boldsymbol{\lambda}_J|\boldsymbol{x})$, Chib's ([8]) approximation to the marginal likelihood is

$$\hat{m}_J(\boldsymbol{x}) = \frac{f_J(\boldsymbol{x}|\boldsymbol{\lambda}_J^*)\,\pi_J(\boldsymbol{\lambda}_J^*)}{\hat{\pi}_J(\boldsymbol{\lambda}_J^*|\boldsymbol{x})}\,. \tag{8.8}$$

In the case of mixtures, a natural approximation to $\pi_J(\boldsymbol{\lambda}_J|\boldsymbol{x})$ is the Rao-Blackwell estimate

$$\hat{\pi}_J(\boldsymbol{\lambda}_J^*|\boldsymbol{x}) = \frac{1}{T}\sum_{t=1}^{T}\pi_J(\boldsymbol{\lambda}_J^*|\boldsymbol{x},\boldsymbol{z}^{(t)})\,,$$

where the $\boldsymbol{z}^{(t)}$'s are the latent variables simulated by the MCMC sampler. To be efficient, this method requires

(a) a good choice of $\boldsymbol{\lambda}_J^*$ but, since in the case of mixtures, the likelihood is computable, $\boldsymbol{\lambda}_J^*$ can be chosen as the MCMC approximation to the MAP estimator and,
(b) a good approximation to $\pi_J(\boldsymbol{\lambda}_J|\boldsymbol{x})$.

This later requirement is the core of Neal's ([36]) criticism: while, at a formal level, $\hat{\pi}_J(\boldsymbol{\lambda}_J^*|\boldsymbol{x})$ is a converging (parametric) approximation to $\pi_J(\boldsymbol{\lambda}_J|\boldsymbol{x})$ by virtue of the ergodic theorem, this obviously requires the chain $(\boldsymbol{z}^{(t)})$ to converge to its stationarity distribution. Unfortunately, as discussed previously, in the case of mixtures, the Gibbs sampler rarely converges because of the label switching phenomenon described in Section 8.3.1, so the approximation $\hat{\pi}_J(\boldsymbol{\lambda}_J^*|\boldsymbol{x})$ is untrustworthy. [36] demonstrated via a numerical experiment that (8.8) is significantly different from the true value $m_J(\boldsymbol{x})$ when label switching does not occur. There is, however, a fix to this problem, also explored by [2], which is to recover the label switching symmetry a posteriori, replacing $\hat{\pi}_J(\boldsymbol{\lambda}_J^*|\boldsymbol{x})$ in (8.8) above with

$$\hat{\pi}_J(\boldsymbol{\lambda}_J^*|\boldsymbol{x}) = \frac{1}{T\,J!}\sum_{\sigma\in\mathfrak{S}_J}\sum_{t=1}^{T}\pi_J(\sigma(\boldsymbol{\lambda}_J^*)|\boldsymbol{x},\boldsymbol{z}^{(t)})\,,$$

where $\mathfrak{S}_J$ denotes the set of all permutations of $\{1, \ldots, J\}$ and $\sigma(\boldsymbol{\lambda}_J^*)$ denotes the transform of $\boldsymbol{\lambda}_J^*$ where components are switched according to the permutation σ. Note that the permutation can equally be applied to $\boldsymbol{\lambda}_J^*$ or to the $\boldsymbol{z}^{(t)}$'s but that the former is usually more efficient from a computational point of view given that the sufficient statistics only have to be computed once. The justification for this modification either stems from a Rao-Blackwellisation argument, namely that the permutations are ancillary for the problem and should be integrated out, or follows from the general framework of [26] where symmetries in the dominating measure should be exploited towards the improvement of the variance of Monte Carlo estimators.

Example 8.12. (Example 8.8 continued) *In the case of the normal mixture case and the galaxy dataset, using Gibbs sampling, label switching does not occur. If we compute* $\log \hat{m}_J(\boldsymbol{x})$ *using only the original estimate of* [8], *(8.8), the [logarithm of the] estimated marginal likelihood is* $\hat{\rho}_J(\boldsymbol{x}) = -105.1396$ *for* $J = 3$ *(based on* 10^3 *simulations), while introducing the permutations leads to* $\hat{\rho}_J(\boldsymbol{x}) = -103.3479$. *As already noted by [36], the difference between the original Chib's* ([8]) *approximation and the true marginal likelihood is close to* $\log(J!)$ *(only) when the Gibbs sampler remains concentrated around a single mode of the posterior distribution. In the current case, we have that* $-116.3747 + \log(2!) = -115.6816$ *exactly! (We also checked this numerical value against a brute-force estimate obtained by simulating from the prior and averaging the likelihood, up to fourth digit agreement.) A similar result holds for* $J = 3$, *with* $-105.1396 + \log(3!) = -103.3479$. *Both* [36] *and* [19] *also pointed out that the* $\log(J!)$ *difference was unlikely to hold for larger values of* J *as the modes became less separated on the posterior surface and thus the Gibbs sampler was more likely to explore incompletely several modes. For* $J = 4$, *we get for instance that the original Chib's* ([8]) *approximation is* -104.1936, *while the average over permutations gives* -102.6642. *Similarly, for* $J = 5$, *the difference between* -103.91 *and* -101.93 *is less than* $\log(5!)$. *The* $\log(J!)$ *difference cannot therefore be used as a direct correction for Chib's* ([8]) *approximation because of this difficulty in controlling the amount of overlap. However, it is unnecessary since using the permutation average resolves the difficulty. Table 8.3 shows that the preferred value of* J *for the galaxy dataset and the current choice of prior distribution is* $J = 5$. ◀

When the number of components J grows too large for all permutations in $\mathfrak{S}_J$ to be considered in the average, a (random) subsample of permuta-

Table 8.3. Estimations of the marginal likelihoods by the symmetrised Chib's approximation (based on 10^5 Gibbs iterations and, for $J > 5$, 100 permutations selected at random in $\mathfrak{S}_J$).

J	2	3	4	5	6	7	8
$\hat{\rho}_J(\boldsymbol{x})$	−115.68	−103.35	−102.66	−101.93	−102.88	−105.48	−108.44

tions can be simulated to keep the computing time to a reasonable level when keeping the identity as one of the permutations, as in Table 8.3 for $J = 6, 7$. (See [2] for another solution.) Note also that the discrepancy between the original Chib's ([8]) approximation and the average over permutations is a good indicator of the mixing properties of the Markov chain, if a further convergence indicator is requested.

Example 8.13. **(Example 8.6 continued)** *For instance, in the setting of Example 8.6 with $a = b = 1$, both the approximation of* [8] *and the symmetrized one are identical. When comparing a single class model with a two class model, the corresponding (log-)marginals are*

$$\hat{\rho}_1(\boldsymbol{x}) = \prod_{i=1}^{4} \frac{\Gamma(1)}{\Gamma(1/2)^2} \frac{\Gamma(n_i + 1/2)\Gamma(n - n_i + 1/2)}{\Gamma(n+1)} = -552.0402$$

and $\hat{\rho}_2(\boldsymbol{x}) \approx -523.2978$, giving a clear preference to the two class model. ◀

Acknowledgments

We are grateful to the editors for the invitation as well as to Gilles Celeux for a careful reading of an earlier draft and for important suggestions related with the latent class model.

References

[1] Antoniak, C. (1974). Mixtures of Dirichlet processes with applications to Bayesian nonparametric problems. *Ann. Statist.* **2** 1152–1174.

[2] Berkhof, J., van Mechelen, I., and Gelman, A. (2003). A Bayesian approach to the selection and testing of mixture models. *Statistica Sinica.* **13** 423–442.

[3] Blei, D. and Jordan, M. (2006). Variational inference for Dirichlet process mixtures. *Bayesian Analysis.* **1** 121–144.

[4] Brooks, S., Giudici, P. and Roberts, G. (2003). Efficient construction of reversible jump Markov chain Monte Carlo proposal distributions (with discussion). *J. Royal Statist. Society Series B.* **65** 3–55.

[5] Cappé, O., Robert, C. and Rydén, T. (2003). Reversible jump, birth-and-death, and more general continuous time MCMC samplers. *J. Royal Statist. Society Series B.* **65** 679–700.

[6] Casella, G., Robert, C. and Wells, M. (2004). Mixture models, latent variables and partitioned importance sampling. *Statistical Methodology.* **1** 1–18.

[7] Celeux, G., Hurn, M. and Robert, C. (2000). Computational and inferential difficulties with mixtures posterior distribution. *J. American Statist. Assoc.* **95** 957–979.

[8] Chib, S. (1995). Marginal likelihood from the Gibbs output. *J. American Statist. Assoc.* **90** 1313–1321.

[9] Chib, S. and Jeliazkov, I. (2001). Marginal likelihood from the Metropolis–Hastings output. *J. American Statist. Assoc.* **96** 270–281.

[10] Chopin, N. and Robert, C. (2007). Contemplating evidence: properties, extensions of, and alternatives to nested sampling. Technical Report 2007-46, CEREMADE, Université Paris Dauphine. arXiv:0801.3887.

[11] Dempster, A., Laird, N. and Rubin, D. (1977). Maximum likelihood from incomplete data via the EM algorithm (with discussion). *J. Royal Statist. Society Series B.* **39** 1–38.

[12] Diebolt, J. and Robert, C. (1990). Bayesian estimation of finite mixture distributions, Part i: Theoretical aspects. Technical Report 110, LSTA, Université Paris VI, Paris.

[13] Diebolt, J. and Robert, C. (1994). Estimation of finite mixture distributions by Bayesian sampling. *J. Royal Statist. Society Series B.* **56** 363–375.

[14] Dupuis, J. and Robert, C. (2003). Model choice in qualitative regression models. *J. Statistical Planning and Inference* **111** 77–94.

[15] Escobar, M. and West, M. (1995). Bayesian density estimation and inference using mixtures. *J. American Statist. Assoc.* **90** 577–588.

[16] Fearnhead, P. (2005). Direct simulation for discrete mixture distributions. *Statistics and Computing.* **15** 125–133.

[17] Feller, W. (1970). *An Introduction to Probability Theory and its Applications*, volume 1. John Wiley, New York.

[18] Ferguson, T. (1973). A Bayesian analysis of some nonparametric problems. *Ann. Statist.* **1** 209–230.

[19] Frühwirth-Schnatter, S. (2004). Estimating marginal likelihoods for mixture and Markov switching models using bridge sampling techniques. *The Econometrics Journal.* **7** 143–167.

[20] Frühwirth-Schnatter, S. (2006). *Finite Mixture and Markov Switching Models.* Springer-Verlag, New York, New York.

[21] Gelfand, A. and Dey, D. (1994). Bayesian model choice: asymptotics and exact calculations. *J. Royal Statist. Society Series B* **56** 501–514.

[22] Geweke, J. (2007). Interpretation and inference in mixture models: Simple MCMC works. *Comput. Statist. Data Analysis.* (To appear).

[23] Goutis, C. and Robert, C. (1998). Model choice in generalized linear models: a Bayesian approach via Kullback–Leibler projections. *Biometrika* **85** 29–37.

[24] Iacobucci, A., Marin, J.-M. and Robert, C. (2008). On variance stabilisation by double Rao-Blackwellisation. Technical report, CEREMADE, Université

Paris Dauphine.

[25] Kass, R. and Raftery, A. (1995). Bayes factors. *J. American Statist. Assoc.* **90** 773–795.

[26] Kong, A., McCullagh, P., Meng, X.-L., Nicolae, D. and Tan, Z. (2003). A theory of statistical models for Monte Carlo integration. *J. Royal Statist. Society Series B.* **65** 585–618. (With discussion.).

[27] Lavine, M. and West, M. (1992). A Bayesian method for classification and discrimination. *Canad. J. Statist.* **20** 451–461.

[28] Liang, F. and Wong, W. (2001). Real-parameter evolutionary Monte Carlo with applications to Bayesian mixture models. *J. American Statist. Assoc.* **96** 653–666.

[29] MacLachlan, G. and Peel, D. (2000). *Finite Mixture Models.* John Wiley, New York.

[30] Magidson, J. and Vermunt, J. (2000). Latent class analysis. In *The Sage Handbook of Quantitative Methodology for the Social Sciences* (Kaplan, D., ed). 175–198, Thousand Oakes. Sage Publications.

[31] Marin, J.-M., Mengersen, K. and Robert, C. (2005). Bayesian modelling and inference on mixtures of distributions. In *Handbook of Statistics*, volume 25 (Rao, C. and Dey, D. eds). Springer-Verlag, New York.

[32] Marin, J.-M. and Robert, C. (2007). *Bayesian Core.* Springer-Verlag, New York.

[33] McCullagh, P. and Yang, J. (2008). How many clusters? *Bayesian Analysis.* **3** 101–120.

[34] Mengersen, K. and Robert, C. (1996). Testing for mixtures: A Bayesian entropic approach (with discussion). In *Bayesian Statistics 5* (Berger, J., Bernardo, J., Dawid, A., Lindley, D., and Smith, A., eds.) 255–276. Oxford University Press, Oxford.

[35] Moreno, E. and Liseo, B. (2003). A default Bayesian test for the number of components in a mixture. *J. Statist. Plann. Inference.* **111** 129–142.

[36] Neal, R. (1999). Erroneous results in "Marginal likelihood from the Gibbs output". Technical report, University of Toronto.

[37] Neal, R. (2000). Markov Chain sampling methods for Dirichlet process mixture models. *J. Comput. Graph. Statist.* **9** 249–265.

[38] Nilsson, E. D. and Kulmala, M. (2006). Aerosol formation over the Boreal forest in Hyytiälä, Finland: monthly frequency and annual cycles — the roles of air mass characteristics and synoptic scale meteorology. *Atmospheric Chemistry and Physics Discussions.* **6** 10425–10462.

[39] Pérez, J. and Berger, J. (2002). Expected-posterior prior distributions for model selection. *Biometrika.* **89** 491–512.

[40] Richardson, S. and Green, P. (1997). On Bayesian analysis of mixtures with an unknown number of components (with discussion). *J. Royal Statist. Society Series B.* **59** 731–792.

[41] Robert, C. (2001). *The Bayesian Choice.* Second edition, Springer-Verlag, New York.

[42] Robert, C. and Casella, G. (2004). *Monte Carlo Statistical Methods.* Second edition, Springer-Verlag, New York.

[43] Robert, C. and Titterington, M. (1998). Reparameterisation strategies for hidden Markov models and Bayesian approaches to maximum likelihood estimation. *Statistics and Computing.* **8** 145–158.

[44] Roeder, K. and Wasserman, L. (1997). Practical Bayesian density estimation using mixtures of normals. *J. American Statist. Assoc.* **92** 894–902.

[45] Sahu, S. and Cheng, R. (2003). A fast distance based approach for determining the number of components in mixtures. *Canadian J. Statistics.* **31** 3–22.

[46] Skilling, J. (2006). Nested sampling for general Bayesian computation. *Bayesian Analysis.* **1** 833–860.

[47] Steele, R., Raftery, A. and Emond, M. (2006). Computing normalizing constants for finite mixture models via incremental mixture importance sampling (IMIS). *Journal of Computational and Graphical Statistics.* **15** 712–734.

[48] Stephens, M. (1997). *Bayesian Methods for Mixtures of Normal Distributions.* PhD thesis, University of Oxford.

[49] Stephens, M. (2000). Bayesian analysis of mixture models with an unknown number of components—an alternative to reversible jump methods. *Ann. Statist.* **28** 40–74.

[50] Stouffer, S. and Toby, J. (1951). Role conflict and personality. *American Journal of Sociology* **56** 395–406.

[51] Tanner, M. and Wong, W. (1987). The calculation of posterior distributions by data augmentation. *J. American Statist. Assoc.* **82** 528–550.

[52] Verdinelli, I. and Wasserman, L. (1992). Bayesian analysis of outliers problems using the Gibbs sampler. *Statist. Comput.* **1** 105–117.

[53] Wasserman, L. (2000). Asymptotic inference for mixture models using data dependent priors. *J. Royal Statist. Society Series B* **62** 159–180.

Chapter 9

Markov Processes Generated by Random Iterates of Monotone Maps: Theory and Applications

Mukul Majumdar

Department of Economics,
Uris Hall, Fourth Floor,
Cornell University,
Ithaca, New York 14853-7601, USA
mkm5@cornell.edu

The paper is a review of results on the asymptotic behavior of Markov processes generated by i.i.d. iterates of monotone maps. Of particular importance is the notion of splitting introduced by [12]. Some extensions to more general frameworks are outlined, and, finally, a number of applications are indicated.

9.1. Introduction

This paper is an impressionistic overview of some results on Markov processes that arise in the study of a particular class of random dynamical systems. A random dynamical system is described by a triplet (S, Γ, Q) where S is the *state space* (for example, a metric space), Γ an appropriate family of maps on S into itself (interpreted as the set of all possible *laws of motion*) and Q is a *probability measure* on (some σ-field of) Γ.

The evolution of the system can be described as follows: initially, the system is in some state x; an element α_1 of Γ is chosen randomly according to the probability measure Q and the system moves to a state $X_1 = \alpha_1(x)$ in period one. Again, independently of α_1, an element α_2 of Γ is chosen according to the probability measure Q and the state of the system in period two is obtained as $X_2 = \alpha_2(\alpha_1(x))$. In general, starting from some x in S, one has

$$X_{n+1}(x) = \alpha_{n+1}(X_n(x)), \tag{1.1}$$

where the maps (α_n) are *independent* with the common distribution Q.

The initial point x can also be chosen (independently of (α_n)) as a random variable X_0. The sequence X_n of states obtained in this manner is a Markov process and has been of particular interest in developing stochastic dynamic models in many disciplines. With specific assumptions on the structure of S and Γ it has been possible to derive strong results on the asymptotic behavior of X_n.

Random dynamical systems have been particularly useful for modeling long run evolution of economic systems subject to exogenous random shocks. The framework (1.1) can be interpreted as a *descriptive model*; but, one may also start with a discounted (stochastic) dynamic programming problem, and directly arrive at a stationary optimal policy function, which together with the exogenously given law of transition describes the *optimal evolution* of the states in the form (1.1). Of particular significance are results on the "inverse optimal problem under uncertainty" due to [20] and [22] which assert that a very broad class of random systems (1.1) can be so interpreted.

The literature exploring (1.1) is already vast and growing. Given the space limitations, this review is *primarily* restricted to the case when S is an interval (non-degenerate) in R, or a closed (nonempty) subset of R^ℓ, and Γ is a family of *monotone* maps from S into S. Some extensions to more general framework and applications are also outlined. Here I touch upon a few of the issues and provide some references to definitive treatments.

(i) *The existence, uniqueness and global stability of a steady state (an invariant distribution) of random dynamical systems:* Significant progress has been achieved when the laws of motion satisfy either some *"splitting"* or *"contraction"* conditions (see, e.g., [12], [11] [6, 7] and the review in [9], Chapter 3). An awkward problem involving the existence question is worthnoting. Consider $S = [0, 1]$ or $S = R_+$ and assume that $\gamma(0) = 0$ for *all* $\gamma \in \Gamma$. This is a natural property of a law of motion in many population or economic models (viewed as a production function, $\gamma(0) = 0$ means that zero input leads to zero output). The point mass at 0 (the measure δ_0) is obviously an invariant distribution. The challenge, then, is to find an invariant distribution with support in $(0, 1)$.

(ii) *The nature of the invariant distribution.* Suppose, for concreteness, that S is an interval, and F is the distribution function on R of the unique invariant measure. Invoking a standard *decomposition* property (see [18], p. 130, 196), let (i) F_d be the *step part* (a *step function*); (ii) F_{ac} be the *absolutely continuous* part (with respect to the Lebesgue measure) and (iii) F_s be the *singular part* of F.

As a first step one would like to know whether (i) F is *continuous* ($F_d \equiv 0$) or whether (ii) F is *absolutely continuous* or whether (iii) F is *singular.* At the next step, one would like to ask questions of *comparative statics*: how does F (or the components (i)–(iii)) change if a parameter in the model is allowed to change? Finally, one would like to compute (or approximate) F but that typically requires more structure on the model.

All the questions are elusive. Take the standard approach of describing a Markov process with state space $S = R$, and a transition function $p(x, A)$. If for each $x \in S$, $p(x, .)$ is *absolutely continuous* with respect to the Lebesgue measure, then if π is invariant under $p(x, A)$, π is also absolutely continuous with respect to the Lebesgue measure [see [9], Proposition 5.2 of Chapter 5]. This result is to be contrasted with those in Section 9.2.

A study of (i.i.d) random iteration of quadratic maps ($S = [0, 1]$, $\Gamma = \{f : f(x) = \theta x(1 - x),\ 0 \leq \theta \leq 4\}$, Q with a two point support) was initiated by [5]. The subsequent literature offers interesting examples on applications of splitting and open questions. For a review of results when Γ is the quadratic family (the typical $\gamma(x) = \theta x(1 - x)$ does *not* satisfy the monotonicity property that is central here but *does* have 'piecewise monotonicity' which has often been used to invoke the splitting conditions: see [1]; further extensions are in [3]).

The processes considered in this article particularly when Γ is finite are *not* in general Harris irreducible (see, e.g., [23] for a definition of Harris irreducibility). Therefore, the standard techniques used for the study of irreducible Markov processes in the literature are not applicable to many of the cases reviewed. This point was explored in detail in [13] who concluded that "it is surprising and unfortunate that the large classical theory based on compactness and/or irreducibility conditions generally give little information about (1.1) as a population model." The reader interested in this issue is referred to [13], Section 5.

(iii) *Applications of the theoretical results to a few topics*:

(a) *turnpike theorems* in the literature on descriptive and optimal growth under uncertainty: when each admissible law of motion is monotone increasing, and satisfies the appropriate Inada-type 'end point' condition, Theorem 9.1 can be applied directly.

(b) *estimation of the invariant distribution*: as noted above, an important implication of the "splitting theorems" is an estimate of the speed of convergence. This estimate is used in Section 9.5 to prove a result on $\sqrt{n}$-consistency of the sample mean as an estimator of the expected long run equilibrium value (i.e., the value of the state variable with respect to the invariant distribution).

9.2. Random Dynamical Systems

We consider random dynamical systems. Let S be a metric space and $\mathcal{S}$ be the Borel σ-field of S. Endow Γ with a σ-field Σ such that the map $(\gamma, x) \to (\gamma(x))$ on $(\Gamma \times S, \Sigma \otimes \mathcal{S}$ into $(S, \mathcal{S})$ is measurable. Let Q be a probability measure on (Γ, Σ). On some probability space $(\Omega, \mathcal{F}, P)$ let $(\alpha_n)_{n=1}^{\infty}$ be a sequence of *independent* random functions from Γ with a common distribution Q. For a given random variable X_0 (with values in S), independent of the sequence $(\alpha_n)_{n=1}^{\infty}$, define

$$X_1 \equiv \alpha_1(X_0) \equiv \alpha_1 X_0 \tag{2.1}$$

$$X_{n+1} = \alpha_{n+1}(X_n) \equiv \alpha_{n+1}\alpha_n \cdots \alpha_1 X_0 \tag{2.2}$$

We write $X_n(x)$ for the case $X_0 = x$; to simplify notation we write $X_n = \alpha_n \cdots \alpha_1 X_0$ for the more general (random) X_0. Then X_n is a Markov process with the stationary transition probability $p(x, dy)$ given as follows: for $x \in S$, $C \in \mathcal{S}$,

$$p(x, C) = Q(\{\gamma \in \Gamma : \gamma(x) \in C\}) \tag{2.3}$$

The stationary transition probability $p(x, dy)$ is said to be *weakly continuous* or to have the *Feller property* if for any sequence x_n converging to x, the sequence of probability measures $p(x_n, \cdot)$ converges weakly to $p(x, \cdot)$. One can show that if Γ consists of a family of continuous maps, $p(x, dy)$ has the Feller property.

9.3. Evolution

To study the evolution of the process (2.2), it is convenient to define the map T^* [on the space $M(S)$ of all finite signed measures on $(S, \mathcal{S})$] by

$$T^*\mu(C) = \int_S p(x, C)\mu(dx) = \int_\Gamma \mu(\gamma^{-1}C)Q(d\gamma), \qquad \mu \in M(S). \tag{3.1}$$

Let $\mathcal{P}(S)$ be the set of all probability measures on $(S, \mathcal{S})$. An element π of $\mathcal{P}(S)$ is *invariant* for $p(x, dy)$ (or for the Markov process X_n) if it is a fixed point of T^*, i.e.,

$$\pi \textit{ is invariant} \quad iff \quad T^*\pi = \pi \tag{3.2}$$

Now write $p^{(n)}(x, dy)$ for the n-step transition probability with $p^{(1)} \equiv p(x, dy)$. Then $p^{(n)}(x, dy)$ is the distribution of $\alpha_n \cdots \alpha_1 x$. Define T^{*n} as the n-th iterate of T^*:

$$T^{*n}\mu = T^{*(n-1)}(T^*\mu) \ (n \geq 2), \ T^{*1} = T^*, \ T^{*0} = Identity \tag{3.3}$$

Then for any $C \in \mathcal{S}$,

$$(T^{*n}\mu)(C) = \int_S p^{(n)}(x, C)\mu(dx), \tag{3.4}$$

so that $T^{*n}\mu$ is the distribution of X_n when X_0 has distribution μ. To express T^{*n} in terms of the common distribution Q of the i.i.d. maps (α_n), let Γ^n denote the usual Cartesian product $\Gamma \times \Gamma \times \cdots \times \Gamma$ (n terms), and let Q^n be the product probability $Q \times Q \times \cdots \times Q$ on $(\Gamma^n, \mathcal{S}^{\otimes n})$ where $\mathcal{S}^{\otimes n}$ is the product σ-field on Γ^n. Thus Q^n is the (joint) distribution of $\alpha = (\alpha_1, \alpha_2, \ldots, \alpha_n)$. For $\gamma = (\gamma_1, \gamma_2, \ldots, \gamma_n) \in \Gamma^n$ let $\widetilde{\gamma}$ denote the composition

$$\widetilde{\gamma} := \gamma_n \gamma_{n-1} \cdots \gamma_1 \tag{3.5}$$

We suppress the dependence of $\widetilde{\gamma}$ on n for notational simplicity. Then, since $T^{*n}\mu$ is the distribution of $X_n = \alpha_n \cdots \alpha_1 X_0$, one has $(T^{*n}\mu)(A) = \mathrm{Prob}(X_0 \in \widetilde{\alpha}^{-1} A)$, where $\widetilde{\alpha} = \alpha_n \alpha_{n-1} \cdots \alpha_1$. Therefore, by the independence of $\widetilde{\alpha}$ and X_0,

$$(T^{*n}\mu)(A) = \int_{\Gamma^n} \mu(\widetilde{\gamma}^{-1} A) Q^n(d\gamma) \qquad (A \in \mathcal{S}, \ \mu \in \mathcal{P}(S)). \tag{3.6}$$

Finally, we come to the definition of *stability.* A Markov process X_n is *stable in distribution* if there is a unique invariant probability measure π such that $X_n(x)$ converges weakly (or, in distribution) to π irrespective of the initial state x, i.e., if $p^{(n)}(x, dy)$ converges weakly to the same probability measure π for all x.

In what follows, if g is a bounded $\mathcal{S}$-measurable real valued function on S, we write

$$Tg(x) = \int_S g(y) \, p(x, dy) \tag{3.7}$$

9.4. Splitting

If S is a (nonempty) compact metric space and Γ consists of a family of *continuous* functions from S into S, then a fixed point argument ensures that there is an invariant probability measure π^*. However, when Γ consists of *monotone* maps on a suitable subset S of R^ℓ (into S), stronger results on uniqueness and stability can be derived by using a 'splitting' condition, first studied by [12].

9.4.1. *Splitting and Monotone Maps*

Let S be a nondegenerate interval (finite or infinite, closed, semiclosed, or open) and Γ a set of *monotone* maps from S into S; i.e., each element of Γ is either a nondecreasing function on S or a nonincreasing function.

We assume the following *splitting condition*:

(H) *There exist $z_0 \in S, \tilde{\chi} > 0$ and a positive N such that*

(1) $P(\alpha_N\alpha_{N-1}\cdots\alpha_1 x \le z_0 \forall x \in S) \ge \tilde{\chi}$,

(2) $P(\alpha_N\alpha_{N-1}\cdots\alpha_1 x \ge z_0 \forall x \in S) \ge \tilde{\chi}$.

Note that conditions (1) and (2) in (**H**) may be expressed, respectively, as

$$Q^N(\{\gamma \in \mathbf{\Gamma}^N : \tilde{\gamma}^{-1}[x \in S : x \le z_0] = S\}) \ge \tilde{\chi}, \tag{4.1}$$

and

$$Q^N(\{\gamma \in \mathbf{\Gamma}^N : \tilde{\gamma}^{-1}[x \in S : x \ge z_0] = S\}) \ge \tilde{\chi}. \tag{4.2}$$

Recall that $\tilde{\gamma} = \gamma_N\gamma_{N-1}\cdots\gamma_1$.

Denote by $d_K(\mu,\nu)$ the Kolmogorov distance on $\mathcal{P}(S)$. That is, if F_μ, F_ν denote the distribution functions (d.f.) of μ and ν, respectively, then

$$\begin{aligned} d_K(\mu,\nu) &:= \sup_{x\in R} |\mu((-\infty,x]\cap S) - \nu(-\infty,x]\cap S)| \\ &\equiv \sup_{x\in R} |F_\mu(x) - F_\nu(x)|, \mu,\nu \in \mathcal{P}((S)). \end{aligned} \tag{4.3}$$

Remark 4.1. First, it should be noted that *convergence in the distance d_K on $\mathcal{P}(S)$ implies weak convergence* in $\mathcal{P}(S)$. Secondly, $(\mathcal{P}(S), d_K)$ is a *complete metric space.*(See [9], Theorem 5.1 and C11.2(d) of Chapter 2). ■

Theorem 9.1. *Assume that the splitting condition* (**H**) *holds. Then*

(a) *the distribution* $T^{*n}\mu$ *of* $X_n := \alpha_n \cdots \alpha_1 X_0$ *converges to a probability measure* π *on* S *in the Kolmogorov distance* d_K *irrespective of* X_0. *Indeed,*

$$d_K(T^{*n}\mu, \pi) \leq (1 - \tilde{\chi})^{[n/N]} \ \forall \mu \in \mathcal{P}(S) \tag{4.4}$$

where $[y]$ *denotes the integer part of* y.

(b) π *in (a) is the unique invariant probability of the Markov process* X_n.

Proof. [Main Steps] Careful calculations using the splitting condition and monotonicity lead to (see [9], Chapter 3, Theorem 5.1):

$$d_K(T^*\mu, T^*\nu) \leq d_K(\mu, \nu) \tag{4.5}$$

and

$$d_K(T^{*N}\mu, T^{*N}\nu) \leq (1 - \tilde{\chi}) d_K(\mu, \nu) \quad (\mu, \nu \in \mathcal{P}(S)). \tag{4.6}$$

That is, T^{*N} is a uniformly strict contraction and T^* is a contraction. As a consequence, $\forall_n > N$, one has

$$\begin{aligned} d_K(T^{*n}\mu, T^{*n}\nu) &= d_K(T^{*N}(T^{*(n-N)}\mu), T^{*N}(T^{*(n-N)}\nu)) \\ &\leq (1 - \tilde{\chi}) d_K(T^{*(n-N)}\mu, T^{*(n-N)}\nu) \leq \cdots \\ &\leq (1 - \tilde{\chi})^{[n/N]} d_K(T^{*(n-[n/N]N)}\mu, T^{*(n-[n/N]N)}\nu) \\ &\leq (1 - \tilde{\chi})^{[n/N]} d_K(\mu, \nu). \end{aligned} \tag{4.7}$$

Now, by appealing to the contraction mapping theorem, T^{*N} has a unique fixed point π in $\mathcal{P}(S)$, and $T^{*N}(T^*\pi) = T^*(T^{*N}\pi) = T^*\pi$. Hence $T^*\pi$ is also a fixed point of T^{*N}. By uniqueness $T^*\pi = \pi$. Hence, π is a fixed point of T^*. Any fixed point of T^* is a fixed point of T^{*N}. Hence π is the unique fixed point of T^*. Now take $\nu = \pi$ in (4.7) to get the desired relation (4.4). □

The following remarks clarify the role of the splitting condition.

Remark 4.2. Let $S = [a, b]$ and $\alpha_n (n \geq 1)$ a sequence of i.i.d. continuous nondecreasing maps on S into S. Suppose that π is the unique invariant distribution of the Markov process. If π is not degenerate, then the splitting condition holds [[12], Theorem 5.17; for relaxing continuity, see [9], Lemma CS.2 of Chapter 3]. ■

Remark 4.3. Suppose that α_n are strictly monotone a.s. Then if the initial distribution μ is nonatomic (i.e., $\mu(\{x\}) = 0\,\forall x$ or, equivalently the d.f. of μ is continuous), $\mu\, o\, \gamma^{-1}$ is nonatomic $\forall \gamma \in \Gamma$ (outside a set of zero Q-probability). It follows that if X_0 has a continuous d.f., then so has X_1 and in turn X_2 has a continuous d.f., and so on. Since, by Theorem 9.1, this sequence of continuous d.f.s (of $X_n (n \geq 1)$) converges uniformly to the d.f. of π, the latter is continuous. Thus *π is nonatomic* if *α_n are strictly monotone* a.s. ■

Example 4.1. Let $S = [0, 1]$ and Γ be a family of monotone nondecreasing functions from S into S. As before, for any $z \in S$, let

$$X_n(z) = \alpha_n \cdots \alpha_1 z.$$

One can verify the following two results:

[R.1] *$P[X_n(0) \leq x]$ is nonincreasing in n and converges for each $x \in S$.*

[R.2] *$P[X_n(1) \leq x]$ is nondecreasing in n and converges for each $x \in S$.*

Write

$$F_0(x) \equiv \lim_{n \longrightarrow \infty} P(X_n(0) \leq x).$$

$$F_1(x) \equiv \lim_{n \longrightarrow \infty} P(X_n(1) \leq x).$$

Note that $F_1(x) \leq F_0(x)$ for all x. Consider the case when $\Gamma \equiv \{f\}$, where

$$f(x) = \begin{cases} \frac{1}{4} + \frac{x}{4} & \text{if } 0 \leq x < \frac{1}{3}, \\ \frac{1}{3} + \frac{x}{3} & \text{if } \frac{1}{3} \leq x \leq \frac{2}{3}, \\ \frac{1}{3} + \frac{x}{2} & \text{if } \frac{2}{3} < x \leq 1. \end{cases}$$

Verify that f is a monotone increasing map from S into S, but f is not continuous. One can calculate that

$$F_0(x) = \begin{cases} 0 & \text{if } 0 \leq x < \frac{1}{3}, \\ 1 & \text{if } \frac{1}{3} \leq x \leq 1. \end{cases} \quad F_1(x) = \begin{cases} 0 & \text{if } 0 \leq x < \frac{2}{3}, \\ 1 & \text{if } \frac{2}{3} \leq x \leq 1. \end{cases}$$

Neither F_0 nor F_1 is a stationary distribution function. ■

Example 4.2. Let $S = [0, 1]$ and $\Gamma = \{f_1, f_2\}$. In each period f_1 is chosen with probability $\frac{1}{2}$. f_1 is the function f defined in Example 4.1, and $f_2(x) = \frac{1}{3} + \frac{x}{3}$, for $x \in S$.

Then

$$F_0(x) = F_1(x) = \begin{cases} 0 & \text{if } 0 \le x < \frac{1}{2}, \\ 1 & \text{if } \frac{1}{2} \le x \le 1. \end{cases}$$

and $F_0(x)$ is the unique stationary distribution. Note that $f_1(\frac{1}{2}) = f_2(\frac{1}{2}) = \frac{1}{2}$, i.e., f_1 and f_2 have a common fixed point. Examples 4.1 and 4.2 are taken from [30]. ■

We now turn to the case where the state space is a subset of $R^{\ell} (\ell \ge 1)$ satisfying the following assumption:

(A.1) *S is a closed subset of R^{ℓ}.*

Let Γ be a set of *monotone maps* γ on S into S, under the *partial order*: $\mathbf{x} \le \mathbf{y}$ if $x_j \le y_j$ for $1 \le j \le \ell$; $\mathbf{x} = (x_1, \ldots, x_{\ell})$, $\mathbf{y} = (y_1, y_2, \ldots, y_{\ell}) \in R^{\ell}$ (or S). That is, either γ is *monotone increasing*: $\gamma(\mathbf{x}) \le \gamma(\mathbf{y})$ if $\mathbf{x} \le \mathbf{y}$, or γ is *monotone decreasing*: $\gamma(\mathbf{y}) \le \gamma(\mathbf{x})$ if $\mathbf{x} \le \mathbf{y}$; $\mathbf{x}, \ \mathbf{y} \in S$.

On the space $\mathcal{P}(S)$, define, for each $a > 0$, the metric

$$d_a(\mu, \nu) = \sup_{g \in \mathcal{G}_a} \left| \int g d\mu - \int g d\nu \right|, \ (\mu, \nu \in \mathcal{P}(S)), \tag{4.8}$$

where $\mathcal{G}_a$ is the *class of all Borel measurable monotone* (increasing or decreasing) *functions g on S into* $[0, a]$. The following result is due to [10], who derived a number of interesting results on the metric space $(\mathcal{P}(S), d_a)$. One can show that convergence in the metric d_a implies weak convergence if (A.1) holds (see [9], pp. 287-288).

Lemma 9.1. *Under the hypothesis (A.1), $(\mathcal{P}(S), d_a)$ is a complete metric space.*

Consider the following *splitting condition* $(\mathbf{H}')$. To state it, let $\tilde{\gamma}$ be as in (3.5), but with $n = N : \tilde{\gamma} = \gamma_N \gamma_{N-1} \cdots \gamma_1$ for $\gamma = (\gamma_1, \gamma_2, \ldots, \gamma_N) \in \Gamma^N$.

(H$'$) *There exist $F_i \in \sum^{\otimes N} (i = 1, 2)$ for some $N \ge 1$, such that*

(i) $\delta_i \equiv Q^N(F_i) > 0 \ (i = 1, 2)$, *and*

(ii) *for some $\mathbf{x}_0 \in S$, one has*

$$\tilde{\gamma}(\mathbf{x}) \le \mathbf{x}_0 \ \forall \mathbf{x} \in S, \ \ \forall \gamma \in F_1,$$

$$\tilde{\gamma}(\mathbf{x}) \ge \mathbf{x}_0 \ \forall \mathbf{x} \in S, \ \ \forall \gamma \in F_2,$$

Also, assume that the set $H_+ = \{\gamma \in \Gamma^N : \tilde{\gamma}$ is monotone increasing$\} \in \sum^{\otimes N}$.

Theorem 9.2. *Let $\{\alpha_n : n \geq 1\}$ be a sequence of i.i.d. measurable monotone maps with a common distribution Q. Assume (A.1) and ($\mathbf{H'}$) hold. Then there exists a unique invariant probability measure for the Markov process (2.1) and*

$$\sup_{\mathbf{x}\epsilon S} d_1(p^{(n)}(\mathbf{x},.), \pi) \leq (1-\delta)^{\left[\frac{n}{N}\right]} (n \geq 1), \tag{4.9}$$

where $\delta = \min\{\delta_1, \delta_2\}$, and $\left[\frac{n}{N}\right]$ is the integer part of $\frac{n}{N}$.

Proof. The proof uses Lemma 9.1 and is spelled out in [9]. As in the case of Theorem 9.1, we prove:
Step 1. T^{*N} *is a uniformly strict contraction on* $(\mathcal{P}(S), d_1)$

In other words,

$$d_1(T^{*N}\mu, T^{*N}\nu) \leq (1-\delta) d_1(\mu,\nu), \; \forall \mu, \nu \epsilon \wp(S). \tag{4.10}$$

Now, Step 2. *Apply the Contraction Mapping Theorem.* □

For earlier related results see [4].

9.4.2. *An Extension and Some Applications*

An extension of Theorems 9.1–9.2 [proved in [6]] is useful for applications. Recall that $\mathcal{S}$ is the Borel σ-field of the state space S. Let $\mathcal{A} \subset \mathcal{S}$, define

$$d(\mu,\nu) := \sup_{A\epsilon\mathcal{A}} |\mu(A) - \nu(A)| \qquad (\mu, \nu \epsilon \mathcal{P}(S)). \tag{4.11}$$

(1) Consider the following hypothesis ($\mathbf{H}_1$) :

$$(\mathcal{P}(S), d) \text{ is a complete metric space;} \tag{4.12}$$

(2) there exists a positive integer N such that for all $\gamma \,\epsilon\, \Gamma^N$, one has

$$d(\mu\,\tilde{\gamma}^{-1}, \nu\tilde{\gamma}^{-1}) \leq d(\mu,\nu) \;\; (\mu, \nu\epsilon\mathcal{P}(S)) \tag{4.13}$$

(3) there exists $\delta > 0$ such that $\forall A \epsilon \mathcal{A}$, and with N as in (2), one has

$$P(\tilde{\alpha}^{-1}(A) = S \text{ or } \phi) \geq \delta > 0 \tag{4.14}$$

Theorem 9.3. *Assume the hypothesis* ($\mathbf{H}_1$). *Then there exists a unique invariant probability π for the Markov process $X_n := \alpha_n \cdots \alpha_1 X_0$, where X_0 is independent of $\{\alpha_n := n \geq 1\}$. Also, one has*

$$d(T^{*n}\mu, \pi) \leq (1-\delta)^{[n/N]} \qquad (\mu\epsilon\mathcal{P}(S)) \tag{4.15}$$

where $T^{*n}\mu$ is the distribution of X_n when X_0 has distribution μ, and $[n/N]$ is the integer part of n/N.

Remark 4.4. For applications of Theorem 9.3 to derive a Doeblin-type convergence theorem, and to the study of non-linear autoregressive processes see [9]. ■

9.4.3. *Extinction and Growth*

Some light has been thrown on the possibilities of growth and extinction. To review these results (see [13] for proofs and other related results), let us assume that $S = [0, \infty)$, and Γ consists of a family of maps $f : S \to S$ satisfying

C.1 $f(x)$ is continuously differentiable and strictly increasing on $[0, \infty)$
C.2 $\frac{d}{dx}[x^{-1}f(x)] < 0$ for $x > 0$ (concavity)
C.3 There is some $K > 0$ such that $f(K) < K$ for all $f \in \Gamma$ (note that K is independent of f)

Then we have the following:

Theorem 9.4. *Suppose $0 < X_0 < K$ with probability one. Then:*

(a) *X_n converges in distribution to a stationary distribution;*
(b) *The stationary distribution is independent of X_0 and its df has $F(0^+) = 0$ or 1 [$F(0^+) = 1$ means that $X_n \xrightarrow{w} 0$, which is extinction of the population].*

It is often useful to study the non-linear stochastic difference equation written informally as:

$$X_{n+1} = f(X_n, \theta_{n+1})$$

where (θ_n) is a sequence of independent, identically distributed random variables taking values in a (nonempty) finite set $\mathbb{A} \subset R_{++}$. Here $f : R_+ \times \mathbb{A} \to R_+$ satisfies, for each $\theta \in \mathbb{A}$ the conditions (C.1)–(C.2). Write $R(x, \theta) = x^{-1}f(x, \theta)$ for $x > 0$.

For each $\theta \in \mathbb{A}$, let

$$R(0, \theta) = \lim_{x \to 0+} R(x, \theta)$$

and

$$R(\infty, \theta) = \lim_{x \to \infty} R(x, \theta) \equiv f'(x, \theta)$$

Define the growth rates

$$v_0 = E[(\log R(0, \theta)]$$

and

$$v_\infty = E[(\log R(\infty, \theta)]$$

By C.2 v_0 and v_∞ are well-defined.

Theorem 9.5. *Under assumptions C.1–C.2 and $0 < X_0 < \infty$ with probability one,*

(a) *if $v_0 \leq 0$, $X_n \to 0$ with probability one;*
(b) *if $v_\infty \geq 0$, $X_n \to \infty$ with probability one;*
(c) *if $v_0 > 0$, $v_\infty < 0$, X_n converges weakly (independently of the distribution of X_0) to a distribution with support in* $(0, \infty)$.

9.5. Invariant Distributions: Computation and Estimation

The problem of deriving analytical properties of invariant distributions has turned out to be difficult and elusive. In this section we provide an example of a class of Markov processes in which the unique invariant distribution can be completely identified.

Let $Z_1, Z_2 \ldots,$ be a sequence of non-negative i.i.d. random variables. Consider the Markov Chain $\{X_n : n = 0, 1, 2 \ldots\}$ on the state space $S = R_{++}$ defined by

$$X_{n+1} = Z_{n+1} + [1/X_n] \qquad n \geq 0$$

where X_0 is a strictly positive random variable independent of the sequence $\{Z_i\}$. We first summarize the dynamic behavior of the sequence $\{X_n\}$.

Theorem 9.6. *Assume that $\{Z_i\}$ are non-degenerate. Then the Markov chain $\{X_n,\ n = 0, 1\}$ on $S = R_{++}$ has a unique invariant probability π, and $d_k(T^{*n}\mu, \pi)$ converges to zero exponentially fast, irrespective of the initial distribution μ and the invariant probability π is non-atomic.*

Proof. The main step in the proof is to represent X_n as

$$X_n = \alpha_n . \alpha_{n-1} \cdots \alpha_1(X_0)$$

where $\alpha_n(x) = Z_n + 1/x$, $n \geq 1$. The maps α_n are monotone decreasing on S. The splitting condition can also be verified (see [15], Theorem 4.1). Hence Theorem 9.1 can be applied directly. □

Suppose that the common distribution of Z_i is a Gamma distribution. Recall that the Gamma distribution with parameters $\lambda > 0$ and $a > 0$ is the distribution on R_{++} given by the density function

$$\gamma_{\lambda,a}(z) = \begin{cases} \frac{a^\lambda}{\Gamma(\lambda)} z^{\lambda-1} e^{-az} & \text{if } z \epsilon R_{++} \\ 0 & \text{otherwise} \end{cases}$$

where $\Gamma(\cdot)$ is the gamma function:

$$\Gamma(\beta) = \int_0^\infty x^{\beta-1} e^{-x} dx$$

Theorem 9.7. *Suppose that the common distribution of the i.i.d. sequence $\{Z_i\}$ is a Gamma distribution with parameters λ and a. Then the invariant probability π on $(0, \infty)$ is absolutely continuous with density function*

$$g_{\lambda,a}(x) = (2K_\lambda(2a))^{-1} x^{\lambda-1} e^{-a(x+\frac{1}{x})}, \; x \epsilon R_{++}$$

where $K_\lambda(\cdot)$ denotes the Bessel function, i.e., $K_\lambda(z) = \frac{1}{2}\int_0^\infty x^{\lambda-1} \times e^{-\frac{1}{2}z(x+\frac{1}{x})} dx$.

Another interesting example corresponds to Bernoulli $Z_i : \mathcal{P}(Z_i = 0) = p$, $\mathcal{P}(Z_i = 1) = 1 - p$ $(0 < p < 1)$. In this case the unique invariant distribution π is singular with respect to Lebesgue measure, and has full support on $S = (0, \infty)$. An explicit computation of the distribution function of π, involving the classical continued fraction expansion of the argument, may be found in [15], Theorem 5.2.

9.5.1. *An Estimation Problem*

Consider a Markov chain X_n with a unique stationary distribution π. Some of the celebrated results on ergodicity and the strong law of large numbers hold for π-almost every initial condition. However, even with $[0, 1]$ as the state space, the invariant distribution π may be hard to compute explicitly when the laws of motion are allowed to be non-linear, and its support may be difficult to characterize or may be a set of zero Lebesgue measure. Moreover, in many economic models, the initial condition may be historically given, and there may be little justification in assuming that it belongs to the support of π.

Consider, then, a random dynamical system with state space $[c, d]$ (without loss of generality for what follows choose $c > 0$). Assume Γ

consists of a family of monotone maps from S with S, and the splitting condition (**H**) hold. The process starts with a given x. There is, by Theorem 9.1, a unique invariant distribution π of the random dynamical system, and (4.4) holds. Suppose we want to estimate the equilibrium mean $\int_S y\pi(dy)$ by *sample means* $\frac{1}{n}\sum_{j=0}^{n-1} X_j$. We say that the estimator $\frac{1}{n}\sum_{j=0} X_j$ is *$\sqrt{n}$-consistent* if

$$\frac{1}{n}\sum_{j=0}^{n-1} X_j = \int y\pi(dy) + O_P(n^{-1/2}) \tag{5.1}$$

where $O_p(n^{-1/2})$ is a random sequence ε_n such that $\left|\varepsilon_n \cdot n^{1/2}\right|$ is bounded in probability. Thus, if the estimator is $\sqrt{n}$-consistent, the fluctuations of the empirical (or sample-) mean around the equilibrium mean is $O_p(n^{-1/2})$. We can establish (5.1) by using (4.4). One can show that (see [7], pp. 217-219) if

$f(z) = z - \int y\pi(dy)$
then

$$\sup_x \sum_{n=m+1}^{\infty} |T^n f(x)| \leq (d-c) \sum_{n=m+1}^{\infty} (1-\delta)^{[n/N]} \to 0 \qquad as\ m \to \infty$$

Hence, $g = -\sum_{n=0}^{\infty} T^N f$ [where T^0 is the identity operator I] is well-defined, and g, and Tg are bounded functions. Also, $(T-I)g = -\sum_{n=1}^{\infty} T^n f + \sum_{n=0}^{\infty} T^N f \equiv f$. Hence,

$$\begin{aligned}
\sum_{j=0}^{n-1} f(X_j) &= \sum_{j=0}^{n-1}(T-I)g(X_j) \\
&= \sum_{j=0}^{n-1}((Tg)(X_j) - g(X_j)) \\
&= \sum_{j=1}^{n}[(Tg)(X_{j-1}) - g(X_j)] + g(X_n) - g(X_0)
\end{aligned}$$

By the Markov property and the definition of Tg it follows that

$$E((Tg)(X_{j-1}) - g(X_j)\,|\mathcal{F}_{j-1}) = 0$$

where $\mathcal{F}_r$ is the σ-field generated by $\{X_j : 0 \leq j \leq r\}$. Hence, $(Tg)(X_{j-1}) - g(X_j)(j \geq 1)$ is a martingale difference sequence, and are uncorrelated, so that

$$E[\sum_{j=1}^{k}(Tg(X_{j-1}) - g(X_j))]^2 = \sum_{j=1}^{n} E((Tg)(X_{j-1}) - g(X_j))^2 \tag{5.2}$$

Given the boundedness of g and Tg, the right side is bounded by $n.\alpha$ for some constant α. It follows that

$$\frac{1}{n}E(\sum_{j=0}^{n-1} f(X_j))^2 \leq \eta' \qquad for\, all\, n$$

where η' is a constant that does not depend on X_0. Thus,

$$E(\frac{1}{n}\sum_{j=0}^{n-1} X_j - \int y\pi(dy))^2 \leq \eta'/n$$

which implies,

$$\frac{1}{n}\sum_{j=0}^{n-1} X_j = \int y\pi(dy) + 0_p(n^{-1/2})$$

For other examples of $\sqrt{n}$-consistent estimation, see [2] [and [8], Chapter 5].

9.6. Growth Under Uncertainty

9.6.1. *A Stochastic Stability Theorem in a Descriptive Model*

Models of descriptive as well as optimal growth under uncertainty have led to random dynamical systems that are stable in distribution. We look at a "canonical" example and show how Theorem 9.1 can be applied. We begin with a descriptive growth model and follow it up with an optimization problem.

As a matter of notation, for any function h on S into S, we write $h^{(n)}$ for the nth iterate of h. Think of 'x' as per capital output of an economy.

Let $S = R_+$; and $\Gamma = \{F_1, F_2, \ldots, F_i, \ldots, F_N\}$ where the distinct laws of motion F_i satisfy:

F.1. *F_i is strictly increasing, continuous, and there is some $r_i > 0$ such that $F_i(x) > x$ on $(0, r_i)$ and $F_i(x) < x$ for $x > r_i$.*

Note that $F_i(r_i) = r_i$ for all $i = 1, \ldots, N$. Next, assume:

F.2. $r_i \neq r_j$ for $i \neq j$.

In other words, the unique positive fixed points r_i of distinct laws of motion are all distinct. We choose the indices $i = 1, 2, \ldots, N$ so that

$$r_1 < r_2 < \cdots < r_N$$

Let Prob $(\alpha_n = F_i) = p_i > 0 (i \leq i \leq N)$.

Consider the Markov process $\{X_n(x)\}$ with the state space $(0, \infty)$. If $y \geq r_1$, then $F_i(y) \geq F_i(r_1) > r_1$ for $i = 2, \ldots N$, and $F_1(r_1) = r_1$, so that $X_n(x) \geq r_1$ for all $n \geq 0$ if $x \geq r_1$. Similarly, if $y \leq r_N$, then $F_i(y) \leq F_i(r_N) < r_N$ for $i = 1, \ldots, N-1$ and $F_N(r_N) = r_N$, so that $X_n(x) \leq r_N$ for all $n \geq 0$ if $x \leq r_N$. Hence, if the initial state x is in $[r_1, r_N]$, then the process $\{X_n(x) : n \geq 0\}$ remains in $[r_1, r_N]$ forever. We shall presently see that for a long run analysis we can consider $[r_1, r_N]$ as the effective state space.

We shall first indicate that on the state space $[r_1, r_N]$ the splitting condition (H) is satisfied. If $x \geq r_1$, $F_1(x) \leq x$, $F_1^{(2)}(x) \leq F_1(x)$ etc. The limit of this decreasing sequence $F_1^{(n)}(x)$ must be a fixed point of F_1, and therefore must be r_1. Similarly, if $x \leq r_N$, then $F_N^n(x)$ increases to r_N. In particular,

$$\lim_{n\to\infty} F_1^{(n)}(r_N) = r_1, \qquad \lim_{n\to\infty} F_N^{(n)}(r_1) = r_N.$$

Thus, there must be a positive integer n_0 such that

$$F_1^{(n_0)}(r_N) < F_N^{(n_0)}(r_1).$$

This means that if $z_0 \,\epsilon\, [F_1^{(n_0)}(r_N), F_1^{(n_0)}(r_1)]$, then

$$\begin{aligned}
&\Pr\text{ob}(X_{n_0}(x) \leq z_0 \;\; \forall x \epsilon [r_1, r_N]) \\
&\qquad\qquad \geq \Pr ob(\alpha_n = F_1 \, for \, 1 \leq n \leq n_0) = p_1^{n_0} > 0 \\
&\Pr\text{ob}(X_{n_0}(x) \geq z_0 \;\; \forall x \epsilon [r_1, r_n]) \\
&\qquad\qquad \geq \Pr ob(\alpha_n = F_N \, for \, 1 \leq n \leq n_0) = p_N^{n_0} > 0
\end{aligned}$$

Hence, considering $[r_1, r_N]$ as the state space, and using Theorem 9.1, there is a unique invariant probability π with the stability property holding for all initial $x \epsilon [r_1, r_N]$.

Now, define $m(x) = \min_{i=1,\ldots,N} F_i(x)$, and fix the initial state $x \epsilon (0, r_1)$.

One can verify that (i) m is continuous; (ii) m is strictly increasing; (iii) $m(r_1) = r_1$ and $m(x) > x$ for $x \epsilon (0, r_1)$, and $m(x) < x$ for $x > r_1$.

Clearly $m^{(n)}(x)$ increases with n, and $m^{(n)}(x) \leq r_1$. The limit of the sequence $m^{(n)}(x)$ must be a fixed point, and is, therefore r_1. Since $F_i(r_1) > r_1$ for $i = 2, \ldots, N$, there exists some $\varepsilon > 0$ such that $F_i(y) > r_1 (2 \leq i \leq N)$ for all $y \epsilon [r_1 - \varepsilon, r_1]$. Clearly there is some n_ε such that $m^{n_\varepsilon}(x) \geq r_1 - \varepsilon$. If $\tau_1 = \inf\{n \geq 1 : X_n(x) > r_1\}$ then it follows that for all $k \geq 1$

$$\text{Prob}(\tau_1 > n_\varepsilon + k) \leq p_1^k.$$

Since p_1^k goes to zero as $k \to \infty$, it follows that τ_1 is finite almost surely. Also, $X_{\tau_1}(x) \leq r_N$, since for $y \leq r_1$, (i) $F_i(y) < F_i(r_N)$ for all i and (ii) $F_i(r_N) < r_N$ for $i = 1, 2, \ldots, N-1$ and $F_N(r_N) = r_N$. (In a single period it is not possible to go from a state less than r_1 to one larger than r_N). By the strong Markov property, and our earlier result, $X_{\tau+m}(x)$ converges in distribution to π as $m \to \infty$ for all $x \epsilon (0, r_1)$. Similarly, one can check that as $n \to \infty$, $X_n(x)$ converges in distribution to π for all $x > r_N$. ∎

Note that in growth models, the condition F.1 is often derived from appropriate "end point" or Uzawa-Inada conditions. It should perhaps be stressed that convexity assumptions have not appeared in the discussion of this section so far. Of course, in models of optimization, F_i is the *optimal transition* of the system from one state into another, and non-convexity may lead to a failure of the splitting condition (see [19] for details).

9.6.2. *One Sector Log-Cobb-Douglas Optimal Growth*

Let us recall the formulation of the one-sector growth model with a Cobb-Douglas production function $G(x) = x^\alpha, 0 < \alpha < 1$, with a representative decision maker's utility given by $u(c) = \ln c$. Following [21], suppose that an exogenous perturbation may reduce production by some parameter $0 < k < 1$ with probability $p > 0$ (the same for all $t = 0, 1, \ldots$). This independent and identically distributed random shock enters multiplicatively into the production process so that output is given by $G_r(x) = rx^\alpha$ where $r \in \{k, 1\}$. The dynamic optimization problem can be explicitly written as follows:

$$\max \mathbb{E}_0 \sum_{t=0}^{\infty} \beta^t \ln c_t$$

where $0 < \beta < 1$ is the discount factor, and the maximization is over all consumption plans $c = (c_0, c_1, \ldots)$ such that for $t = 0, 1, 2, \ldots$

$$c_t = r_t x_t^\alpha - x_{t+1}, \quad c_t \geq 0, \quad x_t \geq 0$$

and x_0, r_0 are given.

It is well known that the optimal transition of x_t is just described is $g(x,r) = \alpha\beta r x^{\alpha}$ i.e., the plan x_t generated recursively by

$$x_{t+1} = g(x_t, r_t) = \alpha\beta r_t x_t^{\alpha}$$

is optimal.

Consider now the random dynamical system obtained by the following logarithmic transformation of x_t:

$$y_t = -\frac{1-\alpha}{\ln k} \ln x_t + 1 + \frac{1n\,\alpha + \ln \beta}{\ln k}.$$

The new variable y_t, associated with x_t evolves according to a linear policy, so that

$$y_{t+1} = \alpha y_t + (1-\alpha)\left(1 - \frac{\ln r_t}{\ln k}\right),$$

which can be rewritten as

$$\begin{cases} y_{t+1} = \alpha y_t & \text{with probability } p \\ y_{t+1} = \alpha y_t + (1-\alpha) & \text{with probability } 1-p \end{cases}$$

Define the maps γ_0, γ_1 from $[0,1]$ to $[0,1]$ by

$$\begin{cases} \gamma_0(y) = \alpha y \\ \gamma_1(y) = \alpha y + (1-\alpha) \end{cases} \tag{6.1}$$

It is useful to note here that the map γ_0 corresponds to the case where the shock, r, takes the value k; and the map γ_1 corresponds to the case where the shock, r, takes the value 1. Denote $(p, 1-p)$ by (p_0, p_1). Then $S = [0,1]$, $\Gamma \equiv \{\gamma_0, \gamma_1\}$, together with $Q \equiv \{p_0, p_1\}$ is a random dynamical system. The maps γ_i, for $i \in \{0,1\}$, are clearly affine.

9.6.2.1. *The Support of the Invariant Distribution*

Let π be the unique invariant distribution, F_π, its distribution function. The graphs of the functions show that for $0 < \alpha < 1/2$, the image sets of the two functions γ_0 and γ_1 are disjoint, a situation which can be described as the "non-overlapping" case. In this case, the "gap" between the two image sets (in the unit interval) will "spread" through the unit interval by successive applications of the maps (6.1). Thus, one would expect the support of the invariant distribution to be "thin" (with zero Lebesgue measure).

On the other hand, for $1/2 \leq \alpha < 1$, the image sets of the functions γ_0 and γ_1 have a non-empty intersection. We can refer to this as the "overlapping" case. Here, the successive iterations of the overlap can be expected to "fill up" the unit interval, so the invariant distribution should have full support.

The above heuristics are actually seen to be valid.

It is important to remark that this result does not depend on the magnitude of the discount factor β nor on the amplitude of the shock k, but only on the technological parameter α. The discount factor β only shifts the support of the invariant distribution of the original model over the real line, while the exogenous shock k affects its amplitude. The stream of research has been striving around the fundamental question on deciding for what values of α, the invariant F_π is absolutely continuous, and for what values of α, F_π is singular. For an exhaustive mathematical survey on the whole history of Bernoulli convolutions, see [24]. It is known, in the symmetric case $p = \frac{1}{2}$, that the distribution function is "pure"; that is, it is either *absolutely continuous* or it is *singular* (Jessen and Wintner [1935]). Further, Kershner and Wintner [1935] have shown that if $0 < \alpha < 1/2$, the support of the distribution function is a Lebesgue-null Cantor set and, therefore, the distribution function is singular. For $\alpha = \frac{1}{2}$, one gets the uniform distribution, which is *not* singular.

For the symmetric case $p = \frac{1}{2}$, denote by $S_\perp$ the set of $\alpha \in (1/2, 1)$ such that F_π is singular. It was conjectured that the distribution function should be absolutely continuous with respect to Lebesgue measure when $1/2 < \alpha < 1$. Wintner [1935] showed that if α is of the form $(1/2)^{1/k}$ where $k \in \{1, 2, 3, \ldots\}$, then the distribution function is absolutely continuous. However, in the other direction, Erdös [1939] showed that when α is the positive solution of the Equation $\alpha^2 + \alpha - 1 = 0$, so that $\alpha = (\sqrt{5} - 1)/2$, then $\alpha \in S_\perp$.

Erdös also showed that $S_\perp \cap (\xi, 1)$ has zero Lebesgue measure for some $\xi < 1$, so that absolute continuity of the invariant distribution obtains for (almost every) α sufficiently close to 1. A conjecture that emerged from these findings is that the set $S_\perp$ itself should have Lebesgue measure zero. In their brief discussion of this problem, [12] state that deciding whether the invariant distribution is singular or absolutely continuous for $\alpha > 1/2$ is a "famous open question".

Solomyak [1995] made a real breakthrough when he showed that $S_\perp$ has zero Lebesgue measure. More precisely, he established that for almost every $\alpha \in (1/2, 1)$, the distribution has density in $L^2(R)$ and for almost every

$\alpha \in (2^{-1/2}, 1)$ the density is bounded and continuous. A simpler proof of the same result was subsequently presented by [25].

More recent contributions to this literature deal with the asymmetric case $p \neq 1/2$. (see, for example, [26]). For example, F_π is singular for values of parameters (α, p) such that $0 < \alpha < p^p(1-p)^{(1-p)}$, while F_π is absolutely continuous for almost every $p^p(1-p)^{(1-p)} < \alpha < 1$ whenever $1/3 \leq p \leq 2/3$. For more details see [21].

Acknowledgments

This paper is primarily based on my research with Professor Rabi Bhattacharya over a number of years. No formal acknowledgement to his influence is adequate. Thanks are due to the referee for a careful reading and suggestions on expository changes.

References

[1] Athreya, K. B. and Bhattacharya, R. N. (2000). Random Iteration of I.I.D. Quadratic Maps. In *Stochastics in Finite and Infinite Dimensions: in Honor of Gopinath Kallianpur.* (Eds. T. Hida et. al.) Birkhauser, Basel, 49–58.

[2] Athreya, K. B. and Majumdar, M. (2002). Estimating the Stationary Distribution of a Markov Chain. *Economic Theory.* **27** 729–742.

[3] Athreya, K. B. (2004). Stationary Measures for Some Markov Chain Models in Ecology and Economics. *Economic Theory.* **23** 107–122.

[4] Bhattacharya, R. N. and Lee, O. (1988). Asymptotics of a Class of Markov Processes, That Are Not in General Reducible. *Annals of Probability.* **16** 1333–47.

[5] Bhattacharya, R. N. and Rao, B. V. (1993). Random Iterations of Two Quadratic Maps. In *Stochastic Processes* (Eds. S. Cambanis, J. K. Ghosh, R. L. Karandikar and P. K. Sen). Springer Verlag, New York, 13–21.

[6] Bhattacharya, R. N. and Majumdar, M. (1999). On a Theorem of Dubins and Freedman. *Journal of Theoretical Probability.* **12** 1067–1087.

[7] Bhattacharya, R. N. and Majumdar, M. (2001). On a Class of Stable Random Dynamical Systems: Theory and Applications. *Journal of Economic Theory.* **96** 208–229.

[8] Bhattacharya, R. N. and Majumdar, M. (2008). Random Iterates of Monotone Maps. Submitted.

[9] Bhattacharya, R. N. and Majumdar, M. (2007) *Random Dynamical Systems: Theory and Applications.* Cambridge University Press, Cambridge.

[10] Chakraborty, S. and Rao, B. V. (1998). Completeness of Bhattacharya Metric on the Space of Probabilities. *Statistics and Probability Letters.* **36** 321–326.

[11] Diaconis, P. and Freedman, D. (1999). Iterated Random Functions. *SIAM Review.* **41** 45–79.

[12] Dubins, L. E. and Freedman, D. (1966). Invariant Probabilities for Certain Markov Processes. *Annals of Mathematical Statistics.* **37** 837–858.

[13] Ellner, S. (1984). Asymptotic Behavior of Some Stochastic Difference Equation Population Models. *Journal of Mathematical Biology.* **19** 169–200.

[14] Erdös, P. (1939). On a Family of Symmetric Bernoulli Convolutions. *American Journal of Mathematics.* **61** 944–975.

[15] Goswami, A. (2004). Random Continued Fractions: A Markov Chain Approach. *Economic Theory.* **23** 85–106.

[16] Jessen, B. and Wintner, A. (1935). Distribution Function and the Riemann Zeta Function. *Transactions of the American Mathematical Society.* **38** 48–58.

[17] Kershner, R. and Wintner, A. (1935). On Symmetric Bernoulli Convolutions. *American Journal of Mathematics.* **57** 541–548.

[18] Loeve, M. (1963). *Probability Theory.* Van Nostrand, Princeton.

[19] Majumdar, M., Mitra, T., and Nyarko, Y. (1989). Dynamic Optimization under Uncertainty: Non-convex Feasible Set.in *Joan Robinson and Modern Economic Theory.* (Ed. G.R. Feiwel). MacMillan, London, 545–590.

[20] Mitra, K. (1998). On Capital Accumulation Paths in Neoclassical Models in Stochastic Growth Models. *Economic Theory.* **11** 457–464.

[21] Mitra, T., Montrucchio, L. and Privileggi, F. (2004). The Nature of the Steady State in Models of Optimal Growth under Uncertainty. *Economic Theory.* **23** 39–71.

[22] Montrucchio, L. and Privileggi, F. (1999). Fractal Steady States in Stochastic Optimal Control Models. *Annals of Operations Research.* **88** 183–197.

[23] Orey, S. (1971). *Limit Theorems for Markov Chain Probabilities.* Van Nostrand, New York.

[24] Peres, Y., Shlag, W. and Solomyak, B. (1999). Sixty Years of Bernoulli Convolutions. In *Fractal Geometry and Stochastics.* **2** (C. Bandt, S. Graf, M. Zahle, Eds.) Birkhauser, Basel, 39–65.

[25] Peres, Y. and Solomyak, B. (1996). Absolute Continuity of Benoulli Convolutions, A Simple Proof. *Mathematical Research Letters.* **3** 231–239.

[26] Peres, Y. and Schlag, W. (2000). Smoothness of Projections. Bernoulli Convolutions and the Dimension of Exception. *Duke Mathematical Journal.* **102** 193–251.

[27] Solomyak, B. (1995). On the Random Series $\sum \pm\lambda^n$: an Erdös Problem. *Annals of Mathematics.* **142** 611–625.

[28] Solow, R. M. (1956). A Contribution of the Theory of Economic Growth. *Quarterly Journal of Economics.* **70** 65–94.

[29] Wintner, A. (1935). On Convergent Poisson Convolutions. *American Journal of Mathematics.* **57** 827–838.

[30] Yahav, J. A. (1975). On a Fixed Point Theorem and Its Stochastic Equivalent. *Journal of Applied Probability.* **12** 605–611.

Chapter 10

An Invitation to Quantum Information Theory

K. R. Parthasarathy

Indian Statistical Institute, Delhi Centre,
7, S. J. S. Sansanwal Marg, New Delhi – 110 016, India
krp@isid.ac.in

10.1. Introduction

In this lecture we shall present a brief account of some of the interesting developments in recent years on quantum information theory. The text is divided into three sections. The first section is devoted to fundamental concepts like events, observables, states, measurements and transformations of states in finite level quantum systems. the second section gives a quick account of the theory of error correcting quantum codes. The last section deals with the theory of testing quantum hypotheses and the quantum version of Shannon's coding theorem for classical-quantum channels. I have liberally used the books by Nielson and Chuang ([12]), Hayashi ([6]) and also [21], [23]. The emphasis is on the description of results and giving a broad perspective. For proofs and detailed bibliography we refer to [12], [6].

10.2. Elements of Finite Dimensional Quantum Probability

The first step in entering the territory of quantum information theory is an acquaintance with quantum probability where the fundamental notions of events, random variables, probability distributions and measurements are formulated in the language of operators in a Hilbert space. In the present exposition all the Hilbert spaces will be complex and finite dimensional and their scalar products will be expressed in the Dirac notation $\langle \cdot | \cdot \rangle$. For any Hilbert space $\mathcal{H}$, denote $\mathcal{B}(\mathcal{H})$ the $*$-algebra of all operators on $\mathcal{H}$ where, for any operator A, its adjoint will be denoted by $A^\dagger$. We write

$$\langle u\,|A|\,v\rangle = \langle u\,|Av\rangle = \langle A^\dagger u\,|v\rangle \ \forall\ u,v\in\mathcal{H},\ A\in\mathcal{B}(\mathcal{H}).$$

$\mathcal{B}(\mathcal{H})$ itself will also be viewed as a Hilbert space, in its own right, with the scalar product

$$\langle A\,|B\rangle = \operatorname{Tr} A^\dagger B \quad \forall \quad A,B\in\mathcal{B}(\mathcal{H}).$$

In the real linear subspace of all hermitian operators we write $A\geq B$ if $A-B$ is nonnegative definite, i.e., $A-B\geq 0$. It is clear that $\geq$ is a partial ordering. By a projection we shall always mean an orthogonal projection operator. Denote by $\mathcal{O}(\mathcal{H})$ and $\mathcal{P}(\mathcal{H})$ respectively the space of all hermitian and projection operators. Any element A in $\mathcal{O}(\mathcal{H})$ is called an *observable* and any element E in $\mathcal{P}(\mathcal{H})$ an *event*. The elements 0 and I in $\mathcal{P}(\mathcal{H})$ are called the *null* and *certain* events. Clearly, $0\leq E\leq I$ for any E in $\mathcal{P}(\mathcal{H})$. If E is an event, $I-E$ denoted $E^\perp$ is the event called *not* E. For two events E,F let $E\wedge F$ and $E\vee F$ denote respectively the maximum and minimum of the pair E,F in the ordering $\leq$. If $E\leq F$ we say that the event E *implies* F. If E,F are arbitrary events we interpret $E\vee F$ as E *or* F and $E\wedge F$ as E *and* F. It is important to note that for three events E,E_1,E_2 the event $E\wedge(E_1\vee E_2)$ may differ from $(E\wedge E_1)\vee(E\wedge E_2)$. (In the quantum world the operation 'and' need not distribute with 'or' but in the logic of propositions that we prove about a quantum system the logical operation 'and' does distribute with the logical operation 'or'!)

For any observable $X\in\mathcal{O}(\mathcal{H})$ and any real-valued function f on $\mathbb{R}$, the observable described by the hermitian operator $f(X)$ is understood as the function f of the observable X. If $E\subseteq\mathbb{R}$ and 1_E denotes the indicator function of E then $1_E(X)$ is a projection which is to be interpreted as the event that the value of the observable X lies in E. Thus, for a singleton set $\{\lambda\}\subseteq\mathbb{R}$, $1_{\{\lambda\}}(X)\neq 0$ if and only if λ is an eigenvalue of X and the corresponding eigenprojection $1_{\{\lambda\}}(X)$ is the event that X assumes the value λ. This shows that the values of the observable X constitute the spectrum of X, denoted by $\operatorname{spec} X$ and

$$X = \sum_{\lambda\in\operatorname{spec} X} \lambda 1_{\{\lambda\}}(X)$$

is the spectral decomposition of X. We have

$$\mathcal{P}(\mathcal{H})\subset\mathcal{O}(\mathcal{H})\subset\mathcal{B}(\mathcal{H})$$

and any $E\in\mathcal{P}(\mathcal{H})$ is a $\{0,1\}$-valued observable.

Any nonnegative operator $\rho \geq 0$ of unit trace on $\mathcal{H}$ is called a *state* of the quantum system described by $\mathcal{H}$. The set of all such states is denoted by $\mathcal{S}(\mathcal{H})$. Then $\mathcal{S}(\mathcal{H})$ is a compact convex set in $\mathcal{B}(\mathcal{H})$ with its extreme points one dimensional projections. One dimensional projections are called *pure* states and thanks to spectral theorem any state can be expressed as

$$\rho = \sum_j p_j \ |u_j\rangle\langle u_j| \tag{10.1}$$

where $\{u_j\}$ is an orthonormal set of vectors in $\mathcal{H}$, $p_j \geq 0 \ \forall \ j$ and $\sum_j p_j = 1$. In other words every stat e can be expressed as a convex combination of pure states which are mutually orthogonal one dimensional projections. For any $X \in \mathcal{O}(\mathcal{H})$, $\rho \in \mathcal{S}(\mathcal{H})$ the real scalar $\mathrm{Tr}\ \rho X$ is called the *expectation* of the observable X in the state ρ. If $X = E \in \mathcal{P}(\mathcal{H})$ its expectation $\mathrm{Tr}\,\rho E$ is the probability of the event E in the state ρ. Thus 'tracing out' in quantum probability is the analogue of 'integration' in classical probability. Note that

$$\mathcal{S}(\mathcal{H}) \subset \mathcal{O}(\mathcal{H}) \subset \mathcal{B}(\mathcal{H}).$$

In a quantum system described by the pair $(\mathcal{H}, \rho)$ where ρ is a state in $\mathcal{H}$, the observable ρ or its equivalent $-\log \rho$ is a fundamental observable in quantum information theory and its expectation $S(\rho) = -\mathrm{Tr}\,\rho \log \rho$ is called the *von Neumann entropy* or, simply, entropy of ρ. Even though $-log\,\rho$ can take infinite values $S(\rho)$ is finite.

It is often useful to consider the linear functional $X \to \mathrm{Tr}\,\rho X$ on $\mathcal{B}(\mathcal{H})$ for any ρ in $\mathcal{S}(\mathcal{H})$ and call it the *expectation map* in the state ρ. In other words we can view expectation as a nonnegative linear functional on the C^* algebra $\mathcal{B}(\mathcal{H})$ satisfying

$$\mathrm{Tr}\,\rho X^\dagger X \geq 0 \ \forall \ X \in \mathcal{B}(\mathcal{H}),$$
$$\mathrm{Tr}\,\rho I = 1.$$

If $u, v \in \mathcal{H}$ we say that $|v\rangle \in \mathcal{H}$ and call it a *ket* vector whereas $\langle u| \in \mathcal{H}^*$, the dual of $\mathcal{H}$ and call it a *bra* vector. The bra vector $\langle u|$ evaluated at the ket vector $|v\rangle$ is the bracket $\langle u|v\rangle$, the scalar product between u and v in $\mathcal{H}$. Thus for u, v in $\mathcal{H}$ we can define the operator $|u\rangle\langle v|$ by

$$|u\rangle\langle v| \ |w\rangle = \langle v|w\rangle \, |u\rangle \quad \forall \ w \in \mathcal{H}.$$

If $u \neq 0$, $v \neq 0$ then $|u\rangle\langle v|$ is a rank one operator with range $\mathbb{C}\,|u\rangle$. In particular,

$$
\begin{aligned}
&|u_1\rangle\langle u_2|\;|u_3\rangle\langle u_4|\cdots|u_{2j+1}\rangle\langle u_{2j+2}| \\
&\quad = (\langle u_2|u_3\rangle\langle u_4|u_5\rangle\cdots\langle u_{2j}|u_{2j+1}\rangle)\;|u_1\rangle\langle u_{2j+2}|, (|u\rangle\langle v|)^\dagger = |v\rangle\langle u|, \\
&\qquad \operatorname{Tr}|u\rangle\langle v| = \langle v|u\rangle.
\end{aligned}
$$

Equation (10.1) anticipates this notation developed by Dirac.

A finite level quantum system A with a state ρ^A in the Hilbert space $\mathcal{H}^A$ is described by the triple $(\mathcal{H}^A, \mathcal{P}(\mathcal{H}^A), \rho^A)$ where the probability of any event $E \in \mathcal{P}(\mathcal{H}^A)$ is given by $\operatorname{Tr}\rho^A E$. If a finite classical probability space is described by the sample space $\Omega^A = \{1, 2, \ldots, N\}$, the Boolean algebra $\mathcal{F}^A$ of all subsets of Ω^A and a probability distribution P^A then the probability space $(\Omega^A, \mathcal{F}^A, P^A)$ can also be described by the triple $(\mathcal{H}^A, \mathcal{P}(\mathcal{H}^A), \rho^A)$ where $\mathcal{H}^A = \mathbb{C}^n$ and the state operator ρ^A is given by the diagonal matrix

$$
\rho^A = \begin{bmatrix} P^A(1) & 0 & 0 & \cdots & 0 \\ 0 & P^A(2) & 0 & \cdots & 0 \\ \cdots & \cdots & \cdots & \cdots & \cdots \\ 0 & 0 & 0 & \cdots & P^A(n) \end{bmatrix}
$$

in the canonical orthonormal basis $\{e_j\}$, e_j being the vector with 1 in the jth position and 0 in all the other positions.

If A_i, $i = 1, 2, \ldots, k$ are k quantum systems and the Hilbert space $\mathcal{H}^{A_i}$ is the one associated with A_i then the Hilbert space for the joint or composite system $A_1 A_2 \ldots A_k$ is the tensor product:

$$
\mathcal{H}^{A_1 A_2 \ldots A_k} = \mathcal{H}^{A_1} \otimes \mathcal{H}^{A_2} \otimes \cdots \otimes \mathcal{H}^{A_k}.
$$

Suppose

$$
\mathcal{H}^{AB} = \mathcal{H}^A \otimes \mathcal{H}^B
$$

is the Hilbert space of the composite system AB consisting of two subsystems A, B where $\{e_i\}$, $\{f_j\}$ are some orthonormal bases in $\mathcal{H}^A$, $\mathcal{H}^B$ respectively. Let X be an operator on $\mathcal{H}^{AB}$. For $u, v \in \mathcal{H}^A$, $u', v' \in \mathcal{H}^B$ define the sesquilinear forms

$$
\beta_X(u, v) = \sum_j \langle u \otimes f_j | X | v \otimes f_j \rangle, \quad u, v \in \mathcal{H}^A,
$$

$$
\beta'_X(u', v') = \sum_i \langle e_i \otimes u' | X | e_i \otimes v' \rangle, \quad u', v' \in \mathcal{H}^B.
$$

Then there exist operators $T \in \mathcal{B}(\mathcal{H}^A)$, $T' \in \mathcal{B}(\mathcal{H}^B)$ such that

$$\langle u|T|v\rangle \equiv \beta_X(u,v), \langle u'|T'|v'\rangle \equiv \beta'_X(u',v').$$

Then T and T' are respectively the *relative trace* of X over $\mathcal{H}^B$ and $\mathcal{H}^A$ and one writes

$$T = \mathrm{Tr}_{\mathcal{H}^B} X, \quad T' = \mathrm{Tr}_{\mathcal{H}^A} X.$$

The relative traces T, T' are independent of the choice of the orthonormal bases $\{e_i\}$, $\{f_j\}$. The linear maps $X \to \mathrm{Tr}_{\mathcal{H}^B} X$, $X \to \mathrm{Tr}_{\mathcal{H}^A} X$ from $\mathcal{B}(\mathcal{H}^{AB})$ onto $\mathcal{B}(\mathcal{H}^A)$, $\mathcal{B}(\mathcal{H}^B)$ respectively are *completely positive* in the sense that the map

$$[X_{ij}] \to [\mathrm{Tr}_{\mathcal{H}^B}(X_{ij})], \quad i,j \in \{1,2,\ldots,d\}, \quad X_{ij} \in \mathcal{B}(\mathcal{H}^{AB})$$

from $\mathcal{B}\left(\mathcal{H}^{AB} \otimes \mathbb{C}^d\right)$ into $\mathcal{B}\left(\mathcal{H}^A \otimes \mathbb{C}^d\right)$ is positive for every $d = 1, 2, \ldots$. Furthermore for all Y in $\mathcal{B}\left(\mathcal{H}^A\right)$ and X in $\mathcal{B}\left(\mathcal{H}^{AB}\right)$

$$\begin{aligned}
\mathrm{Tr}_{\mathcal{H}^B}(Y \otimes I) X &= Y \mathrm{Tr}_{\mathcal{H}^B} X, \\
\mathrm{Tr}_{\mathcal{H}^B} X (Y \otimes I) &= (\mathrm{Tr}_{\mathcal{H}^B} X) Y, \\
\mathrm{Tr}\,(\mathrm{Tr}_{\mathcal{H}^B} X) = \mathrm{Tr}\,(\mathrm{Tr}_{\mathcal{H}^A} X) &= \mathrm{Tr}\, X.
\end{aligned}$$

In other words, if tracing is viewed as integration the first two relations exhibit the properties of conditional expectation and the last one is analogous to Fubini's theorem. If ρ^{AB} is a state of the composite system then $\mathrm{Tr}_{\mathcal{H}^B} \rho^{AB} = \rho^A$ and $\mathrm{Tr}_{\mathcal{H}^A} \rho^{AB} = \rho^B$ are states on $\mathcal{H}^A$ and $\mathcal{H}^B$ respectively. We call ρ^A and ρ^B the marginal states of ρ^{AB}.

In the context of composite systems there arises the following natural question. Suppose ρ_i is a state in the Hilbert space $\mathcal{H}_i$, $i = 1, 2$. Denote by $\mathcal{C}(\rho_1, \rho_2)$ the compact convex set of all states in $\mathcal{H}_1 \otimes \mathcal{H}_2$ whose $\mathcal{H}_i$-marginal is ρ_i for each i. It is desirable to have a good description of the extreme points of $\mathcal{C}(\rho_1, \rho_2)$. As ρ varies over the set of all such extreme points what is the range of its von Neumann entropy? An answer to this question will throw much light on the *entanglement* between quantum systems $(\mathcal{H}_1, \rho_1)$ and $(\mathcal{H}_2, \rho_2)$. When $\mathcal{H}_i = \mathbb{C}^2$ and $\rho_i = \frac{1}{2} I$ for $i = 1, 2$, it is known that the extreme points are all pure and hence have zero entropy. They are the famous EPR states named after Einstein, Podolskii and Rosen. (See [5], [20], [22].) In general, there can exist extremal states which are not pure.

We now introduce the notion of measurement for finite level quantum systems. They play an important role in formulating the problems of quantum information theory and testing multiple hypotheses. Let $\mathcal{H}$ be the

Hilbert space of a finite level system. By a *measurement* with values in an abstract set $\{m_1, m_2, \ldots, m_k\}$ we mean a partition of I into k nonnegative operators $M_1, M_2, \ldots, M_k$ so that $M_1 + M_2 + \cdots + M_k = I$ and in any state ρ the values m_j occur with respective probabilities $\mathrm{Tr}\, \rho M_j$, $j = 1, 2, \ldots, k$. We express such a measurement by

$$\mathbf{M} = \begin{pmatrix} m_1 & m_2 \cdots m_k \\ M_1 & M_2 \cdots M_k \end{pmatrix} \tag{10.2}$$

When each M_j is a projection it is called a *von Neumann measurement.*

Without loss of generality, for the purposes of information theory we may identify the value set $\{m_1, m_2, \ldots, m_k\}$ with $\{1, 2, \ldots, k\}$. Denote by $M_k(\mathcal{H})$ the compact convex set of all measurements in $\mathcal{H}$ with value set $\{1, 2, \ldots, k\}$. We shall now describe the extreme points of $\mathcal{M}_k(\mathcal{H})$ in a concrete form. The following is a sharpening of Naimark's theorem ([16]) for finite dimensional Hilbert spaces.

Theorem 10.1. *Let* $\mathbf{M} \in \mathcal{M}_k(\mathcal{H})$ *be a measurement given by*

$$\mathbf{M} = \begin{pmatrix} 1 & 2 \cdots k \\ M_1 & M_2 \cdots M_k \end{pmatrix}$$

where

$$M_i = \sum_{j=1}^{r_i} |u_{ij}\rangle\langle u_{ij}| \;\; 1 \le i \le k,$$

$r_i > 0$ *is the rank of* M_i, $\{u_{ij},\ 1 \le j \le r_i\}$ *is an orthogonal set of vectors and* $\|u_{ij}\|^2, 1 \le j \le r_i$ *are the nonzero eigenvalues of* M_i *with multiplicity included. Then there exists a Hilbert space* $\mathcal{K}$, *a set* $\{P_1, P_2, \ldots, P_k\}$ *of projections in* $\mathcal{H} \oplus \mathcal{K}$ *and vectors* $v_{ij} \in \mathcal{K}$, $1 \le j \le r_i$, $1 \le i \le k$ *satisfying the following:*

(i) $\dim \mathcal{H} \oplus \mathcal{K} = \sum_{i=1}^{k} r_i$;

(ii) *The set* $\{u_{ij} \oplus v_{ij} \,|\, 1 \le j \le r_i, 1 \le i \le k\}$ *is an orthonormal basis for* $\mathcal{H} \oplus \mathcal{K}$;

(iii) $P_i = \sum_{j=1}^{r_i} |u_{ij} \oplus v_{ij}\rangle\langle u_{ij} \oplus v_{ij}|$

$$= \left[\begin{array}{c|c} M_i & \sum_{j=1}^{r_i} |u_{ij}\rangle\langle v_{ij}| \\ \hline \sum_{j=1}^{r_i} |v_{ij}\rangle\langle u_{ij}| & \sum_{j=1}^{r_i} |v_{ij}\rangle\langle v_{ij}| \end{array}\right], \quad 1 \le i \le k$$

and $\text{rank}\, P_i = \text{rank}\, M_i \,\forall\, i$;

(iv) *The triplet consisting of* $\mathcal{K}$, *the family* $\{v_{ij} | 1 \leq j \leq r_i,\ 1 \leq i \leq k\}$ *of vectors in* $\mathcal{K}$ *and the spectral resolution* $P_1, P_2, \ldots, P_k$ *is unique upto a Hilbert space isomorphism;*

(v) $\mathbf{M}$ *is an extreme point of* $\mathcal{M}_k(\mathcal{H})$ *if and only if the set* $\{|u_{ij}\rangle\langle u_{ij'}|\ |1 \leq j \leq r_i, 1 \leq j' \leq r_i,\ \ 1 \leq i \leq k\}$ *of rank one operators is linearly independent in* $\mathcal{B}(\mathcal{H})$. *In particular, von Neumann measurements are extremal.*

Remark:

(1) If one of the M_j's is zero, the test for extremality reduces to the case of M_{k-1}.

(2) If $\mathbf{M}$ is an extremal element of $\mathcal{M}_k$ where $\text{rank}\, M_i = r_i > 0\ \forall\ 1 \leq i \leq k$ then $r_1^2 + r_2^2 + \cdots + r_k^2 \leq (\dim\, \mathcal{H})^2$.

(3) If $\mathcal{H} = \mathbb{C}^2$, $\omega = \exp 2\pi i/3$ then

$$\mathbf{M} = \left(\begin{array}{ccc} 1 & 2 & 3 \\ \frac{1}{3}\begin{bmatrix} 1 & 1 \\ 1 & 1 \end{bmatrix} & \frac{1}{3}\begin{bmatrix} 1 & \omega \\ \overline{\omega} & 1 \end{bmatrix} & \frac{1}{3}\begin{bmatrix} 1 & \omega^2 \\ \overline{\omega}^2 & 1 \end{bmatrix} \end{array} \right)$$

is an extremal measurement in $\mathcal{M}_3(\mathcal{H})$ but not a von Neumann measurement.

(4) It would be interesting to make a finer classification of extremal measurements modulo permutations of the set $\{1, 2, \ldots, k\}$ and conjugation by unitary operators in $\mathcal{H}$.

Evolutions of quantum states in time due to inherent dynamics which may or may not be reversible and measurements made on the quantum system bring about changes in the state. Reversible changes are of the form : $\rho \to U \rho U^{-1}$ where U is a unitary or antiunitary operator. According to the theory proposed by Kraus ([11]), (see also [24] and [14]), the most relevant transformations of states in a finite dimensional Hilbert space assume the form

$$T(\rho) = \frac{\sum_i L_i \rho L_i^\dagger}{\operatorname{Tr} \rho \sum_i L_i^\dagger L_i}, \quad \rho \in \mathcal{S}(\mathcal{H})$$

where $\{L_1, L_2, \ldots\}$ is a finite set of operators in $\mathcal{H}$. Expressed in this form T is nonlinear in ρ owing to the scalar factor in the denominator. If the L_i's satisfy the condition $\sum_i L_i^\dagger L_i = I$ then T is linear and therefore T defines an affine map on the convex set $\mathcal{S}(\mathcal{H})$. Indeed, T is a trace preserving and completely positive linear map on $\mathcal{B}(\mathcal{H})$ when extended in the obvious manner. Such completely positive maps constitute a compact convex set $\mathcal{F}(\mathcal{H}) \subset \mathcal{B}(\mathcal{H})$. Elements of $\mathcal{F}(\mathcal{H})$ are the natural quantum probabilistic analogues of stochastic matrices or transition probability operators of Markov chains in classical probability theory. Note that $\mathcal{F}(\mathcal{H})$ is also a semigroup. One would like to have a detailed theory of $\mathcal{F}(\mathcal{H})$ both from the points of view of convexity structure as well as semigroup structure.

Theorem 10.2. *Let $T \in \mathcal{F}(\mathcal{H})$ be of the form*

$$T(\rho) = \sum_{i=1}^{k} L_i \rho L_i^\dagger, \rho \in \mathcal{S}(\mathcal{H})$$

where $\sum_{i=1}^{k} L_i^\dagger L_i = I$. then T is an extreme point of $\mathcal{F}(\mathcal{H})$ if and only if the family $\{L_i^\dagger L_j, 1 \leq i \leq k, 1 \leq j \leq k\}$ of operators is linearly independent in $\mathcal{B}(\mathcal{H})$.

Remark: For any $T \in \mathcal{F}(\mathcal{H})$ it is possible to find a linearly independent set $\{L_i\}$ of operators in $\mathcal{B}(\mathcal{H})$ satisfying the relation $\sum_i L_i^\dagger L_i = I$ and

$$T(\rho) = \sum_i L_i \rho L_i^\dagger \quad \forall \quad \rho \in \mathcal{S}(\mathcal{H}).$$

If $\{M_j\}$ is another finite set of linearly independent operators satisfying these conditions then there exists a unitary matrix $((u_{ij}))$ satisfying

$$M_i = \sum_j u_{ij} L_j \quad \forall \ i.$$

Given two elements $T_1, T_2 \in \mathcal{F}(\mathcal{H})$ it will be interest to know when they are equivalent modulo unitary conjugations.

As far as the semigroup $\mathcal{F}(\mathcal{H})$ is concerned, we make the following observation. If U is a unitary operator then the map $\rho \to U\rho U^\dagger$ is an invertible element of $\mathcal{F}(\mathcal{H})$. If P is a projection then the map $\rho \to P\rho P + (1-P)\rho(1-P)$ is an irreversible element of $\mathcal{F}(\mathcal{H})$. Both of these maps preserve not only the trace but also the identity. Such maps are called *bistochastic* and they constitute a subsemigroup $\mathcal{F}_b(\mathcal{H})$. Do these elementary bistochastic maps described above generate $\mathcal{B}(\mathcal{H})$?

Theorem 10.3. *Let $T \in \mathcal{F}(\mathcal{H})$. Then there exists an ancillary Hilbert space $\mathfrak{h}$, a unitary operator U in $\mathcal{H} \otimes \mathfrak{h}$ and a pure state $|\Omega\rangle\langle\Omega|$ in $\mathfrak{h}$ with the property*

$$T(\rho) = Tr_{\mathfrak{h}} U(\rho \otimes |\Omega\rangle\langle\Omega|)U^{\dagger} \quad \forall \quad \rho \in \mathcal{S}(\mathcal{H}).$$

Remark: Theorem 10.3 has the interpretation that the irreversible dynamics described by the trace-preserving completely positive map can always be viewed as a 'coarse-graining' of a finer reversible unitary evolution in an enlarged quantum system. When the single T is replaced by a continuous one parameter semigroup $\{T_t, t \geq 0\}$ it is possible to construct a unitary evolution $\{U_t\}$ in a larger Hilbert space which is described by quantum stochastic differential equations. Such semigroups are analogues of classical semigroups and their study leads to a rich theory of quantum Markov processes. (See [14], [18].)

10.3. Quantum Error-Correcting Codes

The mathematical theory of quantum error-correcting codes is based on the following assumptions: **(1)** Messages can be encoded as states of a finite level quantum system and transmitted through a quantum communication channel; **(2)** For a given input state of the channel the output state can differ from the input state owing to the presence of 'noise' in the channel. Repeated transmission of the same input state can result in different output states; **(3)** There exists a collection of 'good' states which, when transmitted through the noisy channel, lead to output states from which the input can be recovered with no error or a small margin of error. The main goal is to identify a reasonably large collection of such good states for a given model of the channel and construct the decoding or recovery procedure. This can be described in the following pictorial form:

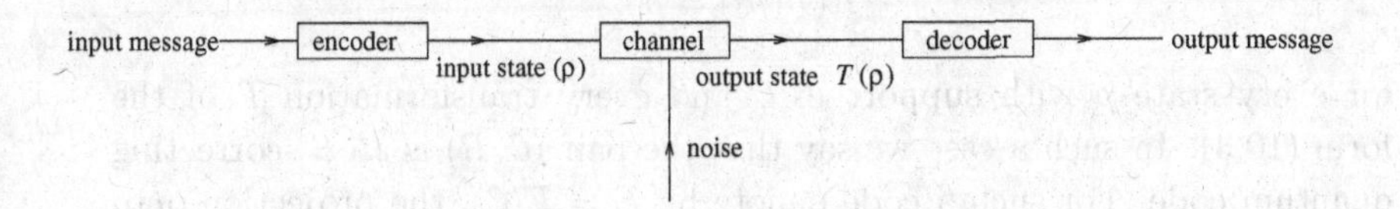

Fig. 10.1. Encoding, transmission and decoding.

Denote by $\mathcal{H}$ the finite dimensional Hilbert space from which the input and output states of the channel appear in the communication system. We assume that there is a linear subspace $\mathcal{E} \subset \mathcal{B}(\mathcal{H})$, called the *error space* such that for any input state ρ of the channel the output state $T(\rho)$ has the form

$$T(\rho) = \frac{\sum_j L_j \rho L_J^\dagger}{\operatorname{Tr} \rho \sum_j L_j^\dagger L_j}, \quad L_j \in \mathcal{E} \quad \forall\, j \tag{10.3}$$

where the summations are over finite sets of indices. If the same state ρ is transmitted repeatedly the corresponding 'corrupting' or 'noise-creating' operators $\{L_j\}$ can differ in different transmissions but they always constitute some finite subsets of $\mathcal{E}$. The L_j's may depend on the input state ρ.

For any subspace $\mathcal{S} \subset \mathcal{H}$ denote by $E(S)$ the projection onto S and by $S^\perp$ the orthogonal complement of S in $\mathcal{H}$. A state ρ is said to have its *support* contained in the subspace S if $\rho\,|u\rangle = 0 \;\forall\; u$ in $S^\perp$. This means that we can choose an orthonormal basis $\{e_i, e_2, \ldots, e_N\}$ for $\mathcal{H}$ such that the subset $\{e_i, e_2, \ldots, e_k\}$ is an orthonormal basis of S and the matrix of ρ in this basis has the form $\left[\begin{array}{c|c} \tilde{\rho} & 0 \\ \hline 0 & 0 \end{array}\right]$ where $\tilde{\rho}$ is a nonnegative matrix of order k and unit trace. To recover the input state ρ from the output state $T(\rho)$ of the channel we look for a recovery operation R of the form

$$R(T(\rho)) = \sum_i M_i T(\rho) M_i^\dagger$$

where $\{M_i\}$ is a finite set of operators depending only on the error space $\mathcal{E} \subset \mathcal{B}(\mathcal{H})$ describing the noise. Whatever be the output we apply the same recovery operation R. The goal is to construct a reasonably large subspace $\mathcal{C} \subset \mathcal{H}$ and a recovery operation R satisfying the requirement

$$R(T(\rho)) = \rho$$

for every state ρ with support in $\mathcal{C}$ and every transformation T of the form (10.3). In such a case we say that the pair $(\mathcal{C}, R)$ is an $\mathcal{E}$ -correcting quantum code. For such a code denote by $E = E(\mathcal{C})$, the projection onto $\mathcal{C}$. It is a theorem of Knill and Laflamme ([12], [21]) that

$$EL^\dagger ME = \lambda(L^\dagger M)E \quad \forall \quad L, M \in \mathcal{E} \tag{10.4}$$

where $\lambda(L^\dagger M)$ is a scalar depending on $L^\dagger M$. Conversely, if E is any projection in $\mathcal{H}$ satisfying (10.4) for all L, M in a subspace $\mathcal{E}$ of $\mathcal{B}(\mathcal{H})$ then there exists a recovery operation R such that the range $\mathcal{C}$ of E and R constitute an $\mathcal{E}$-correcting quantum code. It is important to note that equation (10.4) is sesquilinear in L, M and hence it suffices to verify (10.4) for L, M varying over a basis of $\mathcal{E}$. In view of this property we say that the projection E satisfying (10.4) or its range may be called an $\mathcal{E}$-correcting quantum code.

In order to construct quantum error-correcting codes it is often useful to identify $\mathcal{H}$ with $L^2(A)$ where A is a finite abelian group of order equal to the dimension of $\mathcal{H}$ and consider the so-called Weyl operators as an orthogonal basis for the Hilbert space $\mathcal{B}(\mathcal{H})$. To this end we view A as an additive abelian group with null element 0 and addition operation $+$. Denote by $|a\rangle$ the ket vector equal to the singleton indicator function $1_{\{a\}}$ for every $a \in A$. Then $\{|a\rangle, a \in A\}$ is an orthonormal basis for $\mathcal{H} = L^2(A)$. Choose and fix a symmetric nondegenerate bicharacter $\langle \cdot, \cdot \rangle$ on $A \times A$ satisfying

(i) $\langle x, y\rangle = \langle y, x\rangle$ and $|\langle x, y\rangle| = 1 \ \forall \ x, y \in A$;
(ii) $\langle x, y_1 + y_2\rangle = \langle x, y_1\rangle\langle x, y_2\rangle \ \forall \ x, y_1, y_2 \in A$;
(iii) $\langle x, y\rangle = 1 \ \forall \ y \in A$ if and only if $x = 0$.

Such a choice is always possible. It is then clear there exist unitary operators U_a, V_a, $a \in A$ satisfying

$$U_a |x\rangle = |x + a\rangle, \quad V_a |x\rangle = \langle a, x\rangle |x\rangle \quad \forall \quad x \in A.$$

They satisfy the following relations:

$$U_a U_b = U_{a+b}, \quad V_a V_b = V_{a+b}, \quad V_b U_a = \langle b, a\rangle U_a V_b, \quad a, b \in A.$$

These are analogous to the famous Weyl commutation relations of quantum theory for the unitary groups $U_a = e^{-iaq}$, $V_a = e^{-iap}$, $a \in \mathbb{R}$ where p and q are the momentum and position operators. Here the real line is replaced by a finite abelian group alphabet A with U_a as a location and V_a as a phase operator. We write

$$W(a, b) = U_a V_b, \quad (a, b) \in A \times A$$

and call them *Weyl operators*. They satisfy the relations

$$W(a, b)W(a', b') = \langle b, a'\rangle W(a + a', b + b'),$$
$$W(a, b)W(x, y)W(a, b)^{-1} = \langle b, x\rangle \overline{\langle a, y\rangle} W(x, y).$$

We write

$$\langle\langle (a,b),(x,y)\rangle\rangle = \langle b,x\rangle\overline{\langle a,y\rangle} \quad \forall \quad (a,b),(x,y) \text{ in } A\times A$$

and call $\langle\langle\cdot,\cdot\rangle\rangle$ the symplectic bicharacter on $A\times A$. Two Weyl operators $W(a,b)$ and $W(x,y)$ commute with each other if and only if $\langle\langle (a,b),(x,y)\rangle\rangle = 1$. If $S\subset A\times A$ is a subgroup and $\langle\langle (a,b),(a',b')\rangle\rangle = 1$ for all $(a,b),(a',b')$ in S then it is a theorem that there exists a function $\omega : S\to\mathbb{T}$, $\mathbb{T}$ denoting the multiplicative group of complex scalars of modulus unity, such that the correspondence

$$(a,b)\to\omega(a,b)W(a,b)$$

is a unitary representation of S. If χ is a character on such a subgroup S define the projection

$$E_\chi(S) = \frac{1}{\#S}\sum_{(a,b)\in S}\chi(a,b)\omega(a,b)W(a,b), \quad \chi\in\widehat{S}, \qquad (10.5)$$

$\widehat{S}$ denoting the character group of S. Subgroups $S\subset A\times A$ of the type introduced above are called *selforthogonal* or *Gottesman groups* (See [3]. [1].) Denote

$$S^\perp = \{(x,y)\,|\langle\langle (x,y),(a,b)\rangle\rangle = 1 \quad \forall \quad (a,b)\in S\}\subset A\times A. \qquad (10.6)$$

$S^\perp$ is a subgroup of $A\times A$ and $S^\perp\supset S$. It may be called the *symplectic annihilator* of S. Note that

$$\mathrm{Tr}\, W(a,b) = \begin{cases} 0 & \text{if } (a,b)\neq(0,0) \\ \dim\,\mathcal{H} & \text{otherwise.}\end{cases}$$

This shows that $\left\{(\dim\,\mathcal{H})^{-\frac{1}{2}}W(x,y),(x,y)\in A\times A\right\}$ is an orthonormal basis for the Hilbert space $\mathcal{B}(\mathcal{H})$ introduced earlier. In particular, the Weyl operators consitute an irreducible family and every operator X on $\mathcal{H}$ admits a 'Fourier expansion'

$$X = \frac{1}{\dim\,\mathcal{H}}\sum_{(a,b)\in A\times A}\left\{\mathrm{Tr}\, W(a,b)^\dagger X\right\}W(a,b), \quad X\in\mathcal{B}(\mathcal{H}). \qquad (10.7)$$

Elementary algebra using Schur orthogonality relations for characters shows that

$$E_\chi(S)W(x,y)E_\chi(S) = \begin{cases} 0 & \text{if } \quad (x,y)\notin S^\perp \\ \lambda(x,y)E_\chi(S) & \text{if } \quad (x,y)\in S.\end{cases}$$

where $\lambda(x, y)$ is a scalar. In other words the linear space

$$\mathcal{D}(S) = \text{lin span}\left\{W(x, y), (x, y) \in S \cup (A \times A \backslash S^{\perp})\right\} \tag{10.8}$$

satisfies the property

$$E_{\chi}(S) L E_{\chi}(S) = \lambda(L) E_{\chi}(S) \quad \forall \quad L \in \mathcal{D}(S).$$

Clearly, $\mathcal{D}(S)$ is closed under the adjoint operation. Suppose now that $\mathcal{E} \subset \mathcal{B}(\mathcal{H})$ is a subspace satisfying the property

$$\mathcal{E}^{\dagger}\mathcal{E} = \left\{L^{\dagger} M \,|L, M \in \mathcal{E}\right\} \subset \mathcal{D}(S).$$

Then (10.4) holds when $E = E_{\chi}(S)$ and by Knill-Laflamme theorem the range of $E_{\chi}(S)$ is an $\mathcal{E}$-correcting quantum code for any Gottesman subgroup S and character χ of S. Furthermore,

$$E_{\chi}(S) E_{\chi'}(S) = 0 \quad \text{if} \quad \chi, \chi' \in \widehat{S} \quad \text{and} \quad \chi \neq \chi'.$$

$$\sum_{\chi \in \widehat{S}} E_{\chi}(S) = I.$$

If $\mathcal{C}_{\chi}(S)$ denotes the range of $E_{\chi}(S)$ it follows that $\mathcal{H}$ decomposes into a direct sum of $\mathcal{E}$-correcting or $\mathcal{D}(S)$-detecting quantum codes of dimension $\dim \mathcal{H} \backslash \# S$.

We now discuss the special case when $\mathcal{H}$ is replaced by its n- fold tensor product $\mathcal{H}^{\otimes^n} = L^2(A)^{\otimes^n} \cong L^2(A^n)$. Denoting any point $\mathbf{x} \in A^n$ by $\mathbf{x} = (x_1, x_2, \ldots, x_n)$ where $x_j \in A$ is the j-th coordinate we get a symmetric nondegenerate bicharacter on $A^n \times A^n$ by putting

$$\langle \mathbf{x}, \mathbf{y} \rangle = \prod_{j=1}^{n} \langle x_j, y_j \rangle$$

and the corresponding Weyl operators

$$W(\mathbf{x}, \mathbf{y}) = \bigotimes_{j=1}^{n} W(x_j, y_j)$$

where $W(x_j, y_j)$ is the Weyl operator in $L^2(A)$ associated with (x_j, y_j) for each j. Let $\mathcal{E}_d$ denote the linear span of all operators in $\mathcal{H}^{\otimes^n}$ of the form $X_1 \otimes X_2 \otimes \cdots \otimes X_n$ where X_i is different from I for at most d values of i. Then $\mathcal{E}_d$ has the orthogonal unitary operator basis

$$\left\{W(\mathbf{a}, \mathbf{b}) \,|\# \left\{i \,|(a_i, b_i) \neq (0, 0)\right\} \leq d\right\}.$$

One says that the element $(\mathbf{a},\mathbf{b})$ in $A \times A$ has *weight*

$$w(\mathbf{a},\mathbf{b}) = \#\{i\,|(a_i,b_i) \neq (0,0)\}\,.$$

Let $S \subset A^n \times A^n$ be a Gottesman subgroup of $A^n \times A^n$ such that every element $(\mathbf{a},\mathbf{b})$ in $S^{\perp}\backslash S$ has weight $\geq d$. Then it follows from the preceding discussions and the definition in (10.8) that $\mathcal{E}_{\chi}(S)$, i.e., the range of $E_{\chi}(S)$ is an $\mathcal{E}_t$-correcting quantum code where $t = \left\lfloor \frac{d-1}{2} \right\rfloor$. It is also called an $\mathcal{E}_d$-detecting quantum code. Thus the problem of constructing $\mathcal{E}_d$-detecting quantum codes reduces to the algebraic problem of constructing symplectic self-orthogonal or Gottesman subgroups S of $A^n \times A^n$ satisfying the property that every element $(\mathbf{a},\mathbf{b})$ in $S^{\perp}\backslash S$ has weight $\geq d$.

Choosing $A = \mathbb{Z}_2$ the additive group of the field $GF(2)$ the problem of constructing symplectic selforthogonal subgroups of $A^n \times A^n$ has been reduced to the construction of classical error correcting codes over $GF(4)$ by [3]. See also [1], [21].

We conclude this section with an example of a quantum code based on a Gottesman subgroup which also yields states exhibiting maximal entanglement for multipartite quantum systems. To this end, we introduce in $L^2(A^5)$ a single error correcting, (i.e., $\mathcal{E}_1$-correcting) quantum code as follows. Introduce the cyclic permutation σ in A^5 by

$$\sigma(x_0,x_1,x_2,x_3,x_4) = (x_1,x_2,x_3,x_4,x_0,)$$

and put

$$\tau(\mathbf{x}) = \sigma^2(\mathbf{x}) + \sigma^{-2}(\mathbf{x}), \quad \mathbf{x} = (x_0,x_1,x_2,x_3,x_4) \in A^5.$$

Define

$$\widetilde{W}_{\mathbf{x}} = \langle \mathbf{x}, \sigma^2(\mathbf{x})\rangle W(\mathbf{x},\tau(\mathbf{x})), \quad \mathbf{x} \in A^5.$$

Consider the subgroup $C \subset A^5$ given by

$$C = \{\mathbf{x}\,|x_0 + x_1 + x_2 + x_3 + x_4 = 0\}\,.$$

Then C is a Gottesman subgroup and $\mathbf{x} \to \widetilde{W}_{\mathbf{x}}$ is a unitary representation of C and the operator

$$E(C) = (\#A)^{-4} \sum_{\mathbf{x}\in C} \widetilde{W}_{\mathbf{x}}$$

is a projection on the subspace

$$\mathcal{C} = \left\{\psi \middle| \widetilde{W}_{\mathbf{x}}\psi = \psi \quad \forall \quad \mathbf{x} \in C\right\}.$$

This subspace is an $\mathcal{E}_1$-correcting quantum code of dimension $\#A$. It has the maximum entanglement property in the sense that for any $\psi \in \mathcal{C}$ the pure state $|\psi\rangle\langle\psi|$ satisfies the relation

$$\mathrm{Tr}_{L^2(A^3)} |\psi\rangle\langle\psi| = \frac{I}{(\#A)^2}$$

for any three copies of A occurring in A^5. Here we view $L^2(A^5)$ as $L^2(A) \otimes \cdots \otimes L^2(A)$ with 5 copies and the factorization

$$L^2(A^5) = L^2(A^3) \otimes L^2(A^2).$$

For more details see [19], [22]. It is an interesting problem to construct such examples for products of n copies of A with $n > 5$ and construct subspaces of the form $\mathcal{C}$ with maximal dimension. It is desirable to have a formula for this maximal dimension.

10.4. Testing Quantum Hypotheses

Suppose a quantum system with Hilbert space $\mathcal{H}$ is known to be in one of the states $\{\rho_1, \rho_2, \ldots, \rho_k\}$ and we have to decide the true state by making a measurement $\mathbf{M} = \begin{pmatrix} 1 & 2 & \ldots & k \\ M_1 & M_2 & \ldots & M_k \end{pmatrix}$ as described in (10.2) with the value set $\{1, 2, \ldots, k\}$ and applying the decision rule that the state of the system is ρ_j if the outcome of the measurement is j. In such a case, if the true state is ρ_i then the probability of deciding ρ_j is equal to $\mathrm{Tr}\, \rho_i M_j$. Suppose there is a prior distribution $\pi = (\pi_1, \pi_2, \ldots, \pi_k)$ on the set of states, i.e., the system is in the state ρ_i with probability π_i for each i and there is cost matrix $((c_{ij}))$ according to which c_{ij} is the cost of deciding ρ_j when the true state ρ_i. In such a case the expected cost of the decision rule associated with the measurement is

$$C(\mathbf{M}) = \sum_{i,j} (\pi_i\ \mathrm{Tr}\ \rho_i\ M_j)\ c_{ij}.$$

The natural thing to do is to choose a measurement $\mathbf{M}$ which minimizes the cost $C(\mathbf{M})$. Define

$$\gamma = \inf_{\mathbf{M}}\ C(\mathbf{M})$$

and the hermitian operators

$$A_j = \sum_i \pi_i c_{ij} \rho_i. \qquad (10.9)$$

Then

$$\gamma = \inf_{\mathbf{M}} \text{ Tr} \sum_{j=1}^{k} A_j M_j. \tag{10.10}$$

Since the space of all measurements $\mathbf{M}$ with value set $\{1, 2, \ldots, k\}$ constitute the compact convex set $\mathcal{M}_k(\mathcal{H})$, γ is attained at a measurement $\mathbf{M}^\circ$ which is also an extreme point of $\mathcal{M}_k(\mathcal{H})$. An appropriate application of Lagrange multipliers shows that the following theorem holds.

Theorem 10.4. (*cf.* [26], [9], [16]) *Let* $\mathbf{M}^\circ = \begin{pmatrix} 1 & 2 & \ldots & k \\ M_1^\circ & M_2^\circ & \ldots & M_k^\circ \end{pmatrix}$ *be a measurement satisfying*

$$\gamma = Tr \sum_{j=1}^{k} A_j \, M_j^\circ$$

where A_j *and* γ *are given by* (10.9) *and* (10.10) *and let* $\Gamma = \sum_{j=1}^{k} A_j M_j^\circ$. *Then the following holds:*

(i) Γ *is hermitian and* $\Gamma \leq A_j \; \forall \; j = 1, 2, \ldots, k$;
(ii) $(A_j - \Gamma) M_j^\circ = 0 \quad \forall \quad i$;
(iii) Γ *is independent of the measurement* $\mathbf{M}^\circ$ *at which* γ *is attained.*

Remark: In the context of Theorem 10.4 it is natural to seek a good algorithm for evaluating Γ when the hermitian operators $\{A_j, 1 \leq j \leq k\}$ are known. One may interpret Γ as a 'minimum' of the noncommuting observables A_j, $j = 1, 2, \ldots, k$.

In the discussion preceding Theorem 10.4 note that $\sum_i \pi_i$ Tr ρ_i M_i is the probability of correct decision. It is natural to consider the problem of maximising this probability in the spirit of the method of maximum likelihood. Let

$$\delta = \max_{\mathbf{M}} \text{ Tr} \sum_{i=1}^{k} \pi_i \rho_i M_i.$$

As before δ is attained at some measurement $\mathbf{M}^\circ$. Then one has the following theorem.

Theorem 10.5. *Let* $\mathbf{M}^\circ = \begin{pmatrix} 1 & 2 & \ldots & k \\ M_1^\circ & M_2^\circ & \ldots & M_k^\circ \end{pmatrix}$ *be a measurement satisfying*

$$\delta = Tr \sum_i \pi_i \rho_i M_i^\circ$$

and let $\Delta = \sum \pi_i \rho_i M_i^\circ$. *Then the following holds:*

(i) Δ *is hermitian and* $\Delta \geq \pi_i \rho_i \quad \forall\ i$;
(ii) $(\Delta - \pi_i \rho_i) M_i^\circ = 0 \quad \forall\ i$;
(iii) Δ *is independent of* $\mathbf{M}^\circ$ *at which* δ *is attained.*

Now consider the situation when n copies of the system are available so that ρ_i can be replaced by $\rho_i^{\otimes^n}$ for each i in $\mathcal{H}^{\otimes^n}$. Write

$$\delta_n = \max_{\mathbf{M}} \operatorname{Tr} \sum_{i=1}^{k} \pi_i \rho_i^{\otimes^n} M_i$$

where $\mathbf{M} \in \mathcal{M}_k(\mathcal{H}^{\otimes^n})$ and Δ_n for the corresponding operator in Theorem 10.5. Then the probability of making a wrong decision is equal to $1-\delta_n$. By the elementary arguments in [17] it follows that $1 - \delta_n$ decreases to zero as n increases to ∞ and the rate of decrease is exponential. It is an open problem to determine the quantities $\underline{\lim}_{n\to\infty} \frac{-\log(1-\delta_n)}{n}$ and $\overline{\lim}_{n\to\infty} \frac{-\log(1-\delta_n)}{n}$. When $k = 2$ so that there are only two states the following theorem holds:

Theorem 10.6. (*Quantum Chernoff Bound, see* [2], [13]) *Let* ρ_1, ρ_2 *be two states and let* π_1, π_2 *be a prior probability distribution on* $\{1, 2\}$ *with* $0 < \pi_1 < 1$. *Suppose*

$$\delta_n = \max_{\mathbf{M}}\ Tr\left(\pi_1 \rho_1^{\otimes^n} M_1 + \pi_2 \rho_2^{\otimes^n} M_2\right)$$

where

$$\mathbf{M} = \begin{pmatrix} 1 & 2 \\ M_1 & M_2 \end{pmatrix}$$

varies over all measurements in $\mathcal{H}^{\otimes^n}$ *with value set* $\{1, 2\}$. *Then*

$$\lim_{n\to\infty} -\frac{1}{n} \log(1-\delta_n) = \sup_{0 \leq s \leq 1} -\log\ Tr\, \rho_1^s \rho_2^{1-s}.$$

Remark: An explicit computation shows that in Theorem 10.6

$$\delta_n = \frac{1}{2}\left(1 + \left|\pi_1 \rho_1^{\otimes^n} - \pi_2 \rho_2^{\otimes^n}\right|_1\right)$$

where $|\cdot|_1$ stands for trace norm.

In the two states case we introduce the quantity

$$\beta(\rho_1, \rho_2, \varepsilon) = \inf\left\{\operatorname{Tr} \rho_2 X \,|\, 0 \leq X \leq I, \operatorname{Tr} \rho_1 X \geq 1 - \varepsilon\right\}$$

for any $0 < \varepsilon < 1$ and states ρ_1, ρ_2. It has the following statistical interpretation. For any $0 \le X \le I$ consider the measurement

$$\mathbf{M} = \begin{pmatrix} 1 & 2 \\ X & I - X \end{pmatrix}$$

with value set $\{1, 2\}$. For the decision rule based on $\mathbf{M}$, $\operatorname{Tr} \rho_1 X$ is the probability of correct decision in the state ρ_1. we vary $\mathbf{M}$ so that this probability is kept above $1 - \varepsilon$. Now we look at the probability of a wrong decision in the state ρ_2 and note that the minimum possible value for this probability is $\beta(\rho_1, \rho_2, \varepsilon)$. In the Neyman-Pearson theory of testing hypothesis the quantity $1 - \beta(\rho_1, \rho_2, \varepsilon)$ is called the maximum possible power at a level of significance above $1 - \varepsilon$. Now we state a fundamental theorem concerning $\beta\left(\rho_1^{\otimes^n}, \rho_2^{\otimes^n}, \varepsilon\right)$ due to [8].

Theorem 10.7. (*Quantum Stein's Lemma, see* [6])

$$\lim_{n \to \infty} - \frac{\log \beta\left(\rho_1^{\otimes^n}, \rho_2^{\otimes^n}, \varepsilon\right)}{n} = Tr\, \rho_1 (\log\ \rho_1 - \log \rho_2)\,.$$

Remark: The right hand side of the relation in Theorem 10.7 is known as the *relative entropy* of ρ_2 with respect to ρ_1 or *Kulback-Leibler divergence* of ρ_2 from ρ_1 and denoted as $S(\rho_1 || \rho_2)$.

We shall conclude with a brief description of a quantum version of Shannon's coding theorem for classical-quantum communication channels. A classical-quantum or, simply, a *cq*-channel $\mathcal{C}$ consists of a finite set, called the input alphabet A and a set $\{\rho_x,\ x \in A\}$ of states in a Hilbert space $\mathcal{H}$. A quantum code of size m and error probability $\le \varepsilon$ consists of a subset $C \subset A$ of cardinality m and a C-valued measurement $\mathbf{M}(C) = \{M_x, x \in C\}$ with the property $\operatorname{Tr} \rho_x M_x \ge 1 - \varepsilon \ \forall\ x \in C$. If we have such a code then m classical messages can be encoded as states $\{\rho_x, x \in C\}$ and the measurement $\mathbf{M}(C)$ yields a decoding rule : if $x \in C$ is the value yielded by the measurement $\mathbf{M}(C)$ decide that the message corresponding to the state ρ_x was transmitted. Then the probability of a wrong decoding does not exceed ε. In view of this useful property the following quantity

$N(\mathcal{C}, \varepsilon) = \max\{m |$ a quantum code of size m and error probability $\le \varepsilon$ exists$\}$

is an important parameter concerning the *cq*-channel $\mathcal{C}$.

If $\mathcal{C}$ is a *cq*-channel with input alphabet A and states $\{\rho_x, x \in A\}$ in $\mathcal{H}$ we can define the n-fold product $\mathcal{C}^{\otimes^n}$ of $\mathcal{C}$ by choosing the alphabet A^n and the states

$$\{\rho_{x_1} \otimes \rho_{x_2} \otimes \cdots \otimes \rho_{x_n} \,|\mathbf{x} = (x_1, x_2, \ldots, x_n) \in A^n,\ x_i \in A \,\forall\, i\}.$$

Define

$$N(n, \varepsilon) = N\left(\mathcal{C}^{\otimes^n}, \varepsilon\right).$$

Then we have the following quantum version of Shannon's coding theorem.

Theorem 10.8. (*Shannon's Coding Theorem, see* [6], [23], [25])

$$\lim_{n\to\infty} \frac{\log\, N(n,\varepsilon)}{n} = \sup_{p(\cdot)} \left\{ S\left(\sum_{x\in A} p(x)\,\rho_x\right) - \sum_{x\in A} p(x)\, S(\rho_x) \right\} \quad (10.11)$$

where the supremum on the right hand side is taken over all probability distributions $\{p(x), x \in A\}$ *and* $S(\rho)$ *stands for the von Neumann entropy of the state* ρ.

Remark: The quantity on the right hand side of (10.11) is usually denoted by $C = C(\mathcal{C})$ called the *Shannon-Holevo capacity* of the *cq*-channel $\mathcal{C}$. It is interesting to note that the expression within the supermum on the right hand side of (10.11) can be expressed as the mean relative entropy $\sum_{x\in A} p(x)\, S\left(\rho_x || \overline{\rho}\right)$ where $\overline{\rho}$ denotes the state $\sum_x p(x)\,\rho_x$.

References

[1] Arvind, V., Parthasarathy, K. R. (2003). A family of quantum stabilizer codes based on the Weyl commutation relations over a finite field. In *A Tribute to C. S. Seshadri: Perspectives in Geometry and Representation Theory.* (Eds. Lakshmibhai et al.) Hindustan Book Agency, New Delhi, 133–153.

[2] Audenaert, K. M. R., Calsamiglia, J., Munoz-Tapia, R., Bagan, E., Masanes, L., Acin, A. and Verstraete, F. (2006). The quantum Chernoff bound. arXiv:quant-ph/0610027.

[3] Calderbank, A. R., Rains, M., Shor, P. W. and Sloane, N. J. A. (1998). Quantum error correction via codes over GF(4), *IEEE Trans. Inform. Theory.* **44** 1369–1387.

[4] Choi, M. D. (1975). Completely positive linear maps on complex matrices. *Lin. Alg. Appl.* **10** 285–290.

[5] Einstein, A., Podolski, R. and Rosen, N. (1935). Can quantum mechanical descriptions of physical reality be considered complete? *Phys. Rev.* **47** 777–780.

[6] Hayashi, M. (2006). *Quantum Information, an Introduction.* Springer, Berlin.

[7] Helstrom, C. W. (1976). *Quantum Detection and Estimation Theory.* Mathematics in Science and Engineering, **123** Academic Press, New York.

[8] Hiai, F. and Petz, D. (1991). The proper formula for relative entropy and its asymptotics in quantum probability. *Commun. Math. Phys.* **143** 99–114.

[9] Holevo, A. S. (1974). Remarks on optimal quantum measurements, *Problemi Peradachi Inform.* (in Russian) **10** 51–55.

[10] Holevo, A. S. (1974). *Probabilistic and Statistical Aspects of Quantum Theory.* North-Holland, Amsterdam.

[11] Kraus, K. (1983). *States, Effects and Operations.* Lecture Notes in Physics, **190** Springer, Berlin.

[12] Nielsen, M. A. and Chuang, I. L. (2000). *Quantum Computation and Quantum Information.* Cambridge University Press, Cambridge.

[13] Nussbaum, M. and Szkola, A. (2006). The Chernoff lower bound for symmetric quantum hypothesis testing. arXiv:quant-ph/0607216.

[14] Parthasarathy, K. R. (1992). *An Introduction to Quantum Stochastic Calculus.* Birkhauser Verlag, Basel.

[15] Parthasarathy, K. R. (1998). Extreme points of the convex set of stochastic maps on a C*algebra. *Inf. Dim. Anal. Quant. Probab. Rel. Topics* **1** 599–609.

[16] Parthasarathy, K. R. (1999). Extremal decision rules in quantum hypothesis testing. *Inf. Dim. Anal. Quant. Probab. Rel. Topics* **2** 557–568.

[17] Parthasarathy, K. R. (2001). On the consistency of the maximum likelihood method in testing multiple quantum hypotheses. In *Stochastics in Finite and Infinite Dimensions.* (Eds. T. Hida et al.) Birkhauser Verlag, Basel, 361–377.

[18] Parthasarathy, K. R. (2003). Quantum probability and strong quantum Markov processes. In *Quantum Probability Communications.* (Eds. R. L. Hudson and J. M. Lindsay) **XII** World Scientific, Singapore, 59–138.

[19] Parthasarathy, K. R. (2004). On the maximal dimension of a completely entangled subspace for finite level quantum systems. *Proc. Ind. Acad. Sci.* (*Math.Sci.*) **114** 365–374.

[20] Parthasarathy, K. R. (2005). Extremal quantum states in coupled systems. *Ann. de L'Institut Henri Poincare, Prob. Stat.* **41** 257–268.

[21] Parthasarathy, K. R. (2006). *Quantum Computation, Quantum Error Correcting Codes and Information Theory.* TIFR Lecture Notes, Narosa Publishing House, New Delhi.

[22] Parthasarathy, K. R. (2006). Extremality and entanglements of states in coupled systems. In *Quantum Computing.* (Ed. Debabrata Goswami), AIP Conference Proceedings **864** 54–66.

[23] Parthasarathy, K. R. (2007). *Coding Theorems of Classical and Quantum Information Theory.* Hindustan Book Agency, New Delhi.

[24] Srinivas, M. D. (2001). *Measurements and Quantum Probabilities.* Universities Press, Hyderabad.

[25] Winter, A. (1999). Coding theorem and strong converse for quantum channels. *IEEE Trans. Inform. Theory.* **45** 2481–2485.

[26] Yuen, H. P., Kennedy, R. S. and Lax, M. (1975). Optimum testing of multiple hypotheses in quantum detection theory. *IEEE Trans. Inform. Theory.* **IT 21** 125–134.

Chapter 11

Scaling Limits

S. R. S. Varadhan

Courant Institute of Mathematical Sciences,
New York University, New York, USA

11.1. Introduction

The simplest example of a scaling limit, is to take a sequence $\{X_i\}$ of independent random variables, with each X_i taking values $\{+1, -1\}$ with probability $\{p, 1-p\}$ respectively. Let

$$S_k = X_1 + X_2 + \cdots + X_k \, .$$

If $m = 2p - 1$, by the law of large numbers

$$\frac{1}{n} S_{[nt]} \to mt \, .$$

Here the rescaling is $x = \frac{s}{n}$ and $t = \frac{k}{n}$ and the limit is non-random. If $m = 0$, i.e. $p = \frac{1}{2}$, the convergence of the random walk, with a different scaling, is to Brownian motion. According to the invariance principle of Donsker ([1]), the distribution of

$$X_n(t) = \frac{1}{\sqrt{n}} S_{[nt]}$$

converges to that of Brownian motion. Here the rescaling is different, $x = \frac{s}{\sqrt{n}}$ rescales space and $t = \frac{k}{n}$ rescales time. In the new time scale the exact position of the particle in the microscopic location is not tracked. Only the motion of its macroscopic location $x(t)$ which changes in a reasonable manner in the macroscopic time scale with a non trivial random motion is tracked. This random motion is the Brownian motion.

Many problems in the physical sciences involve modeling a large system at a small scale while observations and predictions are made at a much larger scale. Of particular interest are dynamical models involving two distinct scales, like the random walk model mentioned earlier, a microscopic one and a macroscopic one. The system and its evolution are specified at the microscopic level. For instance the system could be particles in space that interact with each other and move according to specific laws, that can be either deterministic or random. The system can have many conserved quantities. For instance in the classical case of Hamiltonian dynamics, mass, momentum and energy are conserved in the interaction. In the simplest case of interacting particle systems, in the absence of creation or annihilation the number of particles is always conserved.

Usually there are invariant probability measures for the complex system, which may not be unique due to the presence of conserved quantities. They represent the statistics of the system in a steady state. If the evolution is random and is prescribed by a Markov process, then these are the invariant measures. If we fix the size of the system there will be a family of invariant measures corresponding to the different values of the conserved quantities. When the size of the system becomes infinite, there will be limiting measures, on infinite systems and a family of invariant probability measures indexed by the average values of the conserved quantities. In the theory of equilibrium statistical mechanics, these topics are explored in detail ([2]).

11.2. Non-Interacting Case

Let us illustrate this by the most elementary of non interacting random walks. We have a large number n of independent random walks, with $p = \frac{1}{2}$. The walks $\{S_k^j : 1 \leq j \leq n\}$ start from locations $\{z^j\}$ at time $k = 0$. Their locations at time k are S_k^j. We assume that initially there exists a function $\rho_0(x) \geq 0$ on R with $\int_{-\infty}^{\infty} \rho_0(x)dx = \bar{\rho} < \infty$, such that

$$\lim_{n\to\infty} \frac{1}{n}\sum_{j=1}^{n} f(\frac{z^j}{n}) = \int_{-\infty}^{\infty} f(x)\,\rho_0(x)\,dx \tag{2.1}$$

for every bounded continuous function f on R. If we look at it from a distance we do not see the individual particles, but only a cloud, and the density varies over the location x in the macroscopic scale. The statistical nature of the actual locations can be quite general. We could place them deterministically in some systematic way to achieve the requisite density profile, or place them randomly with the number of particles at site z having a Poisson distribution with expectation

$$p_n(z) = n\int_{z-\frac{1}{2n}}^{z+\frac{1}{2n}} \rho_0(x)dx$$

and independently so for different sites. The relation (2.1) is now a consequence of the ergodic theorem. If $\eta(0,z)$ is the number of particles initially at site z, then (2.1) can be rewritten as

$$\lim_{n\to\infty}\frac{1}{n}\sum_z f(\frac{z}{n})\eta(0,z) = \int_{-\infty}^{\infty} f(x)\,\rho_0(x)\,dx\,. \tag{2.2}$$

If the initial distributions of number of particles at different sites are Poisson with the same constant parameter λ, and they are independent then this distribution is preserved and the same holds at any future time. For any constant λ the product measure P_λ of independent Poissons at different sites with the same parameter λ is invariant. But this corresponds to $\rho_0(x)\equiv\lambda$. If the system evolves from an arbitrary initial state and and if we look at it at time n^2t, since the rescaling is with n in space and n^2 in time, we are working with the Brownian or central limit scaling and

$$P[z^j_{[n^2t]} = z] \simeq \frac{1}{n\sqrt{2\pi t}}e^{-\frac{(z-z^j)^2}{2n^2t}}$$

and the combination of a local limit theorem and the Poisson limit theorem implies that the distribution of $\eta(n^2t,z)$, in the limit, as $\frac{z}{n}\to x$, is Poisson with parameter

$$\rho(t,x) = \lim_{n\to\infty}\sum_{z'}\frac{1}{n\sqrt{2\pi t}}e^{-\frac{(z-z')^2}{2n^2t}}\eta(0,z') = \int\frac{1}{\sqrt{2\pi t}}e^{-\frac{(y-x)^2}{2t}}\rho_0(y)dy\,.$$

In other words we see locally an invariant distribution, but with a parameter that changes with macroscopic location (t,x) in space and time. If we look at the picture through a microscope, at the grain level we see the Poisson distribution, with a parameter determined by the location. It is not hard to prove a law of large numbers,

$$\lim_{n\to\infty}\frac{1}{n}\sum_z f(\frac{z}{n})\eta(n^2t,z) = \int_{-\infty}^{\infty} f(x)\,\rho(t,x)\,dx\,. \tag{2.3}$$

These are statements of convergence in probability. Notice that the only conserved quantity here is the number of particles. Therefore one expects the quantity of interest to be the density $\rho(t,x)$. Once we know $\rho(t,x)$ we are aware that locally, microscopically, it looks like Poisson process on the integer lattice with constant intensity $\rho(t,x)$. While in this case we were able to solve explicitly, we should note that $\rho(t,x)$ is the solution of the PDE

$$\frac{\partial \rho}{\partial t} = \frac{1}{2}\frac{\partial^2 \rho}{\partial x^2}; \quad \rho(0,x) = \rho_0(x). \tag{2.4}$$

It is more convenient if we make time continuous and have the particles execute jumps according to Poisson random times with mean 1. Jump rate is $\frac{1}{2}$ for each direction, left or right, for each particle. Speeding up time now just changes the rates to $\frac{n^2}{2}$. If we think of the whole system as undergoing an infinite dimensional Markov process, then the generator $\mathcal{L}_n$ acting on a function $f(\eta)$ of the configuration $\{\eta(z)\}$ of particle numbers at different sites is give by

$$(\mathcal{L}_n F)(\eta) = \sum_{\eta'} [F(\eta') - F(\eta)]\sigma(\eta,\eta').$$

The new configurations η' involved are the results of a single jump form $z \to z \pm 1$ which occur at rate $\sigma(\eta,\eta') = \frac{1}{2}\eta(z)$. In particular with

$$F(\eta) = \frac{1}{n}\sum_z f(\frac{z}{n})\,\eta(z)$$

we get

$$\begin{aligned}(\mathcal{L}_n F)(\eta) &= \frac{n^2}{2n}\sum_z \eta(z)[(f(\frac{z+1}{n}) - f(\frac{z}{n})) + f(\frac{z-1}{n}) - f(\frac{z}{n})] \\ &\simeq \frac{1}{2n}\sum_z \eta(z) f''(\frac{z}{n}) \\ &\simeq \frac{1}{2}\int f''(x)\rho(t,x)dx\end{aligned}$$

which, with just a little bit of work, leads to the heat equation (2.4).

11.3. Simple Exclusion Processes

Let us now introduce some interaction. The simple exclusion rule imposes the limit of one particle per site. If a particle decides to jump and the site it chooses to jump to is occupied, the jump can not be completed, and the particle waits instead for the next Poisson time. Let us look at the simple case of totally asymmetric walk where the only possible jump is to the next site to the right. The generator is given by

$$(\mathcal{L}_n F)(\eta) = \sum_z \eta(z)(1 - \eta(z+1))F(\eta^{z,z+1}).$$

For any configuration $\eta = \{\eta(z)\}$, the new configuration $\eta^{z',z''} = \{\eta^{z',z''}(z)\}$ is defined by

$$\eta^{z',z''}(z) = \begin{cases} \eta(z') & \textit{if } z = z'' \\ \eta(z'') & \textit{if } z = z' \\ \eta(z) & \textit{otherwise}. \end{cases}$$

Note that because of exclusion, the number of particles $\eta(z)$ at z is either 0 or 1. Now we speed up time by n and rescale space by $x = \frac{z}{n}$. With

$$F(\eta) = \frac{1}{n} \sum_z f(\frac{z}{n})\eta(z),$$

$$(\mathcal{L}_n F)(\eta) \simeq \frac{1}{n} \sum_z \eta(z)(1 - \eta(z+1))f'(\frac{z}{n}). \tag{3.1}$$

The invariant distributions in this case are Bernoulli product measures, with density ρ. That plays a role because we need to average $\eta(z)(1-\eta(z+1))$. While averages of $\eta(z)$ over long blocks will only change slowly due to the conservation of the number of particles, quantities like $\eta(z)(1 - \eta(z+1))$ fluctuate rapidly and their averages are computed under the relevant invariant distribution. The average of $\eta(z)(1-\eta(z+1))$ is $\rho(1-\rho)$. Equation (3.1) now leads to

$$\frac{\partial}{\partial t} \int f(x)\rho(t,x)dx = \int f'(x)\rho(t,x)(1 - \rho(t,x))dx$$

which is the weak form of Burgers equation

$$\frac{\partial \rho}{\partial t} + \frac{\partial}{\partial x}[\rho(t,x)(1 - \rho(t,x))] = 0. \tag{3.2}$$

Weak solutions of this equation are not unique, but the limit can be characterized as the unique solution that satisfies certain "entropy conditions". We have used the independence provided by the Bernoulli distribution. What is really needed is that averages of the type $\frac{1}{2k+1} \sum_{|z| \le k} \eta(z)(1 - \eta(z+1))$ over long blocks can be determined with high probability as being nearly equal to $\bar{\eta}(1-\bar{\eta})$ with $\bar{\eta} = \frac{1}{2k+1}\sum_{|z|\le k}\eta(z)$. See [3] for a discussion of the model and additional references.

We will now consider a general class of simple exclusion processes on Z^d, where the jump rate from z to z' is $p(z'-z)$. The site still needs to be free if a jump is to be executed. The generator is given by

$$(\mathcal{L}F)(\eta) = \sum_{z,z'} \eta(z)(1-\eta(z'))p(z'-z)[F(\eta^{z,z'}) - F(\eta)].$$

If the mean $\sum zp(z) \neq 0$, then the situation is not all that different from the one dimensional case discussed earlier. We shall now assume that $\sum_z zp(z) = 0$. There are two cases. If $p(z) = p(-z)$, then we can symmetrize and rewrite

$$\begin{aligned}(\mathcal{L}F)(\eta) &= \frac{1}{2}\sum_{z,z'}[\eta(z)(1-\eta(z')) + \eta(z')(1-\eta(z')]p(z'-z)[F(\eta^{z,z'}) - F(\eta)]\\ &= \frac{1}{2}\sum_{z,z'} p(z'-z)[F(\eta^{z,z'}) - F(\eta)]\end{aligned}$$

because $\eta(z)(1-\eta(z')) + \eta(z')(1-\eta(z') = 1$ if $\eta(z) \neq \eta(z')$. Otherwise $F(\eta^{z,z'}) = F(\eta)$.

In this case, time is speeded up by a factor of n^2 and one can calculate for

$$F(\eta) = n^{-d}\sum_z f(\frac{z}{n})\eta(z),$$

$$\begin{aligned}(\mathcal{L}_nF)(\eta) &= \frac{n^2}{2n^d}\sum_{z.z'} p(z'-z)[f(\frac{z'}{n}) - f(\frac{z}{n})]\\ &\simeq \frac{1}{2n^d}\sum_z\sum_{i,j} a_{i,j}f_{i,j}(\frac{z}{n})\\ &= \frac{1}{n^d}\sum_z(\mathbf{A}f)(\frac{z}{n}).\end{aligned}$$

Here $\mathbf{A}$ is the second order differential operator $\frac{1}{2}\sum_{i.j} a_{i,j}D_iD_j$, and $A = \{a_{i,j}\}$ is the covariance matrix

$$a_{i,j} = \sum_z \langle z, e_i\rangle\langle z, e_j\rangle p(z).$$

It is not very hard to deduce that if initially one has the convergence of the empirical distributions

$$\nu_n = \frac{1}{n^d}\sum_z \delta_z \eta(z)$$

in the weak topology, to a deterministic limit

$$\nu(dx) = \rho_0(x)dx$$

then the empirical distribution

$$\nu_n(t) = \frac{1}{n^d}\sum_z \delta_z \eta(t,z)$$

in the speeded up time scale, (i.e t is really $n^2 t$) will converge in probability to $\rho(t,x)dx$, where $\rho(t,x)$ solves the heat equation

$$\frac{\partial \rho}{\partial t} = \frac{1}{2}\sum_{i,j} a_{i,j} D_i D_j \, \rho \; ; \qquad \rho(0,x) = \rho_0(x)\,. \tag{3.3}$$

If we drop the assumption of symmetry and only assume that $\sum_z zp(z) = 0$, then we have a serious problem. The summation

$$(\mathcal{L}_n F)(\eta) = \frac{n^2}{n^d}\sum_{z,z'} \eta(z)(1-\eta(z'))p(z'-z)[f(\frac{z'}{n}) - f(\frac{z}{n})]$$

does not simplify. The smoothness of f can be used to get rid of one power of n by replacing $f(\frac{z'}{n}) - f(\frac{z}{n})$ by $\frac{1}{2n}\langle (z'-z), [(\nabla f)(\frac{z'}{n}) + (\nabla f)(\frac{z}{n})]\rangle$. If we denote by

$$W(z) = \frac{1}{2}[\eta(z)\sum_{z'}(1-\eta(z'))(z'-z)p(z'-z) + (1-\eta(z))\sum_{z'}\eta(z')(z-z')p(z-z')]$$

then we have

$$(\mathcal{L}_n F)(\eta) \simeq \frac{n}{n^d}\sum_z \langle W(z), (\nabla f)(\frac{z}{n})\rangle\,.$$

We can write $W(z)$ as $\tau_z W(0)$ where τ_z is translation by z in Z^d. $W(0)$ has mean 0 with respect to every Bernoulli measure. In some models the mean 0 function $W(0)$ takes a special form such as $\sum_z c_z[\tau_z G(\eta) - G(\eta)]$ for some local function G. This allows for another summation by parts which gets rid of one more power of n. We end up with an equation of the form

$$\frac{\partial \rho}{\partial t}\int f(x)\rho(t,x)\,dx = \frac{1}{2}\int (\sum_{i,j} a_{i,j} D_i D_j f)(x)\widehat{G}(\rho(t,x))\,dx$$

where $\widehat{G}(\rho)$ is the expectation of $G(\eta)$ with respect to the invariant (Bernoulli) measure with density ρ. In our case, i.e, symmetric simple exclusion, $G(\eta)$ take the very simple form $G(\eta) = \eta(0)$ and $\widehat{G}(\rho) = \rho$. Leads to an equation of the form

$$\frac{\partial \rho}{\partial t} = \frac{1}{2}\sum_{i,j} a_{i,j} D_i D_j \widehat{G}(\rho(t,x)) = \frac{1}{2}\sum_{i,j} a_{i,j} D_i D_j \rho\,.$$

This type of a situation where one can do summation by parts twice are referred to as gradient models. Another example of a gradient model is the zero range process. Here there can be many particles at site. The sites are again some Z^d. Each particle if it jumps is likely to jump from z to z' with probability $p(z'-z)$. However unlike the Poisson case we saw earlier the jump rate $\lambda(t,z)$ for any particle at site z depends on the number $\eta(t,z)$ of particles currently at site z. Here

$$(\mathcal{L}_n F)(\eta) = \sum_{z,z'} \lambda(\eta(z)) p(z'-z)[F(\eta^{z,z'}) - F(\eta)]$$

where $\eta^{z,z'}$ is obtained by moving a particle from z to z' in the configuration η. With the usual diffusive scaling, if

$$F(\eta) = \frac{1}{n^d}\sum_z J(\frac{z}{n})\eta(z)$$

then

$$\begin{aligned}
(\mathcal{L}_n F)(\eta) &= n^2 \sum_{z,z'} \lambda(\eta(z)) p(z'-z)[F(\eta^{z,z'}) - F(\eta)] \\
&= \frac{n^2}{n^d}\sum_{z,z'} \lambda(\eta(z)) p(z'-z)[f(\frac{z'}{n}) - f(\frac{z}{n})] \\
&\simeq \frac{1}{2\,n^d}\sum_{z,z'} \lambda(\eta(z)) p(z'-z)(z'-z)_i(z'-z)_j (D_i D_j f)(\frac{z}{n}) \\
&\simeq \frac{1}{2\,n^d}\sum_{z} \lambda(\eta(z)) (\mathbf{A} f)(\frac{z}{n}).
\end{aligned}$$

There is a one parameter family of invariant distributions which are product measures with the number k of particles at any site having a distribution ν_θ given by

$$\nu_\theta(k) = \frac{1}{Z(\theta)}\frac{\theta^k}{\lambda(1)\lambda(2)\cdots\lambda(k)}$$

where $Z(\theta)$ is the normalization constant and $\theta = \theta(\rho)$ is adjusted so that $\sum_k k\nu_\theta(k) = \rho$. The expectation $\lambda(k)$ at density ρ, i.e.

$$\sum_k \lambda(k)\nu_{\theta(\rho)} = \widehat{\lambda}(\rho)$$

can be computed and

$$(\mathcal{L}_n F)(\eta) \simeq \int \widehat{\lambda}(\rho(x))(\mathbf{A}f)(x)\,dx$$

leading to the nonlinear PDE,

$$\frac{\partial \rho}{\partial t} = \mathbf{A}\widehat{\lambda}(\rho(t,x))$$

with

$$\mathbf{A}f = \frac{1}{2}\sum_{i,j} a_{i,j} D_i D_j f\,.$$

See [3] for a detailed exposition of zero-range processes.

With this kind of limit theorems we can answer questions of certain type; if the initial mass distribution is given by $\rho_0(x)$ how much of the mass will be in a certain set B at a later time t. But we can not answer questions of the form how much of the mass that was in certain set A at time 0 ended up in a set B at a future time t. This requires us to keep track of the identity of the particles. While the particles that were not in A at time 0 do not count, they do affect the motion of particles starting from A.

A natural question to ask is if we start in equilibrium with density ρ and have one particle at 0, and watch that particle what will its motion be? We can expect a diffusive behavior, especially in the symmetric simple exclusion model, and the so called tagged particle will exhibit a Brownian motion under the standard central limit theorem scaling and the dispersion matrix will in general be a function $S(\rho)$ of ρ. If ρ is very small, there are very few particles to affect the motion of the tagged particle and one would expect $S(\rho) \simeq \mathbf{A} = \{a_{i,j}\}$. However if $\rho \simeq 1$, most sites are filled with particles and free sites to jump to are hard to find. The tagged particle will hardly move and one would expect $S(\rho) \simeq 0$. The motion of two distinct tagged particles can be shown to be asymptotically independent.

In equilibrium at density ρ, the answer to the question we asked earlier can be answered by

$$\rho \int_A dx \int_B p_\rho(t, x, y) dy$$

where p_ρ is the transition probability density for the motion of the tagged particle at density ρ. This suggests that if we take A to be all of R^d, then the density evolves according to the heat equation with dispersion $S(\rho)$ and not **A**. But this is no contradiction, since we are in equilibrium and constants are solutions of every heat equation! The scaling limit of the tagged particle can be found in [5].

It is natural to ask what the motion of a tagged particle would be in non-equilibrium. The particle only interacts with other particles in the immediate neighborhood and since the full system is supposed to be locally in equilibrium, the tagged particle at time t and position x, will behave almost like particle in equilibrium with density $\rho(t, x)$, i.e., a Brownian motion with dispersion $S(\rho(t, x))$. In other words the tagged particle will behave like a diffusion with Kolmogorov backward generator

$$\mathcal{L} = \frac{1}{2} \sum_{i,j} S_{i,j}(\rho(t, x)) D_i D_j \,.$$

May be so, but perhaps there is an additional first order term and the generator is

$$\mathcal{L} = \frac{1}{2} \sum_{i,j} S_{i,j}(\rho(t, x)) D_i D_j + \sum_j b_j(t, x) D_j \,.$$

Since we do not know b, it is more convenient to write the generator in divergence form

$$\mathcal{L} = \frac{1}{2} \nabla S(\rho(t, x)) \nabla + b(t, x) \cdot \nabla \,.$$

If this were to describe the motion of a tagged particle then the density r of the tagged particle would evolve according to the forward Kolmogorov equation, i..e

$$\frac{\partial r}{\partial t} = \frac{1}{2} \nabla S(\rho(t, x)) \nabla r - \nabla \cdot (br) \,.$$

But the motion of the particles is the same tagged or otherwise. Hence ρ itself must satisfy the equation

$$\frac{\partial \rho}{\partial t} = \frac{1}{2} \nabla S(\rho(t, x)) \nabla \rho - \nabla \cdot (b\rho) = \frac{1}{2} \nabla \mathbf{A} \nabla \rho \,.$$

Crossing our fingers, and undoing a ∇,

$$b(t,x) = \frac{[S(\rho(t,x)) - \mathbf{A}](\nabla\rho)(t,x)}{2\rho(t,x)}.$$

It is hard to prove this relationship directly. Instead we look at a system where we have particles of k different colors say $j = 1, 2, \ldots, k$. The evolution is color blind and is the symmetric simple exclusion, but we keep track of the colors. We try to derive equations for the evolution of the k densities $\tilde{\rho} = \{\rho_i(t,x)\}$. We let $\eta_j(z) = 1$ if the site z has particle with color j. $\eta(z) = \sum_j \eta_j(z)$. We denote by ζ the entire configuration $\{\eta_j(z)\}$.

With

$$F(\zeta) = \frac{1}{n^d}\sum_z\sum_j \eta_j(z) f_j(\frac{z}{n})$$

we can compute

$$\mathcal{L}_n F = \frac{n^2}{n^d}\sum_j\sum_{z,z'} \eta_j(z)(1-\eta(z'))p(z'-z)[f_j(\frac{z'}{n}) - f_j(\frac{z}{n})]$$

which is a non-gradient system. We can do one summation by parts and we will be left to handle expressions like

$$\frac{n}{2\,n^d}\sum_j\sum_z (D_r f_j)(\frac{z}{n}) W_{j,r}(\tau_z\zeta)$$

where $W_{j,r}(\zeta)$ is an expression with mean 0 under every invariant measure. The invariant measures are product Bernoulli with $P[\eta_j(z) = 1] = \rho_j$. We wish to replace the term $W_{j,r}$ by

$$\widehat{W}_{j,r} = \sum_{j',r'} C^{j,r}_{j',r'}[\eta_j(e'_r) - \eta_j(0)]$$

and show that the difference is negligible. This is an important and difficult step in the analysis of non-gradient models. The negligibility is proved, in equilibrium after averaging in space and time. In other words quantities of the form

$$\frac{n}{n^d}\int_0^t \sum_z J(\frac{z}{n})[W_{j,r}(\tau_z\zeta(s)) - \widehat{W}_{j,r}(\tau_z\zeta(s))]ds$$

are shown to be negligible in equilibrium. The quantities $\{C^{j,r}_{j',r'}\}$ which exist can be explicitly calculated as functions of $\tilde{\rho} = \{\rho_j\}$.

$$B_{j,j'}(\tilde{\rho}) = (\rho_j\delta_{j,j'} - \frac{\rho_j\rho_{j'}}{\rho})\,S(\rho) + \rho_j\rho_{j'}\frac{(1-\rho)}{\rho}A\,,$$

$$\chi_{j,j'}(\tilde{\rho}) = I(\frac{\delta_{j,j'}}{\rho_j} + \frac{1}{1-\rho})$$

and

$$C = B\chi\,.$$

One has to justify substituting for $\tilde{\rho}$ their local empirical values in non-equilibrium. Finally for the densities $\{\rho_j(t,x)\}$ we get a system of coupled partial differential equations

$$\frac{\partial}{\partial t}\sum\int f_j(x)\,\rho_j(t,x)\,dx = \frac{1}{2}\sum_{j,j',r,r'}\int (D_r f_j)(x) C^{j,r}_{j',r'}(\rho(t,x))(D_{r'}\rho_{j'})(t,x)\,dx$$

or

$$\frac{\partial \rho_{\mathbf{j}}}{\partial t} = \frac{1}{2}\sum_{j'} \nabla \mathbf{C}_{j,j'}(\rho)\nabla \rho'_{\mathbf{j}}\,.$$

The sum $\rho = \sum_j \rho_j$ will satisfy the equation

$$\frac{\partial \rho}{\partial t} = \frac{1}{2}\nabla A \nabla \rho$$

and given $\rho(t,x)$, each $\rho_j(t,x)$ is seen to be a solution of

$$\frac{\partial \rho_j(t,x)}{\partial t} = \frac{1}{2}\nabla S(\rho(t,x))\nabla \rho_j(t,x) - \nabla \cdot \left[\frac{(S(\rho(t,x)) - A)\nabla \rho(t,x)}{2\rho(t,x)}\right]\rho_j(t,x)$$

which is the forward equation for the tagged particle motion. These results were first obtained by Quastel ([7]). It requires more work to conclude that the empirical process

$$R_n(d\omega) = \frac{1}{n^d}\sum_j \delta_{x_j(\cdot)}$$

viewed as a random measure on the space $\omega = D[[0,T];R^d]$ of trajectories, converges in probability, in the topology of weak convergence of measures on Ω to the measure Q_0, which is a Markov process on R^d, with backward generator

$$\frac{1}{2}\nabla S(\rho(t,x))\nabla + \frac{(S(\rho(t,x)) - A)\nabla \rho(t,x)}{2\rho(t,x)}\cdot \nabla$$

and initial distribution $\rho_0(x)$. $\rho(t,x)$ itself is the solution of the heat equation (3.3). The proof involves going through the multicolor system, where the colors code past history and the number of colors increase as the coding gets refined. This was carried out by Rezakhanlou in [9].

11.4. Large Deviations

We will now move on to discuss the issues of large deviations. Large deviations arise by changing the rules of the evolution. The amount of change is measured in terms of relative entropy, suitably normalized. The effect is to produce a different limit. The large deviation rate function is the minimum of the relative entropy over all the changes that produce the desired limit. Our goal is to determine the rate function $I(Q)$ that the empirical process is close to Q. We will carry it out in three steps. The process Q has a marginal $q(t,x)$. The first step is to determine the rate function for the initial configuration $q(0,x)$. This is some quantity $I(q(0,\cdot))$. If the initial condition is chosen randomly with a site z getting a particle with probability $\rho(\frac{z}{n})$, then the typical profile will be $\rho(x)$ but we can have any profile $q(x)$ with a large deviation rate function equal to

$$I(q) = \int [q(x)\log\frac{q(x)}{\rho(x)} + (1-q(x))\log\frac{1-q(x)}{1-\rho(x)}]dx\,.$$

Although we state the results for the full Z^d which scales to R^d, the results are often proved for a large periodic lattice that scales to the torus. But we will ignore this fine point.

The next step is to examine how the density of the untagged system evolves. The unperturbed limit as we saw was the solution of the heat equation (3.3). The underlying system is governed by Poisson jump processes $z \to z'$ with rate $p(z'-z)\eta(z)(1-\eta(z'))$. We can perturb this to $p(z-z') + q_n(t,z,z')$ introducing a spatial and temporal variation. Denoting by q the strength of the perturbation the entropy cost is the entropy of one Poisson process with respect to another which is seen to be of order q^2. The number of sites is of the order of n^d and time is of order n^2. If we want the relative entropy to be of the order of magnitude n^d, then q will have to be of order $\frac{1}{n}$. The asymmetry has to be "weak". We introduce the perturbed operator

$$(\tilde{\mathcal{L}}_n F) = n^2 \sum [p(z'-z) + \frac{1}{n}q(t,\frac{z}{n},z'-z)]\eta(z)(1-\eta(z'))[F(\eta^{z,z'}) - F(\eta)]\,.$$

Here $q(t,x,z)$ determines the perturbation. $b(t,x) = \sum_{z'} z'\,q(t,x,z')$ is the local bias. It is not hard to prove that we do have again a scaling limit but the equation is different.

$$\frac{\partial\rho(t,x)}{\partial t} = \frac{1}{2}\nabla A\nabla\rho(t,x) - \nabla\cdot b(t,x)\rho(t,x)(1-\rho(t,x))\,. \qquad (4.1)$$

The interaction shows up in the nonlinearity and the term $(1-\rho(t,x))$ is the effect of exclusion. The entropy cost normalized by n^{-d} is seen to be asymptotically

$$\frac{1}{2}\int_0^T\int\left[\sum_z \frac{q^2(t,x,z)}{p(z)}\right]\rho(t,x)(1-\rho(t,x))dxdt\,.$$

We need to minimize $\sum_z \frac{q^2(z)}{p(z)}$ over q fixing $b=\sum_z zq(z)$. This is seen to equal $\langle b, A^{-1}b\rangle$. The minimal cost of producing a $b(t,x)$ is therefore

$$J(b)=\frac{1}{2}\int_0^T\int\langle b(t,x),A^{-1}b(t,x)\rangle\rho(t,x)(1-\rho(t,x))dtdx\,.$$

If we are only interested in producing a density profile ρ then we need to minimize $J(b)$ over $\mathcal{B}(\rho(\cdot))$ i.e all b such that (4.1) holds. This can be done and the answer is as worked out in [4], is

$$\begin{aligned}I(\rho)&=\inf_{b\in\mathcal{B}(\rho(\cdot))} J(b)\\&=\sup_F\left[\int_0^T\int F(t,x)[D_t\rho(t,x)-\frac{1}{2}\nabla A\nabla\rho(t,x)]dxdt\right.\\&\qquad\left.-\frac{1}{2}\int\langle\nabla F(t,x),A^{-1}\nabla F(t,x)\rangle\rho(t,x)(1-\rho(t,x))dt\,dx\right].\end{aligned}$$

Now, at the third step, we turn to the question of large deviation of the empirical process $R_n(d\omega)$. It will have a large deviation principle on the space of measures on Ω, with a rate function $H(Q)$. The process Q will have a marginal $q(t,x)$. If $H(Q)$ is to be finite then $I(q)$ has to be finite. For any $b\in\mathcal{B}(q(\cdot))$ with $J(b)<\infty$, there is a Markov process Q_b, which is the motion of the tagged particle in the perturbed system. This process has the backward generator

$$\frac{1}{2}\nabla\cdot S(q(t,x))\nabla+\frac{1}{2}(S(q(t,x))-\mathbf{A})\frac{\nabla q(t,x)}{q(t,x)}\cdot\nabla+(1-q(t,x))b\cdot\nabla\,.$$

The process Q must have finite relative entropy $H(Q;Q_b)$ with respect to any Q_b and therefore the stochastic integrals $\int_0^T\langle f(t,x(t)),dx(t)\rangle$ make sense with respect to Q and we pick $c\in\mathcal{B}(q(\cdot))$ such that

$$E^Q\left[\int_0^T\langle g(t,x(t)),dx(t)\rangle\right]=E^{Q_c}\left[\int_0^T\langle g(t,x(t)),dx(t)\rangle\right].$$

The rate function is given by

$$H(Q) = I(\rho_0) + J(c) + H(Q; Q_c).$$

Details Can be found in [8].

We finally conclude with some results concerning the large deviation probabilities for the totally asymmetric simple exclusion process. We saw that the scaling was given as weak solutions of (3.2) that satisfied an entropy condition. The entropy condition can be stated as

$$\xi(t,x) = \frac{\partial h(\rho)}{\partial t} + \frac{\partial g(\rho)}{\partial x} \leq 0 \tag{4.2}$$

in the sense of distribution. Here h is a convex function and $g'(\rho) = h'(\rho)(1-2\rho)$. Clearly for smooth solutions of (3.2), (4.2) will hold with equality. Of special interest for large deviations is the convex function $h(\rho) = \rho \log \rho + (1-\rho)\log(1-\rho)$. It turns out that the large deviation rate function is finite only for weak solutions of (3.2) and is given by the total mass of the positive part of the distribution ξ given in (4.2). These results can be found in [6], [10], and [11].

References

[1] Donsker, M. D. (1951). An invariance principle for certain probability limit theorems. *Mem. Amer. Math. Soc.* **1951** no. 6.

[2] Georgii, H.-O. (1988). *Gibbs measures and phase transitions.* de Gruyter Studies in Mathematics, **9**. Walter de Gruyter & Co., Berlin.

[3] Kipnis, C. and Landim, C. (1999). *Scaling limits of interacting particle systems.* Grundlehren der Mathematischen Wissenschaften [Fundamental Principles of Mathematical Sciences], **320**. Springer-Verlag, Berlin.

[4] Kipnis, C., Olla, S. and Varadhan, S. R. S. (1989). Hydrodynamics and large deviation for simple exclusion processes. *Comm. Pure Appl. Math.* **42** 115–137.

[5] Kipnis, C. and Varadhan, S. R. S. (1986). Central limit theorem for additive functionals of reversible Markov processes and applications to simple exclusions. *Comm. Math. Phys.* **104** 1–19.

[6] Jensen, L. (2000). *Large Deviations of the Asymmetric Simple Exclusion Process.* Ph.D Thesis, New York University.

[7] Quastel, J. (1992). Diffusion of color in the simple exclusion process. *Comm. Pure Appl. Math.* **45** 623–679.

[8] Quastel, J., Rezakhanlou, F. and Varadhan, S. R. S. (1999). Large deviations for the symmetric simple exclusion process in dimensions $d \geq 3$. *Probab. Theory Related Fields* **113** 1–84.

[9] Rezakhanlou, F. (1994). Propagation of chaos for symmetric simple exclusions. *Comm. Pure Appl. Math.* **47** 943–957.

[10] Varadhan, S. R. S. (2004). Large deviations for the asymmetric simple exclusion process. Stochastic analysis on large scale interacting systems, 1–27, *Adv. Stud. Pure Math.* **39** Math. Soc. Japan, Tokyo.

[11] Vilensky, Y. (2008). *Large Deviation Lower Bounds for the Totally Asymmetric Simple Exclusion Process.* Ph.D Thesis, New York University.

Author Index

Subject Index

Contents of Part II